D1084041

Handbook of
Experimental Pharmacology

Volume 171

Basis and Treatment of Cardiac Arrhythmias

Contributors

M.E. Anderson, C. Antzelevitch, J.R. Balser, P. Bennett,
M. Cerrone, C.E. Clancy, I.S. Cohen, J.M. Fish, I.W. Glaaser,
T.J. Hund, M.J. Janse, C. January, R.S. Kass, J. Kurokawa,
J. Lederer, S.O. Marx, A.J. Moss, S. Nattel, C. Napolitano,
S. Priori, G. Robertson, R.B. Robinson, D.M. Roden,
M.R. Rosen, Y. Rudy, A. Shiroshita-Takeshita, K. Sipido,
Y. Tsuji, P.C. Viswanathan, X.H.T. Wehrens, S. Zicha

Editors
Robert S. Kass and Colleen E. Clancy

 Springer

Robert S. Kass Ph. D.
David Hosack Professor and Chairman
Columbia University
Department of Pharmacology
630 W. 168 St.
New York, NY 10032
USA
e-mail: rsk20@columbia.edu

Colleen E. Clancy Ph. D.
Assistant Professor
Department of Physiology and Biophysics
Institute for Computational Biomedicine
Weill Medical College of Cornell University
1300 York Avenue
LC-501E
New York, NY 10021
e-mail: clc7003@med.cornell.edu

With 60 Figures and 11 Tables

ISSN 0171-2004

ISBN-10 3-540-24967-2 Springer Berlin Heidelberg New York

ISBN-13 978-3-540-24967-2 Springer Berlin Heidelberg New York

Library of Congress Control Number: 2005925472

Springer is a part of Springer Science + Business Media
springeronline.com

© Springer-Verlag Berlin Heidelberg 2006
Printed in Germany

Editor: S. Rallison
Editorial Assistant: S. Dathe
Cover design: design&production GmbH, Heidelberg, Germany
Typesetting and production: LE-TEX Jelonek, Schmidt & Vöckler GbR, Leipzig, Germany
Printed on acid-free paper   27/3151-YL - 5 4 3 2 1 0

Preface

In the past decade, major progress has been made in understanding mechanisms of arrhythmias. This progress stems from much-improved experimental, genetic, and computational techniques that have helped to clarify the roles of specific proteins in the cardiac cycle, including ion channels, pumps, exchanger, adaptor proteins, cell-surface receptors, and contractile proteins. The interactions of these components, and their individual potential as therapeutic targets, have also been studied in detail, via an array of new imaging and sophisticated experimental modalities. The past 10 years have also led to the realization that genetics plays a predominant role in the development of lethal arrhythmias.

Many of the topics discussed in this text reflect very recently undertaken research directions including the genetics of arrhythmias, cell signaling molecules as potential therapeutic targets, and trafficking to the membrane. These new approaches and implementations of anti-arrhythmic therapy derive from many decades of research as outlined in the first chapter by the distinguished professors Michael Rosen (Columbia University) and Michiel Janse (University of Amsterdam). The text covers changes in approaches to arrhythmia therapy over time, in multiple cardiac regions, and over many scales, from gene to protein to cell to tissue to organ.

New York, May 2005   Colleen E. Clancy and Robert S. Kass

List of Contents

List of Contributors

(Addresses stated at the beginning of respective chapters)

HEP (2006) 171:1–39
© Springer-Verlag Berlin Heidelberg 2006

History of Arrhythmias

M. J. Janse[1] (✉) · M. R. Rosen[2]

[1]The Experimental and Molecular Cardiology Group, Academic Medical Center, M 051, University of Amsterdam, Meibergdreef 9, 1105 AZ Amsterdam, The Netherlands
m.j.janse@amc.uva.nl

[2]Center for Molecular Therapeutics, Department of Pharmacology, College of Physicians and Surgeons, Columbia University, 630 W 168th Street, PH7West-321, New York NY, 10032, USA

Abstract A historical overview is given on the techniques to record the electrical activity of the heart, some anatomical aspects relevant for the understanding of arrhythmias, general mechanisms of arrhythmias, mechanisms of some specific arrhythmias and non-pharmacological forms of therapy. The unravelling of arrhythmia mechanisms depends, of course, on the ability to record the electrical activity of the heart. It is therefore no surprise that following the construction of the string galvanometer by Einthoven in 1901, which allowed high-fidelity recording of the body surface electrocardiogram, the study of arrhythmias developed in an explosive way. Still, papers from McWilliam (1887), Garrey (1914) and Mines (1913, 1914) in which neither mechanical nor electrical activity was recorded provided crucial insights into re-entry as a mechanism for atrial and ventricular fibrillation, atrioventricular nodal re-entry and atrioventricular re-entrant tachycardia in hearts with an accessory atrioventricular connection. The components of the electrocardiogram, and of extracellular electrograms directly recorded from the heart, could only be well understood by comparing such registrations with recordings of transmembrane potentials. The first intracellular potentials were recorded with microelectrodes in 1949 by Coraboeuf and Weidmann. It is remarkable that the interpretation of extracellular electrograms was still controversial in the 1950s, and it was not until 1962 that Dower showed that the transmembrane action potential upstroke coincided with the steep negative deflection in the electrogram. For many decades, mapping of the spread of activation during an arrhythmia was performed with a "roving" electrode that was subsequently placed on different sites on the cardiac surface with a simultaneous recording of another signal as time reference. This method could only provide reliable information if the arrhythmia was strictly regular. When multiplexing systems became available in the late 1970s, and optical mapping in the 1980s, simultaneous registrations could be made from many sites. The analysis of atrial and ventricular fibrillation then became much more precise. The old question whether an arrhythmia is due to a focal or a re-entrant mechanism could be answered, and for atrial fibrillation, for instance, the answer is that both mechanisms may be operative. The road from understanding the mechanism of an arrhythmia to its successful therapy has been long: the studies of Mines in 1913 and 1914, microelectrode studies in animal preparations in the 1960s and 1970s, experimental and clinical demonstrations of initiation and termination of tachycardias by premature stimuli in the 1960s and 1970s, successful surgery in the 1980s, the development of external and implantable defibrillators in the 1960s and 1980s, and finally catheter ablation at the end of the previous century, with success rates that approach 99% for supraventricular tachycardias.

Keywords Electrocardiogram · Extracellular electrograms · Transmembrane potentials · Re-entry · Focal activity · Tachycardias · Fibrillation

1
Introduction

The diagnosis of cardiac arrhythmias and the elucidation of their mechanisms depend on the recording of the electrical activity of the heart. The study of disorders of the rhythmic activity of the heart started around the fifth century B.C. in China and in Egypt around 3000 B.C. with the examination of the peripheral pulse (for details see Snellen 1984; Acierno 1994; Lüderitz 1995; Ziskind and Halioua 2004). In retrospect, it is easy to recognize atrioventricular (AV) block, represented by the slow pulse rate observed by Gerber in 1717 (see Music et al. 1984), or atrial fibrillation manifested by the irregular pulse described by de Senac (1749). The recording of arterial, apical and venous pulsations, notably by MacKenzie (1902) and Wenckebach (1903), provided a more rational basis for diagnosing many arrhythmias. Still, the concept that disturbances in the electrical activity of the heart were responsible for abnormal arterial and venous pulsations was not universally known at the turn of the nineteenth century. For example, MacKenzie observed that the A wave disappeared from the venous curve during irregular heart action, and wrote, in

1902 under the heading of "The pulse in auricular paralysis", "I have no clear idea of how the stimulus to contraction arises, and so cannot definitely say how the auricle modifies the ventricular rhythm. But as a matter of observation I can with confidence state that the heart has a very great tendency to irregular action when the auricles lose their power of contraction."

The first demonstration of the electrical activity of the heart was made accidentally by Köllicker and Müller in 1856. Following the experiments of Matteuci in 1842, who used the muscle of one nerve-muscle preparation as a stimulus for the nerve of another, thereby causing its muscle to contract (see Snellen 1984), they also studied a nerve-muscle preparation from a frog (sciatic nerve and gastrocnemius muscle). Accidentally, the sciatic nerve was placed in contact with the exposed heart of another frog, and they observed the gastrocnemius muscle contract in synchrony with the heartbeat. They saw immediately before the onset of systole a contraction of the gastrocnemius, and in some preparations a second contraction at the beginning of diastole. Although Marey (1876) first used Lipmann's capillary electrometer to record the electrical activity of the frog's heart, the explanation for this activity was provided by the classic experiments of Burdon-Sanderson and Page (1879, 1883). They also used the capillary electrometer together with photographic equipment to obtain recordings of the electrical activity of frog and tortoise hearts. They placed electrodes on the basal and apical regions of the frog heart and observed two waves of opposite sign during each contraction. The time interval between the two deflections was in the order of 1.5 s. By injuring the tissue under one of the recording sites, they obtained the first monophasic action potentials and showed how, in contrast to nerve and skeletal muscle, there is in the heart a long period between excitation and repolarization ["... if either of the leading-off contacts is injured ... the initial phase is followed by an electrical condition in which the injured surface is more positive, or less negative relatively to the uninjured surface: this condition lasts during the whole of the isoelectric period ..." (Burdon-Sanderson and Page 1879)]. A second important observation was that by partially warming the surface "... the initial phase (i.e. of the electrogram) is unaltered but the terminal phase begins earlier and is strengthened" (Burdon-Sanderson and Page 1879).

Heidenhain introduced the term arrhythmia as the designation for any disturbance of cardiac rhythm in 1872. With the introduction of better techniques to record the electrical activity of the heart, the study of arrhythmias developed in an explosive way. We will limit this brief account to those studies in which the electrical activity was documented, even though we will make an exception for a number of seminal papers on the mechanisms of arrhythmias in which neither mechanical nor electrical activity was recorded (McWilliam 1887a,b, 1889; Garrey 1914; Mines 1913b, 1914). We will pay particular attention to the early studies, nowadays not easily accessible, and will not attempt to give a complete review of all arrhythmias.

2
Methods to Record the Electrical Activity of the Heart

2.1
The Electrocardiogram

In 1887, Waller was the first to record an electrocardiogram from the body surface of dog and man (see Fig. 1). He used Lippmann's capillary electrometer, an instrument in which in a mercury column borders on a weak solution of sulphuric acid in a narrow glass capillary. Whenever a potential difference between the mercury and the acid is applied, changed or removed, this boundary moves (see Snellen 1995). The capillary electrometer was sensitive, but slow. Einthoven constructed his string galvanometer, which was both sensitive and rapid, based on the principle that a thin, short wire of silver-coated quartz placed in a narrow space between the poles of a strong electromagnet will move whenever the magnetic field changes as a consequence of change in the current flowing through the coils. During the construction of the string galvanometer, Einthoven was aware of the fact that Ader in 1897 also had used an instrument with a string in a magnetic field as a receiver of Morse signals transmitted by undersea telegraph cables. In Einthoven's first publication on the string galvanometer, he did quote Ader (Einthoven 1901). It is often suggested that Einthoven merely improved Ader's instrument. However, as argued by Snellen (1984, 1995), Ader's instrument was never used as a galvanometer, i.e. as an instrument for measuring electrical currents, and if it had, its sensitivity would have been 1:100,000 that of the string galvanometer. To quote Snellen (1995): "... the principle of a conducting wire in a magnetic field moving when a current passes through it, had been known from Faraday's time if not earlier, that is three quarters of a century before Ader. Equalizing all possible instruments which use that principle is perhaps just as meaningless as to put a primitive horse cart on a par with a Rolls Royce, because they both ride on wheels."

Figure 1 shows electrocardiograms recorded with the capillary electrometer by Waller and by Einthoven, Einthoven's mathematical correction of his tracing, and the first human electrocardiogram recorded by Einthoven with his string galvanometer (Einthoven 1902, 1903).

Remarkably, Einthoven constructed a cable which connected his physiological laboratory with the Leiden University hospital, over a distance of a mile (Einthoven 1906). This should have created a unique opportunity to collaborate with clinicians and document the electrocardiographic manifestations of a host of arrhythmias. Unfortunately, according to Snellen (1984):

> Occurrence of extrasystoles had the peculiar effect that Einthoven could warn the physician by telephone that he was going to feel an intermission of the pulse at the next moment. It seems that this annoyed the clinician who was poorly co-operative anyway; in fact, after only a few years he cut the connection to the physiological laboratory. This must have been

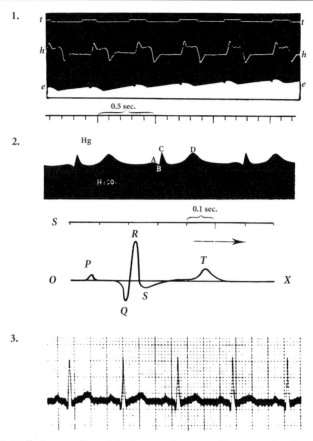

Fig. 1 *Panel 1*: Waller's recording of the human electrocardiogram using the capillary electrometer. *t*, time; *h*, external pulsation of the heart; *e*, electrocardiogram. *Panel 2*: Einthoven's tracing published in 1902 also with the capillary electrometer, with the peaks called *A*, *B*, *C*, and *D*. In the lower tracing, Einthoven corrected the tracing mathematically, and now used the terminology *P*, *Q*, *R*, *S* and *T*. *Panel 3*: One of the first electrocardiograms recorded with the string galvanometer as published in 1902 and 1903 by Einthoven. (Reproduced from Snellen 1995)

a blow to Einthoven, although in 1906 and 1908 he had already collected two impressive series of clinical tracings. Precisely at this time, a young physician and physiologist from London approached him who needed to improve his registration method of the relation between auricular and ventricular contraction in what ultimately proved to be auricular fibrillation. This was Thomas Lewis.

There is no doubt that Lewis was foremost in introducing Einthoven's instrument into clinical practice and in experiments designed to unravel mechanisms of arrhythmias (see later). Einthoven always appreciated Lewis's work. When

Einthoven received the Nobel prize in 1925, he said in his acceptance speech: "It is my conviction that the general interest in electrocardiography would not have risen so high, nowadays, if we had to do without his work and I doubt whether without his valuable contribution I would have the privilege of standing before you today" (Snellen 1995). Others who quickly employed Einthoven's instrument were the Russian physiologist Samojloff, who in 1909 published the first book on electrocardiography, and Kraus and Nicolai who published the second book in 1910 (see Krikler 1987a,b).

Initially, only the three (bipolar) extremity leads were used. Important developments were the introduction of the central terminal and the unipolar precordial leads by Wilson and associates (Wilson et al. 1933a), and of augmented extremity leads by Goldberger (1942). Wilson and Johnston (1938) also paved the way for the development of vectorcardiography.

The first body-surface maps, based on 10 to 20 electrocardiograms recorded from the surface of a human body were published by Waller in 1889. However, the distribution of isopotential lines on the human body surface at different instants of the cardiac cycle took off after the publication by Nahum et al. (1951).

Ambulatory electrocardiography began with Holter's publication in 1957. Further developments in electrocardiography include body surface His bundle electrocardiography, computer analysis of the electrocardiogram, the signal-averaged electrocardiogram, polarcardiography and the magnetocardiogram. For a detailed description of these techniques, the reader is referred to the book *Comprehensive Electrocardiology*, edited by MacFarlane and Lawrie (1989).

A large number of books on the electrocardiography of arrhythmias has been published, and here we will only refer to a few, all written by one or two authors (Samojloff 1909; Kraus and Nicolai 1910; Lewis 1920, 1925; Lepeschkin 1951; Katz and Pick 1956; Spang 1957; Scherf and Cohen 1964; Scherf and Schott 1973; Schamroth 1973; Pick and Langendorf 1973; Josephson and Wellens 1984), and ignore the even greater number of multi-authored books.

2.2
The Interpretation of Extracellular Waveforms

Pruitt (1976) gives a very interesting account of the controversy, confusion and misunderstanding about the interpretation of extracellular electrograms in the 1920s and 1930s. In those days, one generally used the terminology of Lewis (1911), who had written that "the excited point becomes negative relative to all other points of the musculature ... and the wave of negativity travels in all directions from the point of excitation." Burdon-Sanderson and Page (1879) had in fact already written, "Every excited part of the surface of the ventricle is during the excitatory state *negative* to every unexcited part" (their italics). Others interpreted these ideas in the sense that the spread of activation was equal to the propagation of a "wave of negativity". Although Lewis clearly indicated that the excited part of the heart was negative *relative* to the unexcited parts, he never used the terms doublet or dipole. Craib (see Pruitt 1976) was the first to "formulate a concept of myocardial excitation that entailed movement along the fibre not of a wave of negativity, but of an electrical doublet", the latter defined as "intimately related and closely lying foci or loci of raised and lowered potentials". Wilson and associates (1933b) introduced the term bipole, which, much the same as Craib's doublet, represented "two sources of equal but opposite potential lying close together". The word source here may be confusing since Wilson also introduced the terms source and sink, meaning the paired positive and negative charges associated with propagation of the cardiac impulse. In retrospect, the controversy that led to the estrangement of Lewis and Craib (Pruitt 1976) is difficult to understand and seems largely semantic. Why should cardiologists quarrel about the question whether "negativity" could exist on its own, without "positivity" in the immediate neighbourhood?

In addition to the misunderstanding concerning propagation of a wave of "negativity", there is confusion in the early literature regarding the question of which deflection in the extracellular electrogram reflects local excitation. Some of the difficulties in interpreting electrograms directly recorded from the surface of the heart seem to be related to the fact that in the early days only bipolar recordings were used. It took a long time before the concept that a bipolar recording is best understood as the sum of two unipolar recordings became widely accepted among cardiac electrophysiologists. (Strictly speaking, there is of course no such thing as a unipolar recording. We use the term unipolar to indicate that one electrode is positioned directly on the heart, the other electrode, the "indifferent" one, far away. In bipolar recordings, both terminals are close together on the heart's surface.) Lewis introduced the terms "intrinsic" and "extrinsic" deflections, and although we still use these terms today, we do not mean precisely the same thing. Lewis (1915) wrote: "(1) There are deflections which result from arrival of the excitation process immediately beneath the contacts; these we term intrinsic deflections.... (2) There are also deflections which are yielded by the excitation wave, travelling in distant areas of the muscle. To these we apply the term extrinsic deflections." He proves his point by recording a bipolar complex from the atrium. The "usual tall spike" is preceded by a small downward deflection. Crushing the tissue under the electrode pair results in disappearance of the tall spike (the intrinsic deflection), but the small initial deflection (the extrinsic deflection) remains (Lewis 1915). Lewis called this a fundamental observation, and he was right. Still, for us the terminology is somewhat confusing. Today we use bipolar recordings to get rid of extrinsic deflections. The reasoning is that each terminal is affected to (almost) the same degree by extrinsic potentials (far field effects), which are therefore cancelled when one electrode terminal is connected to the negative pole of the amplifier, the other terminal to the positive pole. What then remains is not one single intrinsic deflection, but two intrinsic deflections, one representing the passage of the propagating impulse under one terminal, the other being caused by excitation of the tissue under the other terminal. A unipolar complex has extrinsic compounds, positive when the excitatory wave is travelling towards the electrode, negative when it is moving away, and a single large rapid negative deflection, the intrinsic deflection.

Although Wilson and associates (1933b) introduced unipolar and bipolar recordings, the precise interpretation of the various components of such recordings was not completely clear even in the 1950s. Durrer and van der Tweel began recording unipolar and bipolar electrograms from intramural, multipolar needle electrodes inserted in the left ventricular wall of goats and dogs in the early 1950s. In 1954 they wrote: "In all cases where a fast part of the intrinsic deflection (i.e. in unipolar recordings, MJJ and MRR) could

epicardial surface (dog)

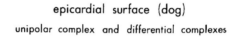

unipolar complex and differential complexes

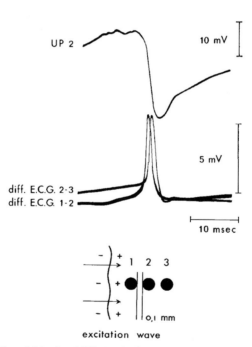

excitation wave

Fig. 2 Unipolar (*UP*) and bipolar (*diff. ECG*) electrograms recorded from the epicardial surface of a canine heart. The direction of the excitation wave and the position of the three electrodes are indicated in the *lower panel*. Bipolar complexes recorded from electrodes *1* and *2* and from electrodes *2* and *3* are shown, together with a unipolar complex from electrode *2*. The intrinsic deflection in the unipolar recording coincides with the intersection of the descending limb from bipolar complex *1–2* and with the ascending limb of bipolar complex *2–3*. Recordings made by Durrer and van der Tweel circa 1960

be detected, the top of the differential spike (i.e. the bipolar recording) was found to coincide with it" (Durrer and van der Tweel 1954a). In other words, the "intrinsic deflection" in bipolar electrograms was thought to be the top of the spike. In a subsequent paper (Durrer et al. 1954b), they found that "the width of the bipolar complex increased proportionally to the distance between the intramural lead points". The implication here is that the bipolar complex has two intrinsic deflections. Figure 2 is an unpublished recording by Durrer and van der Tweel that must have been made in 1960, since a very similar figure was published in 1961 (Durrer et al. 1961). Here it can be seen that the intrinsic deflection in the unipolar recording from terminal 2 coincides with the intersection of the descending limb of the bipolar complex recorded from terminals 1 and 2, and the ascending limb from the bipolar signal from terminals 2 and 3.

That the steep, negative-going downstroke in the unipolar extracellular electrogram coincides with the upstroke of the transmembrane action potential

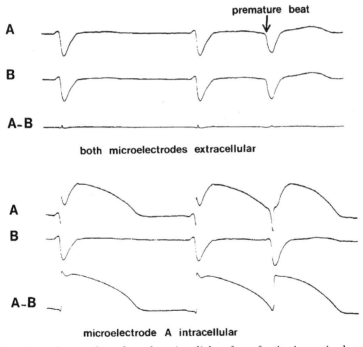

Fig. 3 Microelectrode recordings from the epicardial surface of an in situ canine heart. In the *upper panel*, both microelectrodes *A* and *B* are in the extracellular space as close together as possible, the reference electrode is somewhere in the mediastinum. Note that the "bipolar" electrogram *A–B* is almost a straight line. In the *lower panel*, microelectrode *A* is intracellular, microelectrode *B* extracellular. Note contamination of the unipolar recording of *A* with extrinsic potentials, and how *A–B* gives the true transmembrane potential. (Reproduced from Janse 1993)

was still under debate in the 1950s. Sano et al. (1956) recorded transmembrane potentials together with an extracellular signal from a surrounding ring electrode, and were unable to correlate the action potential upstroke to an extracellular "intrinsic deflection". They concluded that the moment of local excitation could not be detected in extracellular recordings. In a paper, significantly entitled "In Defence of the Intrinsic Deflection", Dower (1962) was finally able to show that the action potential upstroke does indeed coincide with the intrinsic deflection. He was aware of the fact that "to obtain a true transmembrane potential, of course, one electrode should be inside the cell, and the other immediately outside". He attributed Sano's erroneous conclusion to the circumstance that in that study, a leg electrode was used as a second electrode, so that the "transmembrane" potentials were contaminated with the electrocardiogram from the rest of the heart. These effects are illustrated in Fig. 3 (Janse 1993).

2.3
The Recording of Transmembrane Potentials

In 1948, Ling and Gerard managed to pull glass capillaries with a tip diameter in the order of 0.5 μm that were suitable for penetrating the cell membrane (Ling and Gerard 1949). One year later, the first transmembrane potentials from cardiac tissue, in this case the false tendons of canine hearts, were recorded (Coraboeuf and Weidmann 1949). These experiments were made in Cambridge, in the laboratory of A.L. Hodgkin (the future Noble laureate, together with A.F. Huxley), and Weidmann later recalled: "A remark by Hodgkin, 1949, is still in my ears: 'You can now rediscover the whole of cardiac electrophysiology'" (Weidmann 1971). This is indeed what happened. Much of cardiac electrophysiology had previously been studied by either extracellular recordings or measurements of "monophasic action potentials". After Burdon-Sanderson and Page (1879) first recorded the monophasic action potential, quite a number of studies employed this technique by applying suction at the site of recording (for overview of the early studies see Schütz 1936). The monophasic action potential provides a good index to the shape of the action potential as recorded by intracellular microelectrodes (Hoffman et al. 1959), and the technique is especially useful for studies on action potential duration and abnormalities of the repolarization phase of the action potential, such as early and delayed afterdepolarizations. A monophasic action potential can be obtained in humans by suction via an intracardiac catheter (Olsson 1971), or by applying pressure (Franz 1983). However, only by microelectrode recordings can quantitative data be obtained on the various phases of the cardiac action potential.

This is not the place to review in detail the "rediscovery" of cardiac electrophysiology by the use of the microelectrode, and we will refer to some excellent books and reviews summarizing the early studies (Brooks et al. 1955; Weidmann 1956; Hoffman and Cranefield 1960; Noble 1975, 1984).

Important milestones in the elucidation of the mechanisms underlying the action potential were the development of the so-called voltage clamp technique (initially employed in the giant axon of the squid: Marmont 1949; Cole 1949), in which ionic currents flowing through the cell membrane can be measured by keeping the membrane potential constant at a certain level, and the patch-clamp technique, enabling the recording of currents through single ionic channels (Neher and Sakmann 1976, who later shared the Nobel prize). Other chapters in this volume will deal more extensively with the various ionic currents responsible for the action potential, and the molecular biology of ion channels.

2.4
Mapping of the Spread of Activation During Arrhythmias

Some of the early pivotal studies on arrhythmia mechanisms do not contain any recording of either the mechanical or electrical activity of the heart (Mines 1913b, 1914; Garrey 1914). This is remarkable, because in another paper by Mines (1913a), beautiful recordings of extracellular electrograms from the frog's heart, using Einthoven's string galvanometer, are published. These recordings provided important information on changes in the T wave during local warming of the heart, but give no information about arrhythmias. In Rytand's splendid review on the early history of the circus movement hypothesis (Rytand 1966), a copy of page 327 of Mines' paper (1913b) was reproduced on which Mines had added in his own handwriting: "Later I took electrograms of this expt". On page 383 of the same paper, Mines added the following handwritten note: "Cinematographed the ring excn at Toronto, March 1914 ..." He refers to the excitation in a ring-like preparation from the auricle of *Acanthias vulgaris* in which he produced circulating excitations (see the section on arrhythmia mechanisms). Despite strenuous efforts by Rytand in 1964 to retrieve this film, no trace of it could be found. Still, it is fair to consider Mines to be the first to map arrhythmias. A close second is Thomas Lewis, who published the first "real" mapping experiments in 1920 (Lewis et al. 1920).

For decades thereafter, mapping of the spread of activation during arrhythmias was performed with a "roving" extracellular electrode that was subsequently placed on different sites on the cardiac surface, with a simultaneous recording of another lead, usually a peripheral electrocardiogram, as a time reference. This method could only provide reliable information if the arrhythmia was strictly regular and was useless for irregular rhythms such as atrial or ventricular fibrillation. Only when multiplexing systems became available in the late 1970s did simultaneous recordings from many sites in the heart become possible, allowing analysis of excitation patterns during fibrillation.

An important development was the optical mapping technique, in which hearts are loaded with voltage-sensitive fluorescent dyes and the upstroke of the "optical action potential" is rapidly scanned by a laser beam at many sites

(Dillon and Morad 1981; Morad et al. 1986; Rosenbaum and Jalife 2001). This technique provides a high spatial resolution of up to 50 µm.

Clinically, recording of extracellular electrograms by intracardiac catheters led to an explosive development in the diagnosis and treatment of arrhythmias. The first human His bundle electrogram was recorded in Paul Puech's clinic in Montpellier in 1960 (Giraud et al. 1960). Since this paper was published in French, it did not receive the attention it deserved. The real stimulus for the widespread use of His bundle recording in man was provided by the work of Scherlag and colleagues (Scherlag et al. 1976, 1979). The technique—first validated in dogs, and soon applied to man—allowed the localization of the various forms of AV block: proximal and distal to the His bundle, and intra-Hisian block.

Of even greater influence were two papers, simultaneously and independently published from groups in Amsterdam (Durrer et al. 1967) and Paris (Coumel et al. 1967), in which premature stimuli were used to initiate and terminate tachycardias, and intracardiac recordings were made at various sites. This technique, of which Hein J.J. Wellens was the great proponent (Wellens 1971), became known as programmed electrical stimulation. A summary of studies employing this technique can be found in Josephson's book (Josephson 2002).

Josephson developed the technique for endocardial mapping of ventricular tachycardia, which led to the development of surgical techniques, by which pieces of endocardium were resected, based on intra-operative endocardial mapping (Josephson et al. 1978; Harken et al. 1979; De Bakker et al. 1983). "Noncontact" mapping, in which intracavitary potentials are measured from electrodes on an olive-shaped probe introduced in the left ventricle of animals, was first elucidated by Taccardi et al. (1987), and a similar system has been used in humans (Peters et al. 1997). Another mapping system, the nonfluoroscopic mapping system CARTO was first described by Ben-Haim et al. (1996), and is also used to localize arrhythmogenic sites suitable for catheter ablation. For a recent overview of mapping systems, see Shenassa et al. (2003).

3
Some Aspects of Cardiac Anatomy Relevant for Arrhythmias

3.1
Atrioventricular Connections

Around the turn of the twentieth century most aspects of the specialized tissues of the heart were known. Thus, Keith and Flack (1907) described the SA node at the entrance of the superior caval vein into the right atrium, Tawara (1906) demonstrated in the hearts of many species that the AV node is the only structure connecting the atria to the His bundle, already known since 1893

(His 1893), whilst the peripheral Purkinje fibres had been discovered in 1845 (Purkinje 1845). However, at the beginning of the twentieth century, there still was controversy about the AV connections. Kent (1913) wrote:

> Some of the divergent views now held on this question are the following: (A) There is one, and only one, muscular path capable of conveying impulses from auricle to ventricle, viz the atrioventricular bundle.... (B) The muscular path of communication may be multiple. (C) The muscular path of communication is undoubtedly multiple. The view described under A is very generally held.... B. This is a view which has been gradually forced on some of those workers who have been brought into most intimate contact with experimental and clinical evidence. C. This view is held by comparatively few. It is the view put forward by myself in 1892.

Erlanger, the 1944 Noble laureate, wrote in a reminiscence (Erlanger 1964):

> British physiologists, and particularly one Stanley Kent, were steadfastly maintaining, as do some American clinicians to this day..., that there are auriculoventricular conduction paths in addition to the His bundle, which, after a time, can take over when the bundle of His is blocked.... In order to ascertain whether there are such additional conductors, in my experiments the auriculoventricular bundle was crushed aseptically. In the surviving dogs, the block remained complete ... some of them for periods as long as three months. There are no other conducting paths!

It is ironical that Kent's paper, although meant to convey the message that multiple pathways are the rule in the normal heart, is seen as the original paper

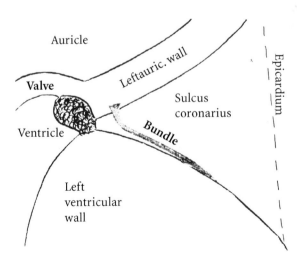

Fig. 4 Öhnell's illustration (1944) of the accessory bundle connecting left atrial myocardium to that of the left ventricle. Note that the bundle does not cross the annulus fibrosis but runs through the subepicardial fat pad

describing (abnormal) accessory AV pathways, which usually are called bundles of Kent (Cobb et al. 1968). In Kent's paper, a "muscular" connection between the right atrium and the right ventricle is described that crosses the annulus fibrosis, and a histological section is shown as well. This observation prompted Mines (1914) to accurately predict the re-entrant pathway of the tachycardia in what we now call the Wolff–Parkinson–White (WPW) syndrome, unknown at the time (see Sect. 5.2). In fact, what Kent described was not an accessory bundle consisting of ordinary muscle, but a node-like structure which is a remnant of an extensive AV ring of specialized tissue present in the embryo. In rare cases, the accessory pathway consists of such specialized cells (Becker et al. 1978). As argued by Anderson and Becker (1981): "... there are indeed good scientific reasons for discontinuing the use of 'Kent bundle' ... the most important being that Kent did not describe connections in terms of morphology we know today.... If an eponym is really necessary, then let us call them nodes of Kent." The true morphology of accessory AV pathways was described by Öhnell in 1944 (see also Sect. 5.2 and Fig. 4).

3.2
Specialized Internodal Atrial Pathways

Controversy regarding the spread of activation of the sinus impulse has existed since the discovery of the sinus node and the AV node. Thorel (1909) claimed to have demonstrated continuity between both nodes via a tract of "Purkinje-like" cells. This possibility was debated during a meeting of the Deutschen Pathologischen Gesellschaft (1910). The consensus of that meeting was that both nodes were connected by simple myocardium. In the 1960s and 1970s, the concept of specialized internodal pathways was again promoted, notably by James (1963). This promotion was so successful that the tracts are denoted in the well-known atlas of Netter (1969) in a fashion analogous to that used to delineate the ventricular specialized conduction system, and in the 1960s and 1970s, paediatric cardiac surgeons took care not to damage the supposed specialized pathways. Moreover, specialized conduction pathways were supposed to be involved in the genesis of atrial flutter (Pastelin et al. 1978). A review of the early literature, together with our own (M.J.J.) experimental and histological data, which concluded that there was no well-defined specialized conduction system connecting sinoatrial (SA) and AV nodes, was presented by Janse and Anderson (1974). In our view, the definitive proof that specialized internodal pathways do not exist was given by Spach and co-workers (1980). They argued that preferential conduction in atrial bundles could either be due to the anisotropic properties of the tissue or to the presence of a specialized tract. If the point of stimulation would be shifted to varying sites of the bundle, isochrones of similar shape would result from stimulating multiple sites if the anisotropic properties primarily influenced the local conduction velocities. On the other hand, isochrones of different shapes would be obtained if there was a fixed position specialized tract in the bundle. Their experimental results provided evidence that preferential conduction in the atria is due to the intrinsic anisotropic properties of cardiac muscle.

4
Mechanisms of Arrhythmias

Table 1 shows the classification of arrhythmia mechanisms as proposed by Hoffman and Rosen (1981).

4.1
Re-entry

Two causes for tachycardia and fibrillation were considered throughout the twentieth century: enhanced impulse formation and re-entrant excitation. McWilliam was the first to suggest that disturbances in impulse propagation could be responsible for tachyarrhythmias, and he clearly envisaged the possibility that myocardial fibres could be re-excited as soon as their refractory

Fig. 5 George Ralph Mines. This photograph was taken by his co-worker Dorothy Dale (later Mrs. Dorothy Thacker) in the Marine Biological Laboratory, Plymouth, in the summer of 1911 and was given to M.J.J. by D.A. Rytand in 1973

Table 1 Classification of mechanisms of arrhythmias by Hoffman and Rosen (1981)

I	II	III
Abnormal impulse generation	Abnormal impulse conduction	Simultaneous abnormalities of impulse generation and conduction
A. Normal automatic mechanism	A. Slowing and block	A. Phase 4 depolarization and impaired conduction
1. Abnormal rate	1. Sinoatrial block	1. Specialized cardiac fibres
a. Tachycardia	2. Atrioventricular block	B. Parasystole
b. Bradycardia	3. His bundle block	
2. Abnormal rhythm	4. Bundle branch block	
a. Premature impulses	B. Unidirectional block and re-entry	
b. Delayed impulses	1. Random re-entry	
c. Absent impulses	a. Atrial muscle	
B. Abnormal automatic mechanism	b. Ventricular muscle	
1. Phase 4 depolarization at low membrane potential	2. Ordered re-entry	
2. Oscillatory depolarizations at low membrane potential preceding upstroke	b. AV node and junction	
C. Triggered activity	c. His-Purkinje system	
1. Early after depolarizations	d. Purkinje fibre-muscle junction	
2. Delayed after depolarizations	e. Abnormal AV connection (WPW)	
3. Oscillatory depolarizations at low membrane potentials following action	3. Summation and inhibition	
	C. Conduction block and reflection	

period had ended (McWilliam 1887a). Yet, it was the work of Mines and Garrey, some 30 years later, that firmly established the role of re-entry as a cause of arrhythmias. Both investigators, working independently, were inspired by the work of Mayer (1906, 1908), who used an unlikely preparation, namely ring-like structures cut from the muscular tissue of the subumbrella of the jellyfish *Scyphomedusa cassiopeia*. Mayer could induce in these rings, by a single stimulus, a contraction wave that continued to circulate: "... upon momentarily stimulating the disk in any manner, it suddenly springs into rapid, rhythmical pulsation so regular and sustained as to recall the movement of clock work" (Mayer 1906). "In one record specimen the pulsation persisted for 11 days during which it travelled 457 miles" (Mayer 1908). Mines (Fig. 5) repeated these experiments on ring-like structures from hearts of different species, and was able to induce circulating excitations by electrical stimulation. As already mentioned, his papers on circulating excitations did not contain records of electrical or mechanical activity (Mines 1913b, 1914). In these papers, written at age 27 and 28 years, Mines formulated the essential characteristics of re-entry:

- For the initiation of re-entry, an area of unidirectional block must be present [Garrey (1914) emphasized this as well, as was acknowledged by Mines (1914)]. Mines describes an experiment on an isolated auricular preparation from a dogfish heart, slit up in such a way as to form a ring. Normally, a stimulus provoked two contraction waves that ran in opposite directions and met on the far side of the ring, where they died out. However: he "repeated the stimulus at diminishing intervals and after several attempts started a wave in one direction and not in the other. The wave ran all the way around the ring and continued to circulate, going around about twice a second. After this had continued for two minutes, extra stimuli were thrown in. After several attempts the wave was stopped" (Mines 1914). Here, Mines not only describes unidirectional block, but also the principles of antitachycardia pacing.

- Mines described the relationship between conduction velocity and refractory period, as illustrated in Fig. 6, and thus can be considered as the first to formulate the "wavelength concept", where re-entrant arrhythmias are more likely to occur when conduction velocity is low and refractory period duration is short. "With increasing frequency of stimulation, each wave of excitation in the heart muscle is propagated more slowly but lasts a shorter time at any point in the muscle. The wave of excitation becomes slower and shorter" (Mines 1913b).

- Mines realized that establishing the activation sequence is not sufficient to prove re-entry: "The chief error to be guarded against is that of mistaking a series of automatic beats originating in one point of the ring and travelling round it in one direction only owing to complete block close to the point origin of the rhythm on one side of this point.... Severance of the ring will obviously prevent the possibility of circulating excitations but will not upset the course of a series of rhythmic spontaneous excitations unless by a rare chance the section should pass through the point actually initiating the spontaneous rhythm" (Mines 1914). Thus, Mines set the stage for catheter ablation.

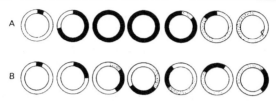

Fig. 6a,b Mines' diagram to explain that re-entry will occur when conduction is slowed and the refractory period is decreased. A stimulated impulse leaves in its wake absolute refractory tissue (*black area*) and relatively refractory tissue (*stippled area*). In both **a** and **b**, the impulse conducts in one direction only. In **a**, because of fast conduction and a long refractory period, the tissue is still absolutely refractory when the impulse has returned to its site of origin. In **b**, because of slow conduction and a short refractory period, the tissue has recovered its excitability by the time the impulse has reached the site of origin and the impulse continues to circulate. (Reproduced from Mines 1913)

Garrey excised pieces of tissue from fibrillating canine ventricles and noted that "... any piece cut from any part of ventricular tissue would cease fibrillating if small enough, e.g. if its surface area was less than four square centimetres" (Garrey 1914). Apart from showing that a minimal tissue mass is required for fibrillation, Garrey also demonstrated that fibrillation is not due to a single, rapidly firing focus, and that re-entry can occur without the involvement of an anatomical obstacle. During ventricular fibrillation there are "blocks of transitory character and shifting location" and "it is in these 'circus contractions', determined by the presence of blocks, that we see the essential phenomena of fibrillation" (Garrey 1914). Again, these striking statements were based on observations made by the naked eye.

Were it not for World War I, Lewis probably would have added to the remarkable clustering of papers on re-entry around 1914. As it was, the war postponed the course of events by a few years, because the first real "mapping" experiments were published in 1920 (Lewis et al. 1920). Initially, Lewis was not convinced about the validity of the circus movement concept, and "leaned to the view that irritable foci in the muscle underlay tachycardia and fibrillation" (Lewis et al. 1920). This view was also expressed in the first edition of the famous book *The Mechanism and Graphic Registration of the Heart Beat* (Lewis 1920). However, in this book an addendum dated May 1920 was added: "In observations recently completed and as yet unpublished, we have observed much direct evidence to show that atrial flutter consists essentially of a single circus movement ... the hypothesis which Mines and Garrey have advocated now definitely holds the field." This definitive statement was based on a study of two dogs (at the time of the addendum), in which atrial flutter was induced by faradic stimulation, or by driving the atria at increasingly faster rates. It is highly unlikely that the original, and important paper (Lewis et al. 1920), would be accepted by present-day reviewers, since none of Mines' criteria for re-entry were met: unidirectional block during the initiation of flutter was not demonstrated; the complete pathway of excitation during flutter was not mapped (no measurements were made from the left side of the caval veins (see Fig. 7); no attempt was made to terminate the arrhythmia by cutting the supposed circuitous pathway. Still, these experiments were the first to document re-entry in the intact heart and were of great influence on later studies.

It took more than 60 years before Allessie and co-workers (1977) provided insight into the nature of Garrey's "block" around which circus movement could occur. In isolated preparations of rabbit atrial muscle, rapid tachycardias were induced by a critically timed premature stimulus (often, but not always, the induction and maintenance of the tachycardia was facilitated by adding carbachol to the superfusing solution, which shortened the refractory period by 40 to 50 ms). A key observation is shown in Fig. 8, where an activation map during stable tachycardia is accompanied by intracellular recordings from seven cells located on a straight line through the zone of functional block. The transmembrane potential from cell 3 shows two responses per tachycardia cycle, where the larger voltage deflection is caused by the wavefront propagating from left to right, and the smaller response is caused by the electrotonic influence of the wave propagating from right to left, half a cycle length later. The same sequence of events occurs on the opposite side of the re-entrant circuit (tracings D, 5 and 4). Allessie and co-workers formulated the "leading circle" concept, where the re-entrant circuit is "the smallest possible pathway in

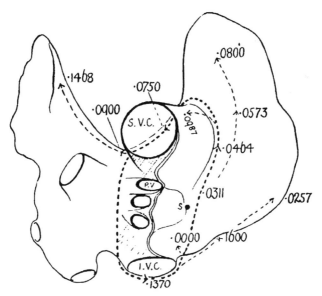

Fig. 7 Lewis' diagram of the canine's atria showing the pathway of excitation during atrial flutter. The *broken line* and *arrows* indicate the course of the excitation wave. *S* is the point originally stimulated to induce flutter; *I.V.C.,* inferior vena cava; *S.V.C.,* superior vena cava; *P.V.,* pulmonary veins. (Reproduced from Lewis et al. 1920)

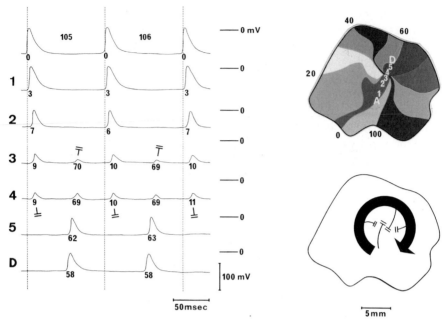

Fig. 8 Functional re-entry and tachycardia. Activation map (*right*) and action potential recordings (*left*) obtained during steady-state tachycardia in an isolated rabbit atrial preparation. Cells in the central area of the re-entrant circuit show double potentials of low amplitude (traces *3* and *4*). *Lower right panel* is the schematic representation of the "leading circle" model, where *double bars* indicate conduction block. See text for further discussion. (Reproduced from Allessie et al. 1977)

which the impulse can continue to circulate ... in which the stimulating efficacy of the circulating wavefront is just enough to excite the tissue ahead, which still is in its relative refractory phase" (Allessie et al. 1977). In other words, there is no fully excitable gap, and maintenance of the leading circle is due to repetitive centripetal wavelets that keep the core in a constant state of refractoriness.

Following the pioneering work of Allessie et al. it became apparent that in addition to excitability, the curvature of the wavefront is an important factor in maintaining functional re-entry (Fast and Kléber 1997). In fact, a curving wavefront may cease to propagate altogether when a critical curvature is reached, despite the presence of excitable tissue, and it is this phenomenon that is at the core of so-called spiral wave re-entry. A study by Athill et al. (1998) highlights the difference between "leading circle" and spiral wave re-entry: in the former the core is kept permanently refractory, in the latter the core is excitable but not excited. These authors used a preparation similar to that of Allessie et al. They also recorded transmembrane potentials from the core of the re-entrant circuit, and in contrast to the findings shown in Fig. 8, cells at the core were sometimes quiescent at almost normal levels of diastolic membrane

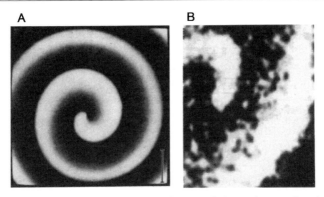

Fig. 9a, b Spiral wave re-entry in chemical Belousov-Zhabotinsky reaction (**a**) and in an isolated preparation of canine epicardial muscle (**b**). (**a**, reproduced from Müller et al. 1985, and **b** from Davidenko et al. 1992)

potential. Thus, the excitable gap was larger near the core than in the periphery of the re-entrant circuit, which is incompatible with the leading circle concept.

Spiral waves (also called vortices or rotors) were initially described for a chemical reaction in which malonic acid is reversibly oxidized by bromate in the presence of ferroin. In this process, ferroin changes in colour from red to blue and then back to red, which allows the visual observation of the reaction. This so-called Belousov-Zhabotinsky reaction is shown in Fig. 9 (Müller et al. 1985). Spiral waves have been implicated in the genesis of arrhythmias for quite some time (Winfree 1987), and can account for both tachycardias (Fig. 9, Davidenko et al. 1992) and fibrillation (Jalife et al. 2002; see sections on atrial and ventricular fibrillation).

The various forms of functional re-entry, including spiral waves, have recently been reviewed by Kléber and Rudy (2004).

4.2
Abnormal Focal Activity

Around the beginning of the twentieth century, it was generally assumed that arrhythmias were caused by rapidly firing, ectopic foci (Rothberger and Winterberg 1909; Lewis et al. 1920), and this view was still held by Scherf and Cohen in 1953. The first studies that shed some light on the mechanisms of abnormal focal activity were those of Segers (1941) and Bozler (1943). (For a more complete account, see Cranefield and Aronson 1988.) Both authors recorded monophasic action potentials in isolated preparations of cardiac muscle exposed to elevated extracellular calcium concentrations and/or adrenaline. They observed single and series of afterpotentials, and in Bozler's words: "oscillatory afterpotentials provide a simple explanation for extrasystoles and paroxysmal tachycardia" (Bozler 1943).

Today, this arrhythmogenic mechanism is called triggered activity, that is impulse generation caused by afterdepolarizations (for an extensive review, see Wit and Rosen 1991). An afterdepolarization is a second depolarization which occurs either during repolarization of a propagated action potential (referred to as an early afterdepolarization), or after repolarization has been completed (a delayed afterdepolarization). Both types of afterdepolarizations may reach threshold and initiate action potentials, either singly or in a repetitive series. Early afterdepolarizations occur when heart rate is slow and action potential duration is long (as for instance in the congenital or acquired long Q-T syndrome), and the triggered rhythm is a "bradycardia-dependent tachycardia". Delayed afterdepolarizations occur in conditions when there is cellular calcium overload, such as during digitalis intoxication, exposure to catecholamines, reperfusion after a period of ischaemia or in heart failure. The amplitude of delayed afterdepolarizations, and hence the chance of initiating a salvo of repetitive responses, is increased at short cycle lengths ("tachycardia-dependent tachycardia").

For a long time, the fact that an arrhythmia can be initiated and terminated by an appropriately timed premature beat has been considered evidence that the arrhythmia is re-entrant in nature. The fact that this can also occur in triggered arrhythmia has caused some problems in identifying the mechanism of clinical arrhythmias. There are, however, some differences in the response of re-entrant and triggered arrhythmias to programmed electrical stimulation. Thus, for a re-entrant arrhythmia initiated by a premature stimulus there is an inverse relationship between the coupling interval of the premature impulse and the interval between this impulse and the first complex of the tachycardia, and this is not the case for triggered arrhythmias (Rosen and Reder 1981). On the basis of this criterion, tachycardias were re-entrant in nature in 417 out of 425 patients (most with ischaemic heart disease) in whom tachycardias could be initiated reproducibly by premature stimuli (Brugada and Wellens 1983). This gives some indication of the importance of re-entry and triggered activity for clinical arrhythmias.

Triggered activity depends on the presence of a propagated action potential, whilst automaticity occurs de novo. The basis for automaticity is a spontaneous, gradual fall in membrane potential during diastole, referred to as diastolic, or phase-four depolarization. Automaticity is a normal property of cardiac cells in the sinus node, in some parts of the atria and AV node, and in the His–Purkinje system. Normally, the sinus node is the dominant pacemaker of the heart over a wide range of frequencies, because diastolic depolarization in latent pacemakers is inhibited by so-called overdrive suppression. When a pacemaker cell is driven at a faster rate than its intrinsic spontaneous rate, the Na^+/K^+ pump is activated, which moves more Na^+ ions out of the cell than K^+ ions into the cell, thus generating a hyperpolarizing current which counteracts spontaneous diastolic depolarization. When overdrive is stopped, a period of quiescence follows until the rate of Na^+/K^+ pumping decreases, allowing latent pacemakers to depolarize spontaneously to threshold. If the quiescent period lasts too long, syncope may occur, causing the well-known Adams–Stokes attacks (Adams 1827; Stokes 1846). A shift in the site of impulse formation to a region other than the sinus node can occur following block of sinus impulses to atria or ventricles. Although sympathetic stimulation can enhance the rate of subsidiary pacemakers, the maximum rates in normal Purkinje fibres seldom will exceed 80 beats/min.

Atrial and ventricular myocardial cells normally do not show automaticity. When, however, diastolic potentials are reduced to less than about −60 mV, spontaneous diastolic depolarization occurs, resulting in repetitive activity. Such so-called abnormal automaticity is not overdrive suppressed and usually occurs at more rapid rates than normal automaticity. Therefore, even transient sinus pauses may permit the abnormal ectopic focus to manifest itself.

5
Some Specific Arrhythmias

5.1
Atrial Fibrillation

Lüderitz (2003) speculated that William Shakespeare may have been the discoverer of atrial fibrillation because he wrote in his 1611 play *The Winter's Tale*: "I have tremor cordis on me: my heart dances; But not for joy, not joy." An older description of paroxysmal atrial fibrillation may have been in the Bible: in the book of Job it is stated that in thinking of God: "My heart trembleth and is moved out of its place."

The first electrocardiograms of atrial fibrillation were recorded by Einthoven (1906, 1908) and Hering (1908). Rothberger and Winterberg (1909) named the arrhythmia auricular fibrillation, replacing older names such as pulsus irregularis and arrhythmia perpetua. Hering (1908) and Lewis (1909a) showed f waves, corresponding to the fibrillatory activity of the atria. Rothberger and Winterberg (1909, 1915) favoured as mechanism for atrial fibrillation a single, rapidly firing focus, whereas Lewis and Schleiter (1912) suggested a multifocal mechanism. As already mentioned, Lewis later changed his mind and supported re-entry as the mechanism (Lewis 1920). The circus movement theory held the field for a long time until Scherf in 1947 revived the theory of the rapidly firing focus. He applied aconitine focally to the atrium and in this way could cause atrial fibrillation. Later experiments by Moe and Abildskov (1959) showed that, after application of aconitine to the atrial appendage, clamping off the appendage resulted in restoration of sinus rhythm, whereas the clamped-off appendage exhibited a rapid, regular tachycardia. Thus, in this case, atrial fibrillation was due to a focus that fired so rapidly that uniform excitation of the rest of the atria was no longer possible. The irregularity of the electrocardiogram was due to "fibrillar conduction" emerging from the focus. When atrial fibrillation was induced by rapid stimulation, or application of faradic shocks to the appendage, and the atrial refractory period was shortened by the administration of acetylcholine or stimulation of the vagal nerves, clamping off the appendage resulted in disappearance of fibrillation in the appendage, whereas it continued in the remainder of the atrium (Moe and Abildskov 1959). This led Moe and Abildskov (Moe and Abildskov 1959; Moe 1962) to formulate the "multiple wavelet hypothesis" in which multiple independent re-entrant wavelets, to maintain fibrillation, "... must be ... changing in position, shape, size and number with each successive excitation" (Moe 1962). A direct test of this hypothesis was performed by Allessie and colleagues (1985), who recorded simultaneous electrograms from 192 atrial sites during atrial fibrillation in isolated, Langendorff-perfused canine hearts. The activation patterns were compatible with the presence of multiple, independent wavelets. The width of the wavelets could be as small as a few millimetres, but broad wavefronts propagating uniformly over large segments of the atria were observed as well. The wavelets were short-lived, being extinguished by collision with another wavelet, by reaching the borders of the atria, or by meeting refractory tissue. The critical number of wavelets in both atria required to maintain fibrillation was estimated to be between three and six, which was much smaller than the number of wavelets in the computer model of Moe. Later studies, both in animals and humans, largely confirmed these findings (Konings et al. 1994; Schuessler et al. 1997). However, several observations revived the "focus" theory combined with fibrillary conduction, even though the "focus" was in itself re-entrant in nature. Thus, in isolated canine atria exposed to large doses of acetylcholine, which shortened the refractory period to about 95 ms, and in which fibrillation became stable, the "focus" consisted of a small, single and stable re-entrant circuit which activated the rest of the atrium by fibrillatory conduction (Schuessler et al. 1997). Also in humans, fibrillation was found to be due to a single re-entrant circuit (Cox et al. 1991; Konings et al. 1994) at "such a high rate that that it cannot be followed in a 1:1 fashion by all parts of the atria" (Konings et al. 1994). Optical mapping studies have shown that atrial fibrillation may indeed be caused by a single rotor in the left atrium with fibrillatory conduction to the right atrium (Jalife et al. 2002). Finally, the studies of Haissaguerre and co-workers (1994, 1998) showed that a rapid focus in the pulmonary vein can be responsible for atrial fibrillation in patients, and that ablation of the focus can be a successful treatment. So, in the final analysis, there is no single answer to the question: focus or re-entry?

5.2
Atrioventricular Re-entrant Tachycardia

After describing circulating excitation in ring-like preparations, Mines (1913b) wrote: "I venture to suggest that a circulating excitation of this type may be

responsible for some cases of paroxysmal tachycardia as observed clinically."
One year later, he repeated this suggestion:

> "... in the light of the new histological demonstration of Stanley Kent ...
> that an extensive muscular connection is to be found at the right hand
> margin of the heart at the junction of the right auricle and right ventricle.
> Supposing that for some reason an impulse from the auricle reached the
> main A-V bundle but failed to reach this 'right lateral' connection, it is
> possible then that the ventricle would excite the ventricular end of this
> right lateral connection, not finding it refractory as it normally would at
> such a time. The wave spreading then to the auricle might be expected to
> circulate around the path indicated." (Mines 1914)

This was written 16 years before Wolff, Parkinson and White (1930) described
the clinical syndrome that now bears their name, 18 years before Holzmann and
Scherf (1932) ascribed the abnormal ECG in these patients to pre-excitation
of the ventricles via an accessory AV bundle, 19 years before Wolferth and
Wood (1933) published diagrams showing the pathway for orthodromic and
antidromic re-entry, and 53 years before the first studies in patients employing
intraoperative mapping and programmed stimulation during cardiac catheter-
ization proved Mines' predictions to be correct (Durrer and Roos 1967; Burchell
et al. 1967; Durrer et al. 1967). It is remarkable that none of these papers quotes
Mines. As already mentioned in Sect. 3.1), Kent did not describe the usual
accessory pathway. For Mines, what was important was that a human heart
had been described with multiple connections between atria and ventricles,
thereby providing and anatomical "substrate" for re-entrant excitation.

Mines had indicated that the best therapy would be "severance of the ring",
and the first attempts to surgically interrupt the accessory pathway were published by Burchell et al. (1967). The initial attempts of antitachycardia surgery,
including those of the Amsterdam group, which began in 1969 (Wellens et al.
1974), consisted of making an incision through the atrium just above the fibrous
annulus at the site of the accessory pathway. These attempts were unsuccessful
in the long run, probably because the physicians involved were unaware of the
findings of Öhnell (1944; see Fig. 4) who had demonstrated that "the accessory
connection skirts through the epicardial fat, being well outside a well-formed
annulus fibrosis" (Becker et al. 1978). Investigators at Duke University were
the first to recognize this, and they developed a "fish hook" which was used to
scrape through the epicardial fat pad, destroying the accessory pathway (Sealy
et al. 1976).

The era of surgical treatment of the WPW syndrome did not last long: in
1983 the first study describing catheter ablation of an accessory pathway was
published (Weber and Schmitz 1983), and after the introduction of radiofrequency current ablation (Borggrefe et al. 1987), this became the therapy of
choice (Kuck et al. 1991; Jackman et al. 1991) from which by now thousands of
patients have benefited. The success rates are over 95%, and the risk for seri-

ous complications is between 2% and 4% (Cappato et al. 2000; Miler and Zipes 2000). Some 70 years elapsed between Mines' description of the mechanism of the arrhythmia and the widespread application of a safe and successful therapy.

5.3
Atrioventricular Nodal Re-entrant Tachycardia

It was again Mines (1913b) who first described re-entry in the AV node, which he called a reciprocating rhythm:

> The connexion between the auricle and ventricle is never a single muscular fibre but always a number of fibres, and although these are ordinarily in physiological continuity, yet it is conceivable that exceptionally, as after too rapid stimulation, different parts of the bundle should lose their intimate connexion.... A slight difference in the rate of recovery of two divisions of the A-V connexion might determine that an extrasystole of the ventricle, provoked by a stimulus applied to the ventricle shortly after activity of the A-V connexion, should spread up to the auricle by that part of the A-V connexion having the quicker recovery process and not by the other part. In such a case, when the auricle became excited by this impulse, the other portion of the A-V connexion would be ready to take up transmission again back to the ventricle. Provided the transmission in each direction was slow, the chamber at either end would be ready to respond (its refractory period being short) and thus the condition once established would tend to continue, unless upset by the interpolation of a premature systole.

One could not wish for a more beautiful description of AV nodal re-entry, and it is sobering to realize that upsetting AV nodal re-entry by "premature systoles" was accomplished in patients 54 years later (Coumel et al. 1967) and 58 years later in isolated rabbit heart preparations (Janse et al. 1971). In both papers, as well as in an earlier paper on AV nodal re-entry (Moe and Mendez 1966), proper credit was given to Mines.

Figure 10 shows microelectrode recordings from an isolated rabbit heart preparation in which AV nodal re-entry could be induced and terminated by premature atrial stimuli. One cell (N2) belongs to the anterograde pathway, the other (N1) to the retrograde pathway. During the tachycardia, the stimulator on the atrium was switched on and accidentally the regular stimuli captured the atrium, resulting in a "premature" atrial impulse. In panel a, the premature impulse excited N2 in the anterograde pathway prematurely (open circle in the diagram) and reached N1 in the retrograde pathway almost simultaneously with the retrograde "tachycardia" wavefront coming up from the node. The tachycardia was merely reset. In panel b, a slightly earlier atrial impulse failed to elicit an action potential in the anterograde path, but entered the retrograde path to collide with the circulating wavefront, terminating the tachycardia.

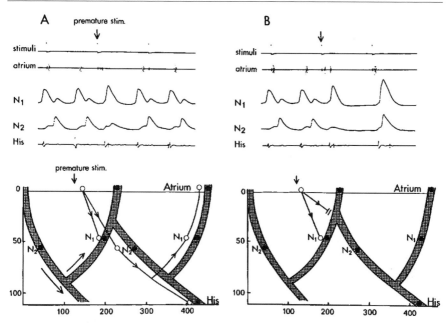

Fig. 10a,b Effects of premature atrial impulses during sustained AV nodal re-entrant tachycardia. **a** "Resetting" of tachycardia by a premature impulse 145 ms after the atrial complex. **b** Termination of tachycardia by a premature impulse 135 ms after the atrial complex. During tachycardia, stimuli were applied to the atrium at a fixed rate, most of them falling in the atrial refractory period and some of them capturing the atrium prematurely. In the diagram, the *ordinate* is a time scale indicating the moment of activation of *N1* and *N2* during a regularly paced beat from the atrium. The *abscissa* is a time scale indicating the moments of activation of *N1* and *N2* during steady-state tachycardia (*hatched band, solid circles*) and during the interpolated atrial premature beats (*open circles*). The time scale of the diagram is not the same as that of the recordings. See text for further discussion. (Reproduced from Janse et al. 1971)

This is what Mines meant by upsetting the reciprocal rhythm by the interpolation of extrasystoles. Figure 10 also illustrates what Mines wrote about the different fibres being "ordinarily in physiological continuity ..." unless "... after too rapid stimulation, different parts of the bundle should lose their intimate connection...." When following termination of the tachycardia in panel b the atrial stimulus causes normal conduction from atrium to His bundle, N1 and N2 are activated almost simultaneously. During the tachycardia, both action potentials show an electrotonic component coinciding with the "active" component of the other action potential, indicating loss of "intimate connection". The fact that during this loss of intimate connection, or, as we would call it now, longitudinal dissociation, there is still electrotonic contact between the two pathways means that the pathways are very close together. It would seem foolish to attempt, at least in this part of the node (the so-called compact node) to cut through one pathway to abolish re-entry, since one would certainly damage the other pathway as well, causing AV nodal block.

Although all authors working on AV nodal re-entry agree that the lower level of the junction between anterograde and retrograde pathway is above the level of the His bundle, controversy has existed regarding the question whether the atrium forms part of the re-entrant circuit, or whether the circuit is entirely confined to the node itself. The fact that it is possible, both by surgery and by catheter ablation, to abolish AV nodal re-entry by destroying tissue far away from the compact node whilst preserving AV conduction seems clear evidence that the atrium must be involved in the circuit (Marquez-Montez et al. 1983; Ross et al. 1985; Cox et al. 1987; Haissaguerre et al. 1989; Epstein et al. 1989). The reason why these therapeutic interventions were attempted was that both in animals and in humans the atrial inputs to the AV node during AV conduction, and the exits during ventriculo-atrial conduction are far apart, superior and inferior to the ostium of the coronary sinus (Janse 1969; Sung et al. 1981).

We therefore seem to have a very satisfactory and logical sequence of milestones on the road from understanding the mechanism of an arrhythmia to its successful therapy: Mines' description in 1913, microelectrode studies in animal preparations in the 1960s and 1970s, experimental and clinical demonstration of termination of the tachycardia by premature stimuli, demonstration of atrial input and exit sites to and from the AV node that are wide apart, successful surgery in the 1980s and finally catheter ablation with success rates that approach 99% and with complication rates well below 1% (Strickberger and Morady 2000). Clearly, this is a success story. Paradoxically, whereas in AV re-entry, understanding of the mechanism of the arrhythmia and therapy go hand in hand, in AV nodal re-entry we still are in doubt about the exact location of the re-entrant circuit. For example, in the canine heart the re-entrant circuit during ventricular and atrial echo beats is confined to the compact AV node, and regions immediately adjacent to it, and atrial tissue is not involved (Loh et al. 2003). It is of course possible that circuits involved in echo beats are not the same as those responsible for sustained tachycardias, but it is also possible that radiofrequency ablation of sites far from the compact node alter input sites and/or innervation of the compact node without actually interrupting parts of the re-entrant circuit. To quote Zipes (2000), who borrowed the words Churchill used to characterize Russia, the AV node is "a riddle wrapped in a mystery inside an enigma".

5.4
Ventricular Tachycardia, Fibrillation and Sudden Death

Although sudden death is mentioned in the Bible, the first studies linking sudden death to coronary artery disease date from the eighteenth century. In 1799, Caleb Parry quoted a letter from a good friend, Edward Jenner, the discoverer of smallpox vaccination. Jenner described an autopsy he had done on a patient with angina pectoris who had died suddenly: "... I was making a transverse section of the heart pretty near its base when my knife struck something so hard and gritty as to notch it. I well remember looking up to the ceiling, which was old and gritty, conceiving that some plaster had fallen down. But upon further scrutiny, the real cause appeared: the coronaries were becoming bony canals" (Parry 1799; see also Friedman and Friedland 1998).

Jenner believed that coronary artery obstruction might be the cause of angina pectoris as well as of the often-associated sudden death. He did not, however, mention ventricular arrhythmias. The first to do so was Erichsen (1842) who ligated a coronary artery in a dog heart and noted that this caused the action of the ventricles to cease, with a "slight tremulous motion alone continuing". Subsequent studies confirmed and expanded these findings (Begold 1867; Porter 1894; Lewis 1909b), and Cohnheim and Schulthess-Rechberg (1881) showed that ventricular fibrillation occurred even more often after reperfusion following a brief ischaemic episode than during the ischaemic period itself. The clinical importance of these findings was not at all recognized, except by McWilliam, who wrote "... sudden syncope from plugging or obstructing some portion of the coronary system (in patients) is very probably determined or ensured by the occurrence of fibrillar contractions in the ventricles. The cardiac pump is thrown out of gear, and the last of its vital energy is dissipated in a violent and prolonged turmoil of fruitless activity in the ventricular walls" (McWilliam 1889).

McWilliam's ideas were largely ignored for many decades. He expressed his disappointment in 1923: "It may be permissible to recall that in the pages of this journal 34 years ago I brought forward a new view as to the causation of sudden death by a previously unrecognized form of failure of the heart's action in man (e.g. ventricular fibrillation)—a view fundamentally different from those entertained up to that time. Little attention was given to the new view for many years" [MacWilliam 1923 (his name in 1923 was spelled MacWilliam rather than McWilliam)]. Little attention was given to his views for many more years. The reason for that was probably that the occurrence of ventricular fibrillation is difficult to document in man, and because ventricular fibrillation could not be treated, a view already expressed by Lewis in 1915. It was not until the 1960s that clinicians began to recognize how often ventricular fibrillation occurs in man. In 1961 Julian noted, "Cardiac arrest due to ventricular fibrillation or asystole is a common mode of death in acute myocardial ischaemia and infarction" (Julian 1961). His recommendations to train all medical, nursing and auxiliary staff in the techniques of closed-chest cardiac massage and mouth-to-mouth breathing, and to monitor the cardiac rhythm, marked the beginning of the coronary care unit. Some milestones are the introduction of the d.c. defibrillator (Lown et al. 1962), and the advent of mobile coronary care units recording ECGs from individuals suffering from cardiac arrest outside the hospital and providing defibrillation (Pantridge and Geddes 1967; Cobb et al. 1980).

In the setting of myocardial ischaemia and infarction, ventricular tachycardia and fibrillation are the causes of cardiac arrest. In heart failure, sudden death is reportedly caused in about 50% of patients by ventricular tachyarrhythmias, in the other half by bradyarrhythmias, asystole or electromechanical dissociation (Luu et al. 1989; Stevenson et al. 1993).

The risk of sudden death in the general population aged 35 years and older is in the order of 1–2 per 1,000 per year. In the presence of coronary artery disease, and other risk factors, the risk increases to 10%–25% per year. In the adolescent and young adult population, the risk is in the order of 0.001% per year, and familial diseases such as hypertrophic cardiomyopathy, the congenital long Q-T syndrome, the Brugada syndrome and right ventricular dysplasia, play a dominant role (Myerburg and Spooner 2001).

McWilliam (1887a) was the first to suggest that ventricular fibrillation is caused by re-entry, a view also held by Mines and Garrey. Mines (1914) described what we now call the vulnerable period. He induced ventricular fibrillation by single induction shocks, applied at various times during the cardiac cycle. "The point of interest is that the stimulus employed would never cause fibrillation unless it was set at a critical instant" (Mines 1914). He showed that a stimulus falling in the refractory period had no effect, "a stimulus coming a little later set up fibrillation" and a stimulus applied "later than the critical instant for the production of fibrillation merely induces an extrasystole" (Mines 1914). As described in detail by Acierno (1994), in the 1920s a considerable number of people were accidentally electrocuted because more and more electrical devices were installed in households. This eventually prompted electricity companies such as Consolidated Edison to provide grants to university departments to investigate the effects of electrical currents on the heart. This led to the introduction of defibrillation by countershock and external cardiac massage (Hooker et al. 1933; Kouwenhoven et al. 1960) and the rediscovery of the vulnerable period by Wiggers and Wegria (1940).

Hoffa and Ludwig (1850) were the first to show that electrical currents can cause fibrillation. This was later confirmed by Prevost and Battelli (1899), who also showed that similar shocks could restore sinus rhythm. It is perhaps somewhat surprising that it took more than half a century before defibrillation by electrical countershock became common clinical practice. Lown, who in the early 1960s introduced d.c. defibrillation and cardioversion for atrial fibrillation (Lown et al. 1962; Lown 1967), wrote recently: "Ignorance of the history of cardiovascular physiology caused me to waste enormous time in attempting to understand a phenomenon long familiar to physiologists" (Lown 2002). He refers to the vulnerable period, and gives full credit to Mines.

As was the case for atrial fibrillation, Moe's multiple wavelet hypothesis also was thought to be valid for ventricular fibrillation, but in recent years, the notion that spiral waves, or rather three-dimensional scroll waves, are responsible for fibrillation gained ground (Winfree 1987; Davidenko 1993; Gray et al. 1995; Jalife et al. 2003). In the setting of acute, regional ischaemia, activation patterns compatible with the multiple wavelet hypothesis have been described during ventricular fibrillation, although non-re-entrant mechanisms, especially the premature beats that initiated re-entry, were demonstrated as well (Janse et al. 1980; Pogwizd and Corr 1987). In human hearts with a healed infarct, monomorphic tachycardias are due to re-entry within the complex network of surviving myocardial fibres within the infarct (De Bakker et al. 1988). To our knowledge, spiral waves or scroll waves have not yet been described in hearts with acute regional myocardial ischaemia, or with a healed infarct.

6
Conclusions

Much has been written of the need to understand history if we are to chart the future. Whether we think of recent world events, or on a minor scale, the diagnosis and treatment of cardiac arrhythmias, we are consistently reminded of the need to learn from the past in coping with the present and preparing for the future. We have reviewed the delays that have occurred in arriving at appropriate diagnosis and therapy by failure to appreciate the work of Mines regarding re-entry (which pushed back the correct conceptualization of WPW syndrome by half a century) as well as similar delays in the appreciation of the potential benefits of electrical defibrillation techniques. We are now in an ever-more reductionist era of research aimed at the appreciation of the molecular root causes of arrhythmias. Yet we must not forget that the present era, as with each preceding one, will likely be followed by even more elemental explorations of the function and structure of the building blocks of cardiac cells in health and disease—charting a new, exciting and uncertain future. And if we can simply remember the lesson that history has given us again and again—that if we look to the past we can chart the future—then it is likely that the fruits born of these new approaches to understanding the workings of the heart will be brought to humanity far more efficiently and more rapidly than if we ignore what has gone before.

References

Acierno LJ (1994) The history of cardiology. The Partenon Publishing Group, London

Adams R (1826) Cases of diseases of the heart, accompanied with pathological observations. Dublin Hosp Rep 4:353–453

Ader C (1897) Sur un nouvel appareil enrégistreur pour cables sous-marins. CR Acad Sci 124:1440–1442

Allessie MA, Bonke FIM, Schopman FJG (1977) Circus movement in rabbit atrial muscle as a mechanism of tachycardia. III. The "leading circle" concept: a new model of circus movement in cardiac tissue without the involvement of an anatomical obstacle. Circ Res 41:9–18

Anderson RH, Becker AE (1981) Stanley Kent and accessory atrioventricular connections. J Thorac Cardiovasc Surg 81:649–658

Athill CA, Ikeda T, Kim Y-H, et al (1998) Transmembrane potential properties at the core of functional reentrant wavefronts in isolated canine right atria. Circulation 98:1556–1567

Becker AE, Anderson RH, Durrer D, et al (1978) The anatomical substrate of Wolff–Parkinson–White syndrome. Circulation 57:870–879

Begold A (1867) Von den Veränderungen des Herzschlages nach Verschliessung der Coronararterien. Unters Physiol Lab Würzburg 2:256–287

Ben-Haim SA, Osadchy D, Schuster I, et al (1996) Nonfluoroscopic, in vivo navigation and mapping technology. Nat Med 2:1393–1395

Borggrefe M, Budde T, Podczeck A, et al (1987) High frequency alternating current ablation of an accessory pathway in humans. J Am Coll Cardiol 10:576–582

Bozler E (1943) The initiation of impulses in cardiac muscle. Am J Physiol 138:273–282

Brooks C, Hoffmann BF, Suckling EE, Orias O (1955) Excitability of the heart. Grune and Stratton, New York

Brugada P, Wellens HJJ (1983) The role of triggered activity in clinical arrhythmias. In: Rosenbaum M, Elizari M (eds) Frontiers of electrocardiography. Martinus Nijhoff, The Hague, pp 195–216

Burchell HB, Frye RB, Anderson M, et al (1967) Atrioventricular and ventriculo-atrial excitation in Wolff–Parkinson–White syndrome (type B). Temporary ablation at surgery. Circulation 36:663–672

Burdon-Sanderson JS, Page FJM (1879) On the time relations of the excitatory process in the ventricle of the heart of the frog. J Physiol (Lond) 2:384–435

Burdon-Sanderson JS, Page FJM (1883) On the electrical phenomena of the excitatory process in the heart of the frog and of the tortoise, as investigated photographically. J Physiol (Lond) 4:327–338

Cappato R, Schlüter M, Kuck KH (2000) Catheter ablation of atrioventricular reentry. In: Zipes DP, Jalife JJ (eds) Cardiac electrophysiology: from cell to bedside, third edn. WB Saunders, Philadelphia, pp 1035–1049

Cobb FR, Blumenschein SD, Sealy WC, et al (1968) Successful interruption of the bundle of Kent in a patient with the Wolff–Parkinson–White syndrome. Circulation 38:1018–1029

Cobb LA, Werner JA, Trobaugh GB (1980) Sudden cardiac death. I. A decade's experience with out-of-hospital resuscitation. Mod Concepts Cardiovasc Dis 49:31–36

Cohnheim J, Schulthess-Rechberg AV (1881) Über die Folgen des Kranzarterienverschliessung fuer das Herz. Virchows Arch 85:503–537

Cole KS (1949) Dynamic electrical characteristics of the squid axon membrane. Arch Sci Physiol (Paris) 3:253–258

Coraboeuf E, Weidmann S (1949) Potentiel de repos et potentiel d'action du muscle cardiaque mesurés à l'aide d 'électrodes intracellulaires. CR Soc Seances Soc Biol Fil 143:1329–1331

Coumel P, Cabrol C, Fabiato A, et al (1967) Tachycardie permanente par rythme réciproque. Arch Mal Coeur Vaiss 60:1830–1864

Cox JL, Holman WL, Cain ME (1987) Cryosurgical treatment of atrioventricular node reentrant tachycardia. Circulation 76:1329–1336

Cox JL, Canavan TE, Schuessler RB, et al (1991) The surgical treatment of atrial fibrillation. II. Intraoperative electrophysiologic mapping and description of the electrophysiological basis of atrial flutter and fibrillation. J Thorac Cardiovasc Surg 101:406–426

Cranefield PF, Aronson RS (1988) Cardiac arrhythmias: the role of triggered activity and other mechanisms. Futura Publishing Company, Mount Kisco

Davidenko JM (1993) Spiral wave activity: a possible common mechanism for polymorphic and monomorphic ventricular tachycardias. J Cardiovasc Electrophysiol 4:730–746

Davidenko JM, Pertsov AV, Salomonsz R, et al (1992) Stationary and drifting spiral waves of excitation in isolated cardiac muscle. Nature 355:349–351

De Bakker JMT, Janse MJ, van Capelle FJL, et al (1983) Endocardial mapping by simultaneous recording of endocardial electrograms during cardiac surgery for ventricular aneurysm. J Am Coll Cardiol 2:947–953

De Bakker JMT, Van Capelle FJL, Janse MJ, et al (1988) Reentry as a cause of ventricular tachycardia in patients with chronic ischemic heart disease: electrophysiologic and anatomic correlation. Circulation 77:589–606

de Senac JB (1749) Traité de la structure du coeur, de son action et de ses maladies, vol. 2. Vincent, Paris

Deutschen pathologischen Gesellschaft (1910) Bericht über die Verhandlungen der XIV Tagung der Deutschen pathologischen Gesellschaft in Erlangen vom 4–6 April 1910. Zbl allg Path path Anat 21:433–496

Dillon S, Morad M (1981) A new laser scanning system for measuring action potential propagation in the heart. Science 214:453–456

Durrer D, Roos JR (1967) Epicardial excitation of the ventricles in a patient with a Wolff–Parkinson–White syndrome (type B). Circulation 35:15–21

Durrer D, van der Tweel LH (1954a) Spread of activation in the left ventricular wall of the dog. II. Activation conditions at the epicardial surface. Am Heart J 47:192–203

Durrer D, van der Tweel LH, Blickman JP (1954b) Spread of activation in the left ventricular wall of the dog. III. Transmural and intramural analysis. Am Heart J 48:13–35

Durrer D, Formijne P, van Dam RTh, et al (1961) The electrocardiogram in normal and some abnormal conditions. Am Heart J 61:303–314

Durrer D, Schoo L, Schuilenburg RM, et al (1967) The role of premature beats in the initiation and termination of supraventricular tachycardia in the Wolff–Parkinson–White Syndrome. Circulation 36:644–662

Einthoven W (1901) Sur un nouveau galvanomètre. Arch néerl des Sciences Exact Nat, série 2, 6:625–633

Einthoven W (1902) [Galvanometric registration of the human electrocardiogram]. In: Rosenstein SS (ed) Herinneringsbundel. Eduard Ijdo, Leiden, pp 101–106

Einthoven W (1903) Die galvanometrische Registrierung des menschlichen Elektrokardio-gramms, zugleichs eine Beurteilung der Anwendung des Capillär-Elektrometers in der Physiologie. Pflugers Arch Gesamte Physiol 99:472–480

Einthoven W (1906) Le télecardiogramme. Arch Int Physiol 4:132–164

Einthoven W (1908) Weiteres über das Elektrokardiogramm. Nach gemeinschaftlich mit Dr. B. Vaandrager angestellten Versuchen mitgeteilt. Pflugers Arch Gesamte Physiol 122:517–584

Epstein LM, Scheinman MM, Langberg JJ, et al (1989) Percutaneous catheter modification of atrioventricular node reentrant tachycardia. Circulation 80:757–768

Erichsen JE (1842) On the influence of the coronary circulation on the action of the heart. Lond Med Gaz 2:561–565

Erlanger J (1964) A physiologist reminisces. Annu Rev Physiol 26:1–14

Fast VG, Kléber AG (1997) Role of wavefront curvature in propagation of cardiac impulse. Cardiovasc Res 33:258–271

Franz MR (1983) Long-term recording of monophasic action potentials from human endo-cardium. Am J Cardiol 51:1629–1634

Friedman M, Friedland GW (1998) Medicine's ten greatest discoveries. Yale University Press, New Haven, pp 76–77

Garrey WE (1914) The nature of fibrillar contractions of the heart. Its relation to tissue mass and form. Am J Physiol 33:397–414

Giraud G, Latour H, Puech P (1960) L'activité du noeud de Tawara et du faisceau de His en electrocardiographie chez l'homme. Arch Mal Coeur Vaiss 33:757–776

Goldberger E (1942) A single indifferent, electrocardiographic electrode of zero potential and a technique of obtaining augmented, unipolar, extremity leads. Am Heart J 23:483–492

Gray RA, Jalife J, Panfilov A, et al (1995) Nonstationary vortex like reentrant activity as a mechanism of polymorphic ventricular tachycardia in the isolated rabbit heart. Circulation 91:2454–2469

Haissaguerre M, Warin J, Lemetayer JJ, et al (1989) Closed-chest ablation of retrograde conduction in patients with atrioventricular nodal reentrant tachycardia. N Engl J Med 320:426–433

Haissaguerre M, Marcus FI, Fischer B, et al (1994) Radiofrequency ablation in unusual mechanisms of atrial fibrillation. A report of three cases. J Cardiovasc Electrophysiol 5:743–751

Haissaguerre M, Jaïs P, Shah DC, et al (1998) Spontaneous initiation of atrial fibrillation by ectopic beats originating in the pulmonary veins. N Engl J Med 339:659–666

Harken AH, Josephson ME, Horowitz LN (1979) Surgical endocardial resection for the treatment of malignant ventricular tachycardia. Ann Surg 190:456–465

Heidenhain R (1872) Ueber arrhytmische Herztätigkeit. Pflugers Arch Gesamte Physiol 5:143–153

Hering HE (1908) Das Elektrokardiogramm des Pulsus irregularis perpetuus. Dtsch Arch Klin Med 94:205–208

His W Jr (1893) Die Tätigkeit des embryonalen Herzens und deren Bedeutung für die Lehre von der Herzbewegung beim Erwachsenen. Arch Med Klin Leipzig 14–49

Hoffa M, Ludwig C (1850) Einige neue Versuche über Herzbewegung. Z Ration Med 9:107–144

Hoffman BF, Cranefield PF (1960) The electrophysiology of the heart. McGraw-Hill, New York

Hoffman BF, Rosen MR (1981) Cellular mechanisms for cardiac arrhythmias. Circ Res 49:1–15

Hofman BF, Cranefield PF, Lepeschkin E, et al (1959) Comparison of cardiac monophasic action potentials recorded by intracellular and suction electrodes. Am J Physiol 196:1297–1306

Holter NJ (1957) Radioelectrocardiography: a new technique for cardiovascular studies. Ann NY Acad Sci 65:913–923

Holzmann M, Scherf D (1932) Über Elektrokardiogrammen mit verkürzter Vorhof-Kammer-Distanz und positiven P-Zacken. Z Klin Med 121:404–423

Hooker DR, Kouwenhoven WB, Langworthy OR (1933) The effect of alternating current on the heart. Am J Physiol 103:444–454

Jackman WM, Wang X, Friday KJ, et al (1991) Catheter ablation of accessory atrioventricular pathways (Wolff–Parkinson–White syndrome) by radiofrequency current. N Engl J Med 334:1605–1611

Jalife J, Berenfeld O, Mansour M (2002) Mother rotors and fibrillatory conduction: a mechanism of atrial fibrillation. Cardiovasc Res 54:204–216

Jalife J, Anumonwo JMB, Berenfeld O (2003) Toward an understanding of the molecular mechanism of ventricular fibrillation. J Interv Card Electrophysiol 9:119–129

James TN (1963) Connecting pathways between the sinus node and A-V node and between the right and left atrium in the human heart. Am Heart J 66:498–508

Janse MJ (1969) Influence of the direction of the atrial wavefront on A-V nodal transmission in isolated hearts of rabbits. Circ Res 25:439–449

Janse MJ (1993) Some historical notes on the mapping of arrhythmias. In: Shenassa M, Borggrefe M, Breithardt G (eds) Cardiac mapping. Futura Publishing, Mount Kisco, pp 3–10

Janse MJ, Anderson RH (1974) Specialized internodal atrial pathways—fact or fiction? Eur J Cardiol 2:117–136

Janse MJ, van Capelle FJL, Freud GE, et al (1971) Circus movement within the A-V node as a basis for supraventricular tachicardia as shown by multiple microelectrode recording in the isolated rabbit heart. Circ Res 28:403–414

Janse MJ, van Capelle FJL, Morsink H, et al (1980) Flow of "injury" current and patterns of activation during early ventricular arrhythmias in acute regional myocardial ischemia in isolated porcine and canine hearts. Evidence for two different arrhythmogenic mechanisms. Circ Res 47:151–165

Josephson ME (2002) Clinical cardiac electrophysiology: techniques and interpretation, third edn. Lippincott, Williams and Wilkins, Philadelphia

Josephson ME, Wellens HJJ (eds) (1984) Tachycardias: mechanism, diagnosis, therapy. Lea and Febiger, Philadelphia

Josephson ME, Horowitz LN, Farshidi A (1978) Continuous local electrical activity: a mechanism of recurrent ventricular tachycardia. Circulation 57:659–665

Julian DG (1961) Treatment of cardiac arrest in acute myocardial ischaemia and infarction. Lancet 14:840–844

Katz LN, Pick A (1956) Clinical electrocardiography. Part I. The arrhythmias. Lea and Febiger, Philadelphia

Keith A, Flack M (1907) The form and nature of the muscular connections between the primary divisions of the vertebrate heart. J Anat Physiol 41:172–189

Kent AFS (1913) Observations on the auriculo-ventricular junction of the mammalian heart. Q J Exp Physiol 7:193–195

Kléber AG, Rudy Y (2004) Basic mechanisms of cardiac impulse propagation and associated arrhythmias. Physiol Rev 84:431–488

Köllicker A, Müller H (1856) Nachweis der negativen Schwankung des Muskelstromes am naturlich sich contrahierenden Muskel. Verh Phys-Med Ges Würzburg 6:528–533

Konings KTS, Kirchhof CJHJ, Smeets JRLM, et al (1994) High-density mapping of electrically induced atrial fibrillation in humans. Circulation 89:1665–1680

Kouwenhoven WB, Jude JR, Knickerbocker GG (1960) Closed-chest cardiac massage. JAMA 173:1064–1067

Kraus F, Nicolai GF (1910) Das Elektrokardiogram des gesunden und kranken Menschen. Verlag von Veit and Co., Leipzig

Krikler DM (1987a) Historical aspects of electrocardiology. Cardiol Clin 5:349–355

Krikler DM (1987b) The search for Samojloff: a Russian physiologist in time of change. Br Med J 295:1624–1627

Kuck KH, Schlüter M, Geiger M, et al (1991) Radiofrequency current catheter ablation therapy for accessory atrioventricular pathways. Lancet 337:1578–1581

Lepeschkin E (1951) Modern electrocardiography. Williams and Wilkins, Baltimore

Lewis T (1909a) Auricular fibrillation: a common clinical condition. Br Med J 2:1528–1548

Lewis T (1909b) The experimental production of paroxysmal tachycardia and the effect of ligation of the coronary arteries. Heart 1:98–137

Lewis T (1911) The mechanism of the heart beat. Shaw and Sons, London

Lewis T (1915) Lectures on the heart. Paul B Hoeber, New York

Lewis T (1920) The mechanism and graphic registration of the heart beat, first edn. Shaw and Sons, London

Lewis T (1925) The mechanism and graphic registration of the heart beat, third edn. Shaw and Sons, London

Lewis T, Schleiter HG (1912) The relation of regular tachycardias of auricular origin to auricular fibrillation. Heart 3:173–193

Lewis T, Feil S, Stroud WD (1920) Observations upon flutter and fibrillation. II. The nature of auricular flutter. Heart 7:191–346

Ling G, Gerard RW (1949) The normal membrane potential. J Cell Comp Physiol 34:383–396

Loh P, Ho SY, Kawara T, et al (2003) Reentrant circuits in the canine atrioventricular node during atrial and ventricular echoes. Electrophysiological and histological correlation. Circulation 108:231–238

Lown B (1967) Electrical reversion of cardiac arrhythmias. Br Heart J 29:469–489

Lown B (2002) The growth of ideas. Defibrillation and cardioversion. Cardiovasc Res 55:220–224

Lown B, Amarasingham R, Neuman J (1962) New method for terminating cardiac arrhythmias. Use of synchronized capacitor discharge. JAMA 182:548–555

Lüderitz B (1995) History of the disorders of cardiac rhythm, second edn. Futura Publishing Company, Armonk

Lüderitz B (2003) The story of atrial fibrillation. In: Capucci A (ed) Atrial fibrillation. Centro Editoriale Pubblicitario Italiano, Rome, pp 101–106

Luu M, Stevenson WG, Stevenson LW, et al (1989) Diverse mechanisms of unexpected cardiac arrest in advanced heart failure. Circulation 80:1675–1680

MacFarlane PW, Lawrie TDV (eds) (1989) Comprehensive electrocardiology. Theory and practice in health and disease, 3 volumes. Pergamon Press, New York

MacWilliam JA (1923) Some applications of physiology to medicine. II. Ventricular fibrillation and sudden death. Br Med J 18:7–43

Marey EJ (1876) Des variations électriques des muscles et du coeur en particulier étudies au moyen de l'électromètre de M.Lipmann. CR Acad Sci 82:975–977

Marmont G (1949) Studies on the axon membrane. I. A new method. J Cell Comp Physiol 34:351–384

Marquez-Montes J, Rufilanchas JJ, Esteve JJ, et al (1983) Paroxysmal nodal reentrant tachycardia. Surgical cure with preservation of atrioventricular conduction. Chest 83:690–693

Mayer AG (1906) Rhythmical pulsation in scyphomedusae. Publication 47 of the Carnegie Institution, Washington, pp 1–62

Mayer AG (1908) Rhythmical pulsation in scyphomedusae II. Papers from the Marine Biological Laboratory at Tortugas; Carnegie Institution, Washington, pp 115–131

McWilliam JA (1887a) Fibrillar contraction of the heart. J Physiol (Lond) 8:296–310

McWilliam JA (1887b) On electrical stimulation of the heart. Trans Int Med Congress, 9th session, Washington, vol III, p 253

McWilliam JA (1889) Cardiac failure and sudden death. Br Med J 1:6–8

Miles WM, Zipes DP (2000) Atrioventricular reentry and variants: mechanisms, clinical features and management. In: Zipes, Jalife J (eds) Cardiac electrophysiology: from cell to bedside, third edn. WB Saunders, Philadelphia, pp 488–504

Mines GR (1913a) On functional analysis by the action of electrolytes. J Physiol (Lond) 46:188–235

Mines GR (1913b) On dynamic equilibrium of the heart. J Physiol (Lond) 46:349–382

Mines GR (1914) On circulating excitations in heart muscles and their possible relation to tachycardia and fibrillation. Trans R Soc Can 4:43–52

Moe GK (1962) On the multiple wavelet hypothesis of atrial fibrillation. Arch Int Pharmacodyn Ther 140:183–188

Moe GK, Abildskov JA (1959) Atrial fibrillation as a self-sustained arrhythmia independent of focal discharge. Am Heart J 58:59–70

Moe GK, Mendez C (1966) The physiological basis of reciprocal rhythm. Prog Cardiovasc Dis 8:461–482

Morad M, Dillon S, Weiss J (1986) An acousto-optically steered laser scanning system for measurement of action potential spread in intact heart. Soc Gen Physiol Ser 40:211–226

Müller SC, Plessr T, Hess B (1985) The structure of the core of the spiral wave in the Belousov-Zhabotinsky reaction. Science 230:661–663

Music D, Rakovec P, Jagodic A, et al (1984) The first description of syncopal attacks in heart block. Pacing Clin Electrophysiol 7:301–303

Myerburg RJ, Spooner PM (2001) Opportunities for sudden death prevention: directions for new clinical and basic research. Cardiovasc Res 50:177–185

Nahum LH, Mauro A, Chernoff HM, et al (1951) Instantaneous equipotential distribution on surface of the human body for various instants in the cardiac cycle. J Appl Physiol 3:454–464

Neher E, Sakmann B (1976) Single-channel currents recorded from membrane of denervated frog muscle fibres. Nature 260:779–802

Noble D (1975) The initiation of the heart beat. Clarendon Press, Oxford

Noble D (1984) The surprising heart: a review of recent progress in cardiac electrophysiology. J Physiol (Lond) 353:1–50

Öhnell RE (1944) Pre-excitation, cardiac abnormality: patho-physiological, patho-anatomical and clinical studies of excitatory spread bearing upon the problem of WPW (Wolff–Parkinson–White) electrocardiogram and paroxysmal tachycardia. Acta Med Scand Suppl 52:1–167

Olsson SB (1971) Monophasic action potentials of right heart (thesis). Elanders Boktryckeri, Göteborg

Pantridge JF, Geddes JS (1967) A mobile intensive-care unit in the management of myocardial infarction. Lancet 2:271–273

Parry CH (1799) An inquiry into the symptoms and causes of the syncope anginosa commonly called angina pectoris. R Crutwell, Bath

Pastelin G, Mendez R, Moe GK (1978) Participation of atrial specialized conduction pathways in atrial flutter. Circ Res 42:386–393

Peters NS, Jackman W, Schilling RJ, et al (1997) Human left ventricular endocardial activation mapping using a novel non-contact catheter. Circulation 95:1658–1660

Pick A, Langendorf R (1979) Interpretation of complex arrhythmias. Lea and Febiger, Philadelphia

Pogwizd SM, Corr PB (1987) Reentrant and nonreentrant mechanisms contribute to arrhythmogenesis during early myocardial ischemia: results using three-dimensional mapping. Circ Res 61:352–371

Porter WT (1894) On the results of ligation of the coronary arteries. J Physiol (Lond) 15:121–138

Prevost JL, Battelli F (1899) Sur quelques effets des décharges électriques sur le coeur des Mammifères. CR Seances Acad Sci 129:1267–1268

Pruitt RD (1976) Doublets, dipoles, and the negativity hypothesis: an historical note on W.H. Craib and his relationship with F.N. Wilson and Thomas Lewis. Johns Hopkins Med J 138:279–288

Purkinje JE (1845) Mikroskopisch-neurologische Beobachtungen. Arch Anat Physiol Med II/III:281–295

Rosen MR, Reder RF (1981) Does triggered activity have a role in the genesis of clinical arrhythmias? Ann Intern Med 94:794–801

Rosenbaum DS, Jalife J (2001) Optical mapping of cardiac excitation and arrhythmias. Futura Press, Armonk

Ross DL, Johnson DC, Denniss AR, et al (1985) Curative surgery for atrioventricular junctional ("AV nodal") reentrant tachycardia. J Am Coll Cardiol 6:1383–1392

Rothberger CJ, Winterberg H (1909) Vorhofflimmern und Arrhythmia perpetua. Wiener Klin Wochenschr 22:839–844

Rothberger CJ, Winterberg H (1915) Über Vorhofflimmern und Vorhofflattern. Pflugers Arch Gesamte Physiol 160:42–90

Rytand DA (1966) The circus movement (entrapped circuit wave) hypothesis and atrial flutter. Ann Intern Med 65:125–159

Samojloff A (1909) Elektrokardiogramme. Verlag von Gustav Fischer, Jena

Schamroth L (1973) The disorders of cardiac rhythm. Blackwell Scientific Publications, Oxford

Scherf D (1947) Studies on auricular tachycardia caused by aconitine administration. Proc Soc Exp Biol Med 64:233–239

Scherf D, Cohen J (1964) The atrioventricular node and selected arrhythmias. Grune and Stratton, New York

Scherf D, Schott A (1953) Extrasystoles and allied arrhythmias. Heinemann, London

Scherlag BJ, Kosowsky BD, Damato AN (1976) Technique for ventricular pacing from the His bundle of the intact heart. J Appl Physiol 22:584–587

Scherlag BJ, Lau SH, Helfant RH, et al (1979) Catheter technique for recording His bundle activity in man. Circulation 39:13–18

Schuessler RB, Grayson TM, Bromberg BI, et al (1997) Cholinergically mediated tachyarrhythmias induced by a single extrastimulus in the isolated canine right atrium. Circ Res 71:1254–1267

Schütz E (1936) Elektrophysiologie des Herzens bei einphasischer Ableitung. Ergebn Physiol 38:493–620

Sealy WC, Gallagher JJ, Wallace AG (1976) The surgical treatment of Wolff–Parkinson–White syndrome: evolution of improved methods for identification and interruption of the Kent bundle. Ann Thorac Surg 22:443–457

Segers M (1941) Le rôle des potentiels tardifs du coeur. Ac Roy Med Belg, Mémoires, Série I:1–30

Shenassa M, Borggrefe M, Briethardt G (eds) (2003) Cardiac Mapping. Blackwell Publishing/Futura Division, Elmsford

Snellen HA (1984) History of cardiology. Donkers Academic Publications, Rotterdam

Snellen HA (1995) Willem Einthoven (1860–1927) Father of electrocardiography. Life and work, ancestors and contemporaries. Kluwer Academic Publishers, Dordrecht

Spach MS, Miller WT III, Barr RC, et al (1980) Electrophysiology of the internodal pathways: determining the difference between anisotropic cardiac muscle and a specialized tract system. In: Little RD (ed) Physiology of atrial pacemakers and conductive tissues. Futura Publishing Company, Mount Kisco, pp 367–380

Spang K (1957) Rhythmusstörungen des Herzens. Georg Thieme Verlag, Stuttgart

Stevenson WG, Stevenson LW, Middlekauf HR, et al (1993) Sudden death prevention in patients with advanced ventricular dysfunction. Circulation 88:2953–2961

Stokes W (1846) Observations on some cases of permanently slow pulse. Dublin Q J Med Sci 2:73–85

Strickberger AS, Morady F (2000) Catheter ablation of atrioventricular nodal reentrant tachycardia. In: Zipes DP, Jalife J (eds) Cardiac electrophysiology: from cell to bedside, third edn. WB Saunders, Philadelphia, pp 1028–1035

Sung RJ, Waxman HL, Saksena S, et al (1981) Sequence of retrograde atrial activation in patients with dual atrioventricular nodal pathways. Circulation 64:1059–1067

Taccardi B, Arisi G, Marchi E, et al (1987) A new intracavitary probe for detecting the site of origin of ectopic ventricular beats during one cardiac cycle. Circulation 75:272–281

Tawara S (1906) Das Reizleitungssystem des Säugetierherzens. Gustav Fischer, Jena

Thorel C (1908) Vorläufige Mitteilung über eine besondere Muskelverbindung zwischen dem Cava superior und die Hisschen Bundel. Münch med Wschr 56:2159–2164

Waller AD (1887) A demonstration on man of electromotive changes accompanying the heart's beat. J Physiol 8:229–234

Waller AD (1889) On the electromotive changes connected with the beat of the mammalian heart, and of the human heart in particular. Philos Trans R Soc Lond B Biol Sci 180:169–194

Weber H, Schmitz L (1983) Catheter technique for closed-chest ablation of an accessory pathway. N Engl J Med 308:653–654

Weidmann S (1956) Elektrophysiologie des Herzmuskelfaser. Huber, Bern

Weidmann S (1971) The microelectrode and the heart. 1950–1970. In: Kao FF, Koizumi K, Vassalle M (eds) Research in physiology. A liber memorialis in honor of Prof. Chandler McCuskey Brooks. Aulo Gaggi Publishers, Bologna, pp 3–25

Wellens HJJ (1971) Electrical stimulation of the heart in the study and treatment of tachycardias. Stenfert Kroese, Leiden

Wellens HJJ, Janse MJ, van Dam RTh, et al (1974) Epicardial mapping and surgical treatment in Wolff–Parkinson–White syndrome. Am Heart J 88:69–78

Wenckebach KF (1903) Die Arhythmie. Engelmann, Leipzig

Wiggers CJ, Wegria R (1940) Ventricular fibrillation due to a single, localised induction and condenser shocks applied during the vulnerable phase of ventricular systole. Am J Physiol 128:500–505

Wilson FN, Johnston FD (1938) The vectorcardiogram. Am Heart J 16:14–28

Wilson FN, Johnston FD, MacLeod AG, et al (1933a) Electrocardiograms that represent the potential variations of a single electrode. Am Heart J 9:447–458

Wilson FN, MacLeod AG, Barker PS (1933b) The distribution of the action currents produced by heart muscle and other excitable tissues immersed in extensive conducting media. J Gen Physiol 16:423–456

Winfree AT (1987) When time breaks down. The three-dimensional dynamics of electrochemical waves and cardiac arrhythmias. Princeton University Press, Princeton

Wit AL, Rosen MR (1991) After depolarizations and triggered activity: distinction from automaticity as an arrhythmogenic mechanism. In: Fozzard HA, Haber E, Jennings RB, Katz AM, Morgan HE (eds) The heart and cardiovascular system. Scientific Foundations, second edn. Raven Press, New York, pp 2113–2164

Wolferth CC, Wood FC (1933) The mechanism of production of short PR intervals and prolonged QRS complexes in patients with presumably undamaged hearts. Hypothesis of an accessory pathway of atrioventricular conduction (bundle of Kent). Am Heart J 8:297–308

Wolff L, Parkinson J, White PD (1930) Bundle-branch block with short P-R interval in healthy young patients prone to paroxysmal tachycardia. Am Heart J 5:685–704

Zipes DP (2000) Introduction. The atrioventricular node: a riddle wrapped in a mystery inside an enigma. In: Mazgalev TN, Tchou PJ (eds) Atrial A–V nodal Electrophysiology. A view from the millennium. Futura Publishing, Armonk, pp XI–XIV

Ziskind B, Halioua B (2004) Contribution de l'Égypte pharaonique à la médecine cardiovasculaire. Arch Mal Coeur Vaiss 97:370–374

HEP (2006) 171:41–71
© Springer-Verlag Berlin Heidelberg 2006

Pacemaker Current and Automatic Rhythms: Toward a Molecular Understanding

I. S. Cohen[1] · R. B. Robinson[2] (✉)

[1]Department of Physiology and Biophysics, Stony Brook University,
Room 150 Basic Science Tower, Stony Brook NY, 11794-8661, USA

[2]Department of Pharmacology, Columbia University, 630 W. 168th St.,
Room PH7W-318, New York NY, 10032, USA
rbr1@columbia.edu

Abstract The ionic basis of automaticity in the sinoatrial node and His–Purkinje system, the primary and secondary cardiac pacemaking regions, is discussed. Consideration is given to potential targets for pharmacologic or genetic therapies of rhythm disorders. An ideal target would be an ion channel that functions only during diastole, so that action potential repolarization is not affected, and one that exhibits regional differences in expression and/or function so that the primary and secondary pacemakers can be selectively targeted. The so-called pacemaker current, I_f, generated by the HCN gene family, best fits these criteria. The biophysical and molecular characteristics of this current are reviewed, and progress to date in developing selective pharmacologic agents targeting I_f and in using gene and cell-based therapies to modulate the current are reviewed.

Keywords Sino-atrial node · Automaticity · Pacemaker · HCN · Arrhythmia · Pharmacologic selectivity

1
Introduction

The sino-atrial (SA) node is the specialized region of the cardiac right atrium where the heartbeat originates. As such, it is capable of spontaneously generating action potentials but also is heavily innervated and thus susceptible to autonomic regulation of its spontaneous rate. The action potential spreads from the SA node into the surrounding atrial muscle, and the anatomical and functional details of that connectivity are critical to proper propagation of the signal. The signal then traverses the atrio-ventricular (AV) node into the His–Purkinje conducting system, which is responsible for rapidly transmitting the electrical signal to the working ventricular myocardium so that an organized and efficient contraction results. However, while the His–Purkinje system normally serves merely to transmit the signal originating at the SA node, cells in this region of the heart also are capable of firing spontaneously. Since their intrinsic rate is slower than that of the SA node they normally do not generate spontaneous action potentials, but when the signal from the SA node is delayed or fails (e.g., in the case of AV block) cells of the His-Purkinje system can serve as subsidiary pacemakers.

This chapter will first discuss the ionic basis of automaticity, autonomic regulation of automaticity, and cytoarchitecture in the SA node, and compare these features to corresponding features in Purkinje fibers. It will then consider appropriate targets for selective modification of automaticity, with particular emphasis on the pacemaker current, since this current has unique characteristics that favor its selective impact on automaticity. Present understanding of the molecular and biophysical characteristics of the pacemaker current will be reviewed, and recent pharmacologic and genetic advances in targeting or employing this current in cardiac therapies discussed.

2
Ionic Basis of Pacemaker Activity

2.1
Sino-atrial Node

2.1.1
Inward Currents of the Sino-atrial Node

The SA node action potential is characterized by a progressive diastolic depolarization (the pacemaker potential) in the voltage range −65 to −40 mV and a relatively slowly rising action potential (Fig. 1a), with a typical maximal rate of depolarization ($\dot{V}_{max}$) of less than 20 V/s, compared to values of several hundred V/s for atrial and ventricular muscle and approaching 1,000 V/s for Purkinje fibers. The ionic basic of the pacemaker potential has been in dispute for decades, with evidence in the literature supporting the contribution of a number of different currents. Almost certainly multiple currents contribute to the net inward current flowing during diastole, with the relative contribution of these individual currents varying with region (central or peripheral node), species, and age. The low $\dot{V}_{max}$ is because the SA node action potential upstroke is largely dependent on Ca channels, with little or no contribution from the Na channels typical of other cardiac regions. However, evidence exists for regional

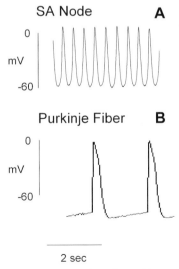

Fig. 1a,b Representative canine SA node and Purkinje fiber action potentials, illustrating the distinct voltage ranges of the diastolic potential. Note the difference in diastolic depolarization voltage range between the primary SA node pacemaker (**a**) and the secondary Purkinje fiber pacemaker (**b**). (SA node recording courtesy of Dr. Lev Protas; Purkinje fiber recording reprinted from Pinto et al. 1999)

and developmental diversity here as well. The major inward currents proposed to flow during diastole include the pacemaker current (I_f), The T-type ($I_{Ca,T}$) and L-type ($I_{Ca,L}$) Ca^{2+} currents, a sustained inward Na^+ current (I_{st}) and the Na^+/Ca^{2+} exchanger current (I_{NaCa}). In certain circumstances, a TTX-sensitive Na^+ current (I_{Na}) also may contribute. Each is discussed in detail below.

2.1.1.1
Pacemaker Current

The diastolic depolarization reflects a period of progressively increasing inward current. This can arise from either a time-dependent and increasing inward current or from a constant (or background) inward current combined with a progressively decreasing outward current. The latter was originally thought to be the case, with the contributing outward current in Purkinje fibers referred to as I_{K2} (McAllister et al. 1975). In 1980, two reports appeared suggesting the existence of a time-dependent inward current in the SA node (Brown and DiFrancesco 1980; Yamagihara and Irisawa 1980). The report by Brown and DiFrancesco included a detailed characterization that argued for the contribution of this new current to SA node automaticity. DiFrancesco termed the current I_f (the nomenclature we will use here), while others have referred to this same current as I_h or I_q. DiFrancesco subsequently provided additional evidence for the contribution of I_f to SA node automaticity (DiFrancesco 1991).

The most distinctive characteristic of I_f is that it is hyperpolarization activated, as opposed to all other voltage-gated channels in the heart, which are activated on depolarization. It is this unique characteristic that makes it particularly suited to serve as a pacemaker current, since it activates at the end of the action potential, when the cell repolarizes, and it deactivates rapidly upon depolarization during the action potential upstroke. Thus, current through these channels flows almost entirely during diastole. The current has mixed Na^+/K^+ selectivity with a reversal potential of approximately -35 mV in normal saline (DiFrancesco 1981a,b). As such, it is inward and largely carried by Na^+ at typical diastolic potentials. It is slowly activating, often with an initial delay resulting in a somewhat sigmoidal shaped time course. The delay is reduced at more negative voltages so that greater hyperpolarization results in greater and more rapid depolarizing current flow (DiFrancesco and Ferroni 1983).

I_f is highly sensitive to modulation by autonomic agonists, and in fact responds to significantly lower concentrations of acetylcholine than the acetylcholine-activated K^+ current $I_{K,Ach}$ (DiFrancesco et al. 1989). Acetylcholine reduces the contribution of the current during diastole by shifting the voltage dependence of activation negative, such that less current flows at a given potential (DiFrancesco and Tromba 1988a,b). Similarly, adrenergic agonists increase the contribution of the current by shifting its voltage dependence positive. These agonists act by reducing or increasing the concentration of cyclic AMP (cAMP), respectively. Unlike many other cAMP-responsive channels, however, the primary cAMP-dependent modulation of I_f in the SA node is independent of protein kinase A and phosphorylation. Rather, as discussed in greater detail below (see Sects. 3.1 and 3.2), cAMP binds directly to the channel to alter voltage dependence of gating (DiFrancesco and Tortora 1991).

Debate on the relative contribution of I_f to SA node automaticity revolves around quantitative issues. That is, whether the current is activated rapidly enough in the appropriate voltage range to make a significant contribution to the pacemaker potential (DiFrancesco 1995; Vassalle 1995). Resolution of these issues has been complicated by the fact that the current magnitude required to achieve SA node automaticity is exceedingly small, and often within the range of leakage currents associated with the employed recording techniques. In addition, the current is highly sensitive to experimental conditions, which can cause a negative shift in its voltage dependence and make it appear less likely to contribute physiologically. The lack of selective I_f blocking agents in the early years of this debate further complicated the problem, since for many years the only available blocker was Cs^+. While Cs^+ is a relatively effective blocker of I_f, the block is voltage dependent and therefore differs with membrane voltage, and Cs^+ also blocks some K^+ currents and is thus not specific for I_f. Newer, more selective blockers have helped to demonstrate that I_f does indeed contribute to SA node automaticity but that the SA node is able to function even without this current (i.e., these blockers slow but do not stop SA node automaticity) (Bois et al. 1996; Thollon et al. 1994). In addition, the recent identification of the HCN gene family as the molecular correlate of I_f (Biel et al. 1999; Santoro and Tibbs 1999) allows the use of transgenic technology to suppress expression of specific HCN isoforms (Stieber et al. 2003) as another approach to elucidating the contribution of this channel to SA node automaticity.

2.1.1.2
T-Type and L-Type Ca Currents

SA node myocytes express both T-type and L-type Ca^{2+} currents (Hagiwara et al. 1988). T-type currents, also referred to as low voltage activated (LVA) channels, activate and inactivate at more negative potentials than L-type currents, which are referred to as high voltage activated (HVA) channels. Given the relative activation voltages, one might anticipate that $I_{Ca,T}$ would contribute only to mid and late diastole and $I_{Ca,L}$ would contribute near the action potential threshold (take off potential) and/or upstroke.

In fact, inhibition of $I_{Ca,T}$ is associated only with suppression of late diastole (Satoh 1995), similar to what is observed when $I_{Ca,L}$ is inhibited (Satoh 1995; Zaza et al. 1996). However, accurate interpretation of the contribution of $I_{Ca,T}$ to automaticity is further complicated by the less-than-ideal selectivity of available inhibitors. Ni^{2+} at micromolar concentrations is reported to be selective for $I_{Ca,T}$ versus $I_{Ca,L}$ (Hagiwara et al. 1988), but the block is voltage dependent and the sensitivity of both LVA and HVA channels to Ni^{2+} is isoform specific (Perez-Reyes 2003). For these reasons measuring the effect of Ni^{2+} on SA node automaticity is not a definitive indicator of the contribution of $I_{Ca,T}$ to pacemaking. Similarly, the T-type blocker mibefradil was found to also block $I_{Ca,L}$ in SA node myocytes (Protas and Robinson 2000). In addition, the SA node expresses two T-type isoforms, $Ca_V3.1$ and $Ca_V3.2$, with the relative predominance varying with species (Perez-Reyes 2003).

Evidence also suggests that two L-type isoforms are present in the SA node, $Ca_V1.2$ and $Ca_V1.3$. The former is the predominant isoform in the working myocardium, but a $Ca_V1.3$ knock-out mouse exhibits sinus bradycardia (Platzer et al. 2000; Mangoni et al. 2003), supporting a role of this isoform in SA node excitability. Typically, $Ca_V1.3$ activates somewhat more negatively than $Ca_V1.2$, and this may be important for its contribution to SA node automaticity. One study found that the $Ca_V1.3$ knock-out SA node myocytes exhibit an L-type current that activates 5 mV more positive with a reduced rate of late diastolic depolarization (Zhang et al. 2002), while another study reported a shift in activation of greater than 20 mV (Mangoni et al. 2003). Consistent with these observations, in the presence of tetrodotoxin (TTX), SA node myocytes from newborn rabbit hearts exhibit a much slower spontaneous rate than those from adult hearts, and L-type current activates 5 mV more positive in the newborn cells, with no difference in T-type current (Protas et al. 2001). However, it is not known whether or not this reflects an age-dependent isoform switch. L-type currents are also responsive to autonomic agonists (unlike T-type) and so may contribute to the autonomic modulation of SA node automaticity. However, muscarinic inhibition of SA node $I_{Ca,L}$ occurs at significantly higher (>1,000×) concentrations than does inhibition of I_f; in contrast, I_f and $I_{Ca,L}$ are enhanced by adrenergic agonists over a similar concentration range (Zaza et al. 1996). This study also suggested that $I_{Ca,L}$ and its adrenergic modulation contribute only to late diastole, although another study argued for a contribution of $I_{Ca,L}$ throughout diastole (Verheijck et al. 1999).

2.1.1.3
Na^+/Ca^{2+} Exchanger Current

The Na^+/Ca^{2+} exchanger operates in forward mode to transport one Ca^{2+} ion out of the cell in exchange for the transport of three Na^+ ions into the cell, generating a net inward current (although it also can operate in the reverse mode). Initial evidence for the contribution of I_{NaCa} to automaticity came from studies in toad SA node cells using calcium chelators and ryanodine (to disrupt sarcoplasmic reticulum stores of calcium), both of which slowed the spontaneous rate (Ju and Allen 1998). More recent studies from several laboratories have confirmed that ryanodine also slows the spontaneous rate of mammalian SA node cells (Rigg et al. 2000; Bogdanov et al. 2001; Bucchi et al. 2003), although the latter study indicated that this involved a positive shift in action potential threshold rather than a reduction in the slope of early diastolic potential. Since the Ca^{2+} uptake mechanism into the sarcoplasmic reticulum is modulated by adrenergic agonists, it also has been argued that I_{NaCa} accounts for the autonomic modulation of rate. Supporting evidence includes the observation that, in the presence of ryanodine, adrenergic modulation of SA node chronotropy was markedly reduced (Rigg et al. 2000; Vinogradova et al. 2002). However, in the presence of ryanodine adrenergic modulation of I_f also is suppressed, while direct cAMP modulation of both I_f and rate are unaffected (Bucchi et al. 2003), suggesting that ryanodine impacts a proximal element of the adrenergic signaling cascade. Thus, while I_{NaCa} clearly contributes to basal automaticity and may also contribute to autonomic modulation of rate, it is not the sole factor in either case.

2.1.1.4
Sustained and TTX-Sensitive Na Currents

I_{st}, a sustained inward current carried by Na^+ ions, was first observed in rabbit SA node myocytes (Guo et al. 1995) and later also reported in other species (Guo et al. 1997; Shinagawa et al. 2000). I_{st} has characteristics consistent with the monovalent cation conductance of L-type Ca^{2+} channels, and so it was originally suggested to represent a novel L-type subtype (Guo et al. 1995). However, single channel analysis reveals distinct unitary conductance and gating kinetics (Mitsuiye et al. 1999). The similar pharmacologic sensitivity to $I_{Ca,L}$ complicates definitive demonstration of a contribution to automaticity. The main evidence in support of such a role is the association of the current with spontaneous activity; it is reported to be present in SA node cells that are spontaneously active and not in cells that are quiescent (Mitsuiye et al. 2000).

Sinus rhythm in the adult rabbit heart, the prototypical SA node preparation for animal studies, is relatively insensitive to TTX, a highly specific blocker of the rapid inward Na^+ current. In addition, to the extent that sinus rhythm is affected in the intact heart by TTX, this may reflect an action on atrial muscle and subsequent inability for the impulse to propagate from the SA node to atrial muscle (exit block), rather than a direct action on SA node myocytes. If I_{Na} exists in SA node myocytes of this preparation, it seems largely restricted to the peripheral rather than central node (Denyer and Brown 1990; Kodama et al. 1997). Further, given the voltage dependence of inactivation of the cardiac isoform of I_{Na}, the channel may be functionally silent in the adult SA node. However, the newborn rabbit SA node exhibits a pronounced I_{Na} that contributes importantly to automaticity (Baruscotti et al. 1996; Baruscotti et al. 1997). This current is not inactivated at the diastolic potentials present in the SA node because it represents a neuronal isoform ($Na_v1.1$) rather than the typical cardiac isoform ($Na_v1.5$, also referred to as SCN5a) (Goldin et al. 2000) and therefore has an inactivation relation that is positively shifted on the voltage axis. It also is more sensitive to TTX than the cardiac isoform. This current is slowly inactivating, so that it flows during diastole (Baruscotti et al. 2000; Baruscotti et al. 2001). Recently, it has been reported that $Na_v1.1$ also is expressed in the adult SA node of mice and rats (Maier et al. 2003; Lei et al. 2004), suggesting that I_{Na} in at least some species may functionally contribute to SA node automaticity in the adult heart. If a neuronal Na channel isoform is expressed in the human SA node, this would have implications for pharmacologic targeting of sinus rhythm. In addition, since I_{Na} is less responsive to adrenergic agonists than $I_{Ca,L}$, and since the different Na channel isoforms exhibit distinct adrenergic responsiveness (Catterall 1992), this also has implications for the autonomic regulation of sinus rate.

2.1.2
Outward Currents of the Sino-atrial Node

Here we address only those outward currents that may contribute to the diastolic potential and automaticity in the SA node. Other currents that flow only during the plateau phase of the action potential, such as the transient outward current, do not significantly impact SA node automaticity.

One of the hallmarks of SA node cells is the relatively low density of inward rectifier current, I_{K1}. This is critical to the ability of the SA node to generate spontaneous action potentials, as it sets membrane potential less negative, reduces membrane stability and greatly decreases the magnitude of the inward currents required during diastole to drive the cell to threshold. Indeed, reduction of I_{K1} in non-spontaneous cells in other cardiac regions is sufficient to induce spontaneous activity (Miake et al. 2002). Similarly, the acetylcholine sensitive K^+ current, $I_{K,Ach}$, contributes to the slowing of sinus rhythm by muscarinic agonists. However, this effect occurs at higher concentrations than those associated with inhibition of I_f and the slowing of rate in single SA node myocytes (DiFrancesco et al. 1989).

Both the rapid (I_{Kr}) and slow (I_{Ks}) components of the delayed rectifier are present in the SA node, but the relative expression varies with species (Anumonwo et al. 1992; Lei and Brown 1996; Cho et al. 2003) as well as with region within the node (Lei et al. 2001). Irisawa and colleagues have summarized the arguments for how a decaying I_K upon repolarization, in the presence of a constant background inward current, can lead to diastolic depolarization (Irisawa et al. 1993). It should be noted that, given that I_{Ks} is susceptible to adrenergic modulation (Marx et al. 2002), to the extent that I_{Ks} contributes to the SA node delayed rectifier and to diastolic depolarization, this provides another mechanism for autonomic regulation of rate. On the other hand, DiFrancesco has provided evidence that net background current in the diastolic potential range is outward rather than inward (DiFrancesco 1991) and therefore has argued that, while I_K decay can contribute to diastolic depolarization, this depolarization is not initiated by the unmasking of a background inward current during I_K decay (DiFrancesco 1993).

2.1.3
Regional Heterogeneity and Coupling to Atrial Tissue

Regional heterogeneity in ionic current expression within the SA node has been extensively studied by the Boyett laboratory (Boyett et al. 2000; Honjo et al. 1996; Kodama et al. 1997; Lei et al. 2001; Zhang et al. 2000). They report differences between the central and peripheral node in Na^+, Ca^{2+}, pacemaker, and delayed rectifier currents, as well as in the spontaneous rate of isolated cells. They also report differences in calcium handling, with a contribution to automaticity being apparent only in the peripheral cells (Lancaster et al. 2004). Finally, they identified regional differences in connexin expression (Honjo et al. 2002), which may have implications for coupling of the SA node to surrounding atrial tissue. This latter observation is consistent with the idea that the way the SA node couples to the atrial myocardium is a critical aspect of the ability of this small region to drive the heart. There is poor coupling between the primary pacemaker cells of the SA node and more peripheral areas, including atrium. This allows for isolation of the SA node from the hyperpolarizing influence of the surrounding atrial muscle with its higher expression of I_{K1}, but also makes it more susceptible to exit block (Boyett et al. 2003). Recently, it has been suggested that one component of that coupling may be represented by fibroblasts, which are able to transmit both electrical signals and small dye molecules between myocytes (Gaudesius et al. 2003; Kohl et al. 2004).

2.2
Bundle of His, Right and Left Bundle Branches and Purkinje Fibers

2.2.1
Specialized Conducting Tissue Serves Two Roles

In larger animals, the speed of conduction in ventricular muscle (up to 0.3 m/s) is insufficient to guarantee almost synchronous activation of the ventricles. To solve this problem, specialized conducting tissue evolved. The cardiac impulse enters the ventricle in the bundle of His and propagates through the left and right bundle branches into the peripheral Purkinje fibers. All of this tissue is rapidly conducting (roughly 3–8 times faster than ventricular muscle) and has a distinctive cellular morphology. The conducting tissue has two major characteristics that allow for this rapid conduction: (1) a high density of Na channels, and (2) a low intercellular resistance. Both these characteristics are in contradistinction to the SA node.

The maximum upstroke velocity of the Purkinje fiber action potential approaches 1,000 V/s, at least threefold greater than that observed for ventricular muscle (Hoffman and Cranefield 1976). This larger upstroke velocity is due to a greater density of Na channels which generate larger local circuit currents bringing adjacent regions more rapidly to threshold. The intercellular resistance of Purkinje fibers, depending on the species, can be as low as 100 Ωcm, which is barely above the resistivity of the intracellular milieu (Hoffman and Cranefield 1976). The low resistance of the intercellular pathway is based on a high density of gap junctions mostly placed at regions of intimate cell-to-cell contact at the ends of cells to facilitate conduction along the longitudinal pathway (Mobley and Page 1972).

Since conduction in the AV node is slow, and prone to block, it is not surprising that the specialized conducting tissue has also evolved as an important secondary pacemaker. Typically if not driven from above, it will assume a spontaneous rate of 25 to 40 beats per minute, appreciably slower than the primary pacemaker, but a rate sufficient to maintain a viable cardiac output. The mechanism of pacemaking is similar but not identical to that in the SA node. It is these differences that might serve as the basis of a selective pharmacology (see Sect. 3.4).

2.2.2
Membrane Currents in Purkinje Fibers and Myocytes that Flow During the Action Potential

As stated above, there is a very high density of TTX-sensitive Na^+ current in Purkinje fibers. This is generated through a cardiac-specific Na channel gene, *SCN5a*, which generates a channel that has a lower sensitivity to TTX ($K_D = 1$ μM) than that found in nerve Na channels ($K_D = 1$ nM) (Cohen et al. 1981). After reaching its peak, the action potential experiences an initial rapid repolarization to the plateau due to transient outward current. This current is thought to have both a Ca-independent and Ca-dependent component (Coraboeuf and Carmeliet 1982). Following this initial rapid repolarization, there is a several-hundred-millisecond period of virtually no change in membrane potential called the action potential plateau. It is a period of very high membrane resistance with relatively small and almost equal inward and outward membrane currents. The inward current is generated through L-type Ca channels (McAllister et al. 1975), as well as slowly inactivating (Gintant et al. 1984) and steady-state TTX-sensitive Na^+ current (Attwell et al. 1979). The outward current is generated by a large rapid and smaller slow component of the delayed rectifier (I_{Kr} and I_{Ks}) (Varro et al. 2000). Final repolarization occurs when the delayed rectifier currents (which continue to activate) exceed in magnitude the L-type Ca^{2+} and TTX-sensitive currents (which inactivate).

2.2.3
Pacemaker Activity in Purkinje Fibers

Following the action potential, Purkinje fibers exhibit a period of diastolic depolarization or pacemaker activity. This automaticity occurs in the voltage range −90 mV to −60 mV [which is much more negative than pacemaker activity in the SA node (see Sect. 2.1.1; Fig. 1b)], and allows Purkinje fibers to function as a subsidiary pacemaking tissue. The basis of this pacemaker activity has been studied for almost 40 years and is still not entirely resolved. It is known that a substantial inwardly rectifying time-independent background current called I_{K1} is present at diastolic potentials in Purkinje fibers (Shah et al. 1987) (and largely absent in SA node). This large outward current makes pacemaker depolarization difficult in this voltage range. Initially it was thought that I_{K2} (yet another outward potassium current, apart from the delayed rectifiers) deactivated, allowing an inward background current to drive the membrane to the threshold potential for firing (Noble and Tsien 1968). However, DiFrancesco (1981a,b) demonstrated that when all diastolic potassium currents in Purkinje fibers were blocked by barium, I_f was revealed. This inward current which activates on hyperpolarization contributes to and may even initiate spontaneous depolarization. More recently, molecular studies have identified the HCN family as the molecular correlate of the α-subunit of the I_f channel (Santoro et al. 1998; Ludwig et al. 1998) and demonstrated the presence of both HCN4 (the predominant HCN isoform present in SA node) and HCN2 (the dominant ventricular isoform) in Purkinje fibers (Shi et al. 1999). However, biophysical studies by Vassalle et al. (1995) have demonstrated that a K current distinct from the delayed rectifiers called I_{Kdd} deactivates at more positive diastolic potentials than I_f in Purkinje fibers and could contribute to diastolic depolarization. Its molecular origin is unknown, which has hampered further study.

In summary, present data support the notion that following repolarization by delayed rectifiers I_{Kdd} deactivates and I_f activates driving the membrane towards threshold. Purkinje fibers also contain an inward background current partly generated by steady-state current flow through TTX-sensitive channels (Attwell et al. 1979). T-type calcium current is also present in Purkinje myocytes (Tseng and Boyden 1989) and may contribute inward current towards the end of diastole to help drive the membrane to threshold.

2.3
Targets for Selective Intervention

From the descriptions of primary pacemaker activity in the SA node and secondary pacemaker activity in Purkinje fibers, it is clear that a number of currents participate, and each must be considered a potential target for selective intervention. In the paragraphs that follow, we consider the diastolic membrane currents and their potential as therapeutic targets.

The ideal target should have the following three characteristics:

– It would flow at diastolic potentials only (since changes in action potential duration can be arrhythmogenic).

– It would have distinctive characteristics in different cardiac regions so that primary or secondary pacemaker activity can be independently altered.

– Its molecular basis should be known (since this dramatically facilitates rational drug design).

Based on these criteria, a number of diastolic membrane currents can be eliminated as potential targets. I_{K1} is selectively expressed in secondary rather than primary pacemaker tissues, but altering its magnitude affects the action potential duration (Miake et al. 2002). It is unknown whether I_{Kdd} is present in the SA node [although it is present in atrium (Wu et al. 1999)]. Unfortunately, as stated above, nothing is known about its molecular origins. The T-type calcium current is expressed in both primary and secondary pacemaker regions, as is the Na/Ca exchanger. Neither provides current selectively at diastolic potentials. The cardiac isoform of the TTX-sensitive sodium current is absent in primary pacemaker cells and present in Purkinje fibers; however, its contribution is not selectively limited to diastole. It contributes to the upstroke of the action potential and helps determine the action potential duration and conduction velocity. I_f is activated selectively at diastolic potentials and so does not influence the action potential duration. Although it is expressed in both primary and secondary pacemakers, the prevalence of individual family members and HCN-associated β-subunits differs in the two cardiac regions. Finally, its molecular basis is known and much is reported already about its structure–function relationship. It is for these reasons that we believe I_f provides a unique target for therapeutic intervention for automatic arrhythmias.

3
The Pacemaker Current I_f

3.1
Biophysical Description

In this section we will review what is known about the biophysical properties of I_f. We will begin by describing its properties in the SA node and then consider differences in other cardiac regions.

I_f activates on hyperpolarization (DiFrancesco 1981a,b). This means that channel open probability increases as the membrane is hyperpolarized. This voltage dependence is distinctly opposite to that observed for other voltage-gated ion channels in the heart which open on depolarization. In SA node myocytes, almost all f channels are closed at −40 mV and open at −100 mV (Wu et al. 2000). Upon hyperpolarization the channels open after a delay. This delay is voltage dependent, lasting only a few milliseconds at −130 mV and hundreds of milliseconds at −65 mV (DiFrancesco and Ferroni 1983). After this initial pause, the current increases along an exponential time course. The kinetics of activation are also voltage dependent, being slowest near the mid-point of activation where the time constant can be several seconds. At extremes of potential, the activation or deactivation can occur with a time constant of tens of milliseconds (DiFrancesco 1981a,b). Activation of β-adrenergic receptors raises cAMP levels and results in a positive voltage shift of all channel properties (Hauswirth et al. 1968). At saturating concentrations of agonist, the shift can approach 15 mV. Activation of muscarinic receptors has the opposite effect, lowering cAMP levels and inducing a negative shift in the voltage dependence of I_f (DiFrancesco and Tromba 1988a,b). The current is carried by both Na^+ and K^+, having a reversal potential roughly midway between E_{Na} and E_K. Increases in extracellular K^+ increase current magnitude in addition to the changes they induce in the current's reversal potential. Changes in extracellular Na^+ affect only the reversal potential (DiFrancesco 1981a,b). DiFrancesco and colleagues (DiFrancesco and Tortora 1991; DiFrancesco 1986) have successfully recorded single f channels simultaneously with whole-cell I_f relaxations, confirming their identity as the basis of the macroscopic current. These channels have an extremely small single channel conductance of about 1 pS in 100 mM extracellular K^+, or an estimated conductance of one fifth that value in physiologic saline. In the SA node, phosphorylation by either serine–threonine or tyrosine kinases increase I_f magnitude without inducing a shift in voltage dependence (Accili et al. 1997).

I_f is present in all cardiac regions studied. As a general rule, as one moves more distal from the primary pacemaker in the conduction pathway, the voltage dependence of activation becomes progressively more negative (Yu et al. 1995). The ubiquitous presence of I_f suggests that all cardiac tissues have the capacity to pace. The progressively more negative threshold as one moves away from the primary pacemaker guarantees its dominance in the pacing hierarchy. The threshold for activation is often negative to -80 mV in Purkinje myocytes and more negative than -100 mV in ventricular tissues (Yu et al. 1993a; Vassalle et al. 1995). The basis of these region-specific differences in voltage dependence is at present unknown. There is also a difference in autonomic responsiveness in Purkinje fibers and ventricle as compared to the SA node. While β-adrenergic agonists cause a positive shift in activation of similar magnitude to the SA node, muscarinic agonists have no direct effect, but can reverse the actions of β-adrenergic agonists (Chang et al. 1990). Other differences also exist in the effects of phosphorylation between SA node and ventricular tissues. In the SA node the shift in voltage dependence induced by β-adrenergic agonists is the direct result of cAMP binding, independent of PKA-mediated phosphorylation. Chang et al. (1991) demonstrated that, in Purkinje fibers, PKA mediated phosphorylation is necessary for the voltage shift to occur. Further, inhibition of serine–threonine phosphatases causes an increase in the amplitude of I_f in the SA node but causes a shift in voltage dependence in Purkinje and ventricular myocytes (Yu et al. 1993b). Inhibiting tyrosine kinases reduces I_f in SA node myocytes (Wu and Cohen 1997), but in ventricular myocytes this reduction in amplitude is accompanied by a negative shift in voltage dependence (Yu et al. 2004).

3.2
Molecular Description

The molecular correlate of the pacemaker current is the HCN (hyperpolarization-activated, cyclic nucleotide-gated) gene family, which was first cloned by Santoro and colleagues as a result of a yeast two-hybrid screen of N-src binding proteins (Santoro et al. 1997). They identified a protein with the typical characteristics of a K channel, but with some modification of the canonical K channel pore motif, and with a cyclic nucleotide-binding domain (CNBD) in the C-terminus. Based both on these features and its localization within specific brain regions, they suggested that this protein, later termed HCN1, might contribute to the current I_q. Subsequent reports by this and another group identified more members of this family and demonstrated that expression resulted in a current with characteristics typical of pacemaker current (Ludwig et al. 1998; Santoro et al. 1998).

There are four known mammalian isoforms of the HCN gene family, three of which (*HCN1, HCN2, HCN4*) are expressed in the heart. They have been the subject of numerous review articles (Clapham 1998; Santoro and Tibbs 1999; Biel et al. 1999; Kaupp and Seifert 2001; Accili et al. 2002; Biel et al. 2002; Robinson and Siegelbaum 2003; Baruscotti and DiFrancesco 2004). The isoforms are 80%–90% identical in the transmembrane and CNBD domains, but show significant diversity elsewhere. They consist of six transmembrane domains (S1–S6), with S4 being positively charged and serving as the putative voltage sensor (Fig. 2). Four subunits are assumed to assemble to form

HCN Channel Domains

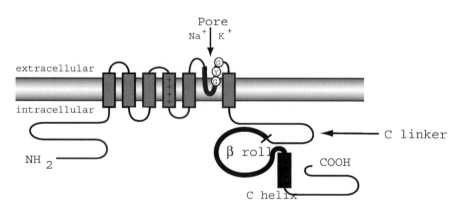

Fig. 2 Transmembrane topology of a single HCN subunit, illustrating the 6 transmembrane domains, including the positively charged S4 domain, and the cyclic nucleotide-binding domain in the C-terminus. (Reprinted from Robinson and Siegelbaum 2003)

a functional channel. The pore-forming region, between S4 and S5, contains the signature GYG sequence of a K^+-selective channel, but exhibits differences immediately outside this region from other K channels, which may account for the mixed Na^+/K^+ selectivity of these channels. All isoforms possess a 120 amino acid CNBD in the C-terminus consisting of a β-roll and C-helix structure; the CNBD is connected to S6 by a region referred to as the C-linker. X-ray crystallographic analysis of the C-terminus, including the C-linker and CNBD, reveal a tetramerization domain with most of the subunit–subunit interactions residing in the C-linker (Zagotta et al. 2003).

A number of laboratories have conducted mutagenesis studies to further elucidate the structure–function relation of HCN channels. Investigation of the C-terminus and S4–S5 linker suggest that interactions between these domains, coupled by the C-linker region, serve to normally inhibit channel gating (Wainger et al. 2001; Chen et al. 2001; Decher et al. 2004). The binding of cAMP to the CNBD relieves this inhibition and allows channel activation at less negative voltages. Investigators also have attempted to determine the basis for the HCN channel's unique activation on hyperpolarization, rather than depolarization typical of other voltage-gated K channels. Studies have confirmed that the S4 region serves as the voltage sensor, and they suggest that its movement is conserved between HCN channels and voltage-gated K channels (Vaca et al. 2000; Chen et al. 2000; Mannikko et al. 2002; Vemana et al. 2004; Bell et al. 2004). The precise molecular explanation for the hyperpolarization-activated mechanism remains to be determined.

3.3
Regional Distribution of HCN Isoforms and MiRP1

If a selective pharmacology is to be developed, it is useful to know not only regional differences in biophysical properties but also the regional distribution of the HCN family members and all auxiliary subunits in the relevant cardiac regions, as well as the potential functional differences that these regional distributions would induce. A number of studies employing either RNase protection assays (RPAs) or quantitative RT-PCR have examined the distribution of HCN isoforms in various cardiac regions in canine and rabbit heart (Shi et al. 2000; Shi et al. 1999; Han et al. 2002). A number of generalizations can be made. First, the SA node has the highest levels of HCN transcripts, and the dominant isoform (>80%) is HCN4. It also contains some HCN1. Ventricular muscle contains predominantly HCN2 and has a much lower expression of HCN transcripts in general. Midway between these two extremes is the Purkinje fiber with about one-third the expression level observed in the SA node, about 40% of which is HCN4, with the other two cardiac isoforms also being expressed. The only β-subunit so far reported to affect expression and biophysical properties of HCN subunits is MinK related peptide 1 (MiRP1) (Yu et al. 2001). This β-subunit is highly expressed in the SA node and much more poorly expressed in ventricular tissues. Within the ventricle, Purkinje fibers have the highest level of expression (Pourrier et al. 2003).

Since the discovery of the HCN ion channel subunit family there has been extensive investigation of the structure–function relationship of the individual family members. In Sect. 3.2 we described the properties common to all family members, here we focus on the differences.

– Voltage dependence of activation: There is only a small difference in activation voltage dependence between the three cardiac isoforms (Moosmang et al. 2001; Altomare et al. 2001).

– Kinetics of activation: There are large differences in the kinetics of activation. HCN1 activates most rapidly. HCN2 activates almost an order of magnitude more slowly than HCN1. HCN4 is the slowest activating of the isoforms, activating almost threefold slower than HCN2 (Moosmang et al. 2001; Altomare et al. 2001).

– cAMP effects: As described above, cAMP directly binds to HCN channels, and this binding results in a shift in the voltage dependence of activation. The shift is only a couple of millivolts with HCN1, but can be 15 mV for the other two isoforms (Fig. 3a) (Wainger et al. 2001; Ludwig et al. 1999).

– Effects of tyrosine phosphorylation: Inhibiting tyrosine kinases has no effect on HCN1. It reduces the magnitude of current flow through HCN4 channels with no effect on either the voltage dependence or kinetics of activation. It also reduces the amplitude of current flowing through HCN2 channels and induces a negative shift in both the voltage dependence and kinetics of activation (Fig. 3b) (Yu et al. 2004).

– Effects of the β-subunit MiRP1: MiRP1 has been shown to increase the amplitude of current flow through all three HCN isoforms expressed in heart (Yu et al. 2001; Decher et al. 2003). However, it has no effect on the voltage dependence of activation of HCN1 and HCN2 but shifts the activation of HCN4 to more negative potentials. It accelerates the activation of HCN1 and HCN2, but slows the activation of HCN4.

3.4
Approaches for Region-Specific Modification of I_f

To change the properties of I_f in a region-specific manner one requires one of the following:
– A change in expression or function of the isoforms that are expressed differentially

– An agent that has a voltage-dependent effect, since diastolic membrane voltages are much more negative in Purkinje fibers and ventricular muscle than in the SA node

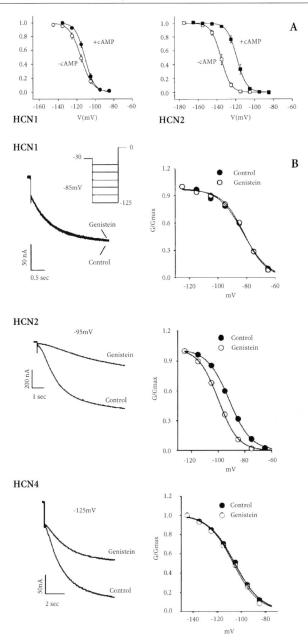

Fig. 3a,b HCN isoform selective pharmacologic modulation. **a** Differential cAMP responsiveness of HCN1 and HCN2. Note the more pronounced shift in activation voltage of HCN2 by cAMP. (Reprinted from Wang et al. 2001.) **b** Distinct modulation of HCN1, HCN2, and HCN4 by the tyrosine kinase inhibitor genistein. Note the reduction by genistein of current magnitude of HCN2 and HCN4, and the shift in voltage dependence only of HCN4. (Reprinted from Yu et al. 2004)

 – An agent that has a current-dependent effect, since the magnitude of current
   flow through f channels will depend on the driving force which will be larger
   at diastolic potentials in Purkinje or ventricular myocytes than in SA node
   myocytes

Current knowledge has already provided evidence that all three approaches to
regional modification of I_f may be feasible.

We begin by considering those agents that act through the first proposed
mechanism. Inhibition of tyrosine kinases (whose effects are described above)
will affect those regions expressing high levels of HCN2 (ventricular myocytes)
more than those regions expressing either HCN4 or HCN1 (SA node). Inhibition of protein kinase A has little or no effect on the effects of cAMP in SA node
(Accili et al. 1997), but decreases or eliminates the effects of cAMP in Purkinje
fibers (Chang et al. 1991). Acetylcholine reduces I_f in SA node myocytes but is
without a direct effect in Purkinje fibers. Nitric oxide increases I_f in SA node
myocytes from guinea pig, but has no effect on I_f in ventricular myocytes from
spontaneously hypertensive rats (Herring et al. 2001; Bryant et al. 2001). One
should remember, however, that selective targeting will require knowledge of
more downstream elements in the signaling cascade.

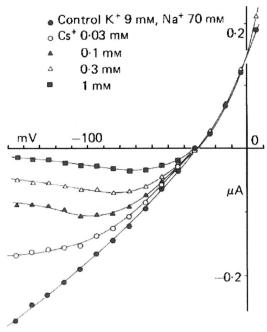

Fig. 4 Voltage-dependent block of I_f by cesium, illustrating the greater block at higher
concentrations of cesium and at more negative potentials. Note the difference in the efficacy
of block that would be experienced at diastolic membrane potentials in the SA node (−65 mV
to −40 mV) and Purkinje fibers (−90 mV to −60 mV) (Reprinted from DiFrancesco 1982)

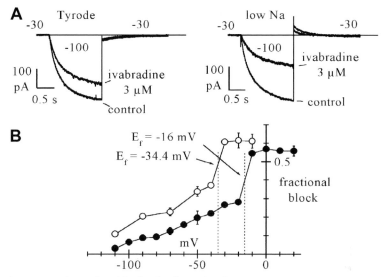

Fig. 5a,b Current-dependent block of I_f by ivabradine. **a** Representative current tracings illustrating block of I_f by ivabradine in normal extracellular solution (*left*) and low Na^+ solution (*right*) to shift the reversal potential. **b** Fractional block by ivabradine as a function of voltage. Note the dependence of fractional block on the calculated reversal potential. (Reprinted from Bucchi et al. 2002)

Now consider mechanism two, in which agents have a voltage-dependent action on pacemaker channels. A number of univalent cations including Cs^+ and Rb^+ can block I_f in a voltage-dependent manner, having a much greater effect at hyperpolarized potentials (Fig. 4) (DiFrancesco 1982). This result suggests that pacemaker activity should be more effectively blocked by lower concentrations of these ions in Purkinje fibers than in SA node myocytes.

Finally, some bradycardic agents have been reported to work by the current dependent mechanism described as alternative No. 3. Bucchi et al. (2002) and van Bogaert and Pittoors (2003) have reported that ivabradine, zatebradine, and cilobradine block I_f in a current-dependent manner (Fig. 5). The drugs enter the open channel from the inside of the cell when the inward current flow is minimal (during channel deactivation) and are displaced from the channel by inward current flow when the driving force on inward current flow is greatest (during activation). Given that diastolic membrane potentials in SA node cells are much closer to the I_f reversal potential than either Purkinje or ventricular myocytes, these results predict a higher affinity for block in the former preparations (that is, a higher affinity in the SA node preparations). However, the difference in the distribution and prevalence of the three HCN cardiac isoforms (which have differing kinetics of activation and deactivation) will also have an effect on the apparent association and dissociation rate constants.

4
Role of I_f in Generating Cardiac Rhythms and Arrhythmias

4.1
Animal Models

Neonatal rat ventricle cell cultures have been employed to demonstrate both the ability of HCN isoforms to enhance automaticity and the contribution of HCN channels to normal automaticity. Qu and colleagues first demonstrated that over-expressing HCN2 in these cultures significantly increased spontaneous rate (Fig. 6) (Qu et al. 2001). Er and colleagues subsequently confirmed this observation and extended it to HCN4 over-expression. They also demonstrated that expression of a dominant negative form of HCN2 markedly reduced native pacemaker current and entirely suppressed spontaneous activity (Er et al. 2003). Other recent studies have employed mouse genetics to produce animals lacking expression of HCN2 or HCN4. The HCN2 knockout was viable, and associated with absence epilepsy and sinus dysrhythmia (Ludwig et al. 2003). A cardiac-specific knockout also exhibited sinus dysrhythmia. I_f was reduced ~30% and activated more slowly in SA node cells from the HCN2 knockout animals, and isolated atria exhibited significantly greater beat-to-

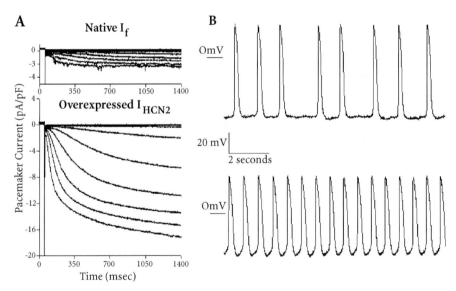

Fig. 6a,b HCN over-expression increases spontaneous rate of neonatal rat ventricular myocyte culture. **a** Representative family of current traces of pacemaker current from a control myocyte expressing only green fluorescent protein (GFP) (*top*) and from a myocyte over-expressing the HCN2 isoform (*bottom*). **b** Representative recordings of spontaneous action potentials from a control monolayer culture expressing only GFP (*top*) and from a monolayer culture over-expressing HCN2 (*bottom*). (Portions reprinted from Qu et al. 2001)

beat variability. In comparison to the HCN2 knockout, a cardiac-specific HCN4 knockout was embryonic lethal, but embryonic myocytes exhibited markedly reduced I_f and a slower spontaneous rate that was unresponsiveness to cAMP (Stieber et al. 2003). Taken together, these data demonstrate an important contribution of I_f and specific HCN isoforms to normal SA node automaticity.

Pacemaker currents and HCN transcripts also have been studied in several animal models of disease. The focus in these studies has tended to be on the appearance or magnitude of I_f current and HCN isoform expression in ventricular tissue. As stated earlier (Sect. 3.1), I_f activates outside the physiologic range in the healthy adult ventricle. However, it appears at physiologically relevant voltages in several disease models. Cerbai and colleagues reported increased I_f density in ventricular myocytes from aged spontaneously hypertensive rats compared to age-matched control animals, and the current density correlated with the severity of hypertrophy (Cerbai et al. 1996). Two other groups have studied an aortic banding model of hypertrophy in rat (Hiramatsu et al. 2002; Fernandez-Velasco et al. 2003). Both found increases in HCN2 and HCN4 message levels at 8 weeks (although the first group also reported a decrease in message levels at earlier times). The latter group also observed an increase in pacemaker current density in septal and left ventricular myocytes.

Finally, hyperthyroidism is associated with sinus tachycardia, and several laboratories have investigated the regulation of HCN subunits by thyroid hormone. Pachucki and colleagues identified a thyroid hormone consensus site in the HCN2 promoter and demonstrated that thyroid hormone administration to hypothyroid rats increased HCN2 message levels (Pachucki et al. 1999). Another group also found a correlation between whole-heart HCN2 message level and thyroid state, as well as HCN4 level, but when they studied atrial tissue, only HCN2 levels decreased with hypothyroidism (Gloss et al. 2001). This same study also reported that knocking out one of two isoforms of the thyroid hormone receptor in the mouse heart resulted in sinus bradycardia and reduced message levels of HCN2 and HCN4.

4.2
Human Disease

There have been two recent reports of human HCN4 mutations associated with sinus node dysfunction (SND). The first study investigated the HCN4 gene in ten patients with idiopathic SND, and found a 1-bp deletion in one patient, resulting in a C-terminal truncated HCN4 protein that lacked the CNBD, HCN4-573X (Schulze-Bahr et al. 2003). The patient exhibited sinus bradycardia with episodes of atrial fibrillation, and during exercise a maximal heart rate significantly less than that predicted for gender and age. When studied in a heterologous expression system, the truncated HCN4 channel produced currents that were relatively similar to those of wildtype HCN4, but lacked cAMP responsiveness. When mutated and wildtype HCN4 were co-expressed, the resulting current also lacked cAMP responsiveness. The second report identified an HCN4 mutation in one of six patients with SND that involved a single amino acid substitution, D553N, in the C-linker region (Ueda et al. 2004). The index patient exhibited severe bradycardia, cardiac arrest, and polymorphic ventricular tachycardia, and the HCN4 mutation co-segregated with the phenotype within the patient's family. The equivalent amino acid mutation in the rabbit HCN4 protein resulted in loss of membrane trafficking and also reduced membrane trafficking of wildtype HCN4 and pacemaker current when both were co-expressed.

Studies in patients with cardiac disease support the idea that HCN subunits are regulated by disease state. I_f density is increased in human ventricle during failure. This was found to be more pronounced in patients with ischemic than dilated cardiomyopathy, and not to correlate with the degree of hypertrophy (Cerbai et al. 2001). Another laboratory reported that HCN4 expression was increased in patients with end-stage heart failure, while HCN2 expression was below the level of detection (Borlak and Thum 2003).

4.3
HCN as a Biological Pacemaker

In sick sinus syndrome, the primary pacemaker oscillates between excessively high and excessively low heart rates. In AV block, life is sustained by an excessively low spontaneous rate which originates in the ventricular conducting system. In both these cases the current approach to therapy is the implantation of an electronic pacemaker to guarantee a constant physiologic rate consistent with a normal lifestyle. However, even the most up-to-date electronic versions are not immediately sensitive to changes in physiologic state induced by activation of the sympathetic and parasympathetic nervous systems. It is for this reason that current pacemaker therapy represents palliation rather than cure.

Subsequent to the cloning of the members of the HCN gene family in the latter part of the 1990s, it became possible to consider replacing the palliation of the electronic pacemakers with a biological cure (Rosen et al. 2004). If HCN family members could be delivered to the appropriate target regions of the heart (atrium for sick sinus syndrome, and the conducting system for AV block) then it would be possible to create a new primary pacemaker responsive to changes in input from the autonomic nervous system.

The initial approaches to create an HCN-based biological pacemaker used an adenoviral delivery system. The adenoviral vector contained the murine HCN2 gene. The virus was delivered either by injection into the wall of canine left atrium (Qu et al. 2003) or via catheter into the ventricular conducting system (Plotnikov et al. 2004). Evidence for a new biological pacemaker was demonstrated several days after injection by anesthetizing the animal and stimulating both vagi to cause sinus standstill. Vagal escape rhythms could be demonstrated in both studies originating near the sites of injection. After the in vivo studies were completed, cells were isolated from the injection sites and studied with the whole-cell patch-clamp technique. The transfected cells [which could be identified by green fluorescent protein (GFP), which was used as a reporter gene], expressed an I_f-like current which was between one and three orders of magnitude larger than I_f recorded from non-transfected myocytes.

As encouraging as these initial successes were, there are significant drawbacks with adenoviral gene delivery. Since the plasmid is not incorporated into the genome, the effect is transient, rarely lasting more than 6 weeks. Second, there is also the risk of allergic reaction, which has limited adenoviral approaches in prior clinical trials for other disease processes. Alternative viruses from the retroviral family are incorporated into the genome but come with an increased associated risk of neoplasia. In 2004 Potapova et al. (2004) employed an altogether different approach to deliver the HCN genes to the cardiac syncytium. They took human mesenchymal stem cells (hMSCs) and incorporated the same HCN2 gene by electroporation. After demonstrating high levels of expression of an I_f-like current in the transfected hMSCs, they injected one million of these transfected cells into the left epicardium and studied the animals 3–10 days later in the same manner described for viral delivery. Spontaneous vagal escape rhythms at roughly 60 beats per minute were observed which mapped to the site of the injection. This rate was significantly higher than those observed in sham animals which received hMSCs containing only enhanced GFP. Although the injected cells were not a stable cell line, separate experiments taking advantage of a neomycin resistance cassette in the original plasmid suggested that the electroporated cells could be selected by growth on antibiotic and continue to express the transfected proteins for at least 3 months.

As encouraging as these initial attempts appear to be, both the long-term reliability and safety of these pacemakers must be demonstrated before they can be considered a clinical tool.

5
Conclusions

Molecular cloning and structure–function studies have allowed the identification and characterization of a multi-gene family that contributes substantially to pacing in all regions of the heart, without contributing to other portions of the action potential. The prevalence of specific members of this family and their biophysical properties varies with cardiac region, raising the possibility of a selective pharmacology, although isoform-selective blockers have to date not been demonstrated. In addition, diastolic depolarization occurs over different voltage ranges in primary and secondary pacemakers. This provides the further possibility of using voltage-dependent or current-dependent block as the means to selectively target individual cardiac regions, and, in fact, existing I_f blockers do exhibit current-dependent block. Finally, recent studies have begun using gene and cell-based therapies to over-express an individual HCN isoform in a localized cardiac region, thereby creating a focal biological pacemaker that may be able to substitute for a dysfunctional natural pacemaker and one day obviate the need of implanting an electronic device.

References

Accili EA, Redaelli G, DiFrancesco D (1997) Differential control of the hyperpolarization-activated current (if) by cAMP gating and phosphatase inhibition in rabbit sino-atrial node myocytes. J Physiol (Lond) 500:643–651

Accili EA, Proenza C, Baruscotti M, DiFrancesco D (2002) From funny current to HCN channels: 20 years of excitation. News Physiol Sci 17:32–37

Altomare C, Bucchi A, Camatini E, Baruscotti M, Viscomi C, Moroni A, DiFrancesco D (2001) Integrated allosteric model of voltage gating of hcn channels. J Gen Physiol 117:519–532

Anumonwo JM, Freeman LC, Kwok WM, Kass RS (1992) Delayed rectification in single cells isolated from guinea pig sinoatrial node. Am J Physiol 262:H921–H925

Attwell D, Cohen I, Eisner DA, Ohba M, Ojeda C (1979) The steady state TTX-sensitive ("window") sodium current in cardiac Purkinje fibres. Pflugers Arch 379:137–142

Baruscotti M, DiFrancesco D (2004) Pacemaker channels. Ann N Y Acad Sci 1015:111–121

Baruscotti M, DiFrancesco D, Robinson RB (1996) A TTX-sensitive inward sodium current contributes to spontaneous activity in newborn rabbit sino-atrial node cells. J Physiol (Lond) 492:21–30

Baruscotti M, Westenbroek R, Catterall WA, DiFrancesco D, Robinson RB (1997) The newborn rabbit sino-atrial node expresses a neuronal type I—like Na$^+$ channel. J Physiol (Lond) 498:641–648

Baruscotti M, DiFrancesco D, Robinson RB (2000) Na(+) current contribution to the diastolic depolarization in newborn rabbit SA node cells. Am J Physiol Heart Circ Physiol 279:H2303–H2309

Baruscotti M, DiFrancesco D, Robinson RB (2001) Single-channel properties of the sinoatrial node Na$^+$ current in the newborn rabbit. Pflugers Arch 442:192–196

Bell DC, Yao H, Saenger RC, Riley JH, Siegelbaum SA (2004) Changes in local S4 environment provide a voltage-sensing mechanism for mammalian hyperpolarization-activated HCN channels. J Gen Physiol 123:5–20

Biel M, Ludwig A, Zong X, Hofmann F (1999) Hyperpolarization-activated cation channels: a multi-gene family. Rev Physiol Biochem Pharmacol 136:165–181

Biel M, Schneider A, Wahl C (2002) Cardiac HCN channels. structure, function, and modulation. Trends Cardiovasc Med 12:206–212

Bogdanov KY, Vinogradova TM, Lakatta EG (2001) Sinoatrial nodal cell ryanodine receptor and Na(+)-Ca(2+) exchanger: molecular partners in pacemaker regulation. Circ Res 88:1254–1258

Bois P, Bescond J, Renaudon B, Lenfant J (1996) Mode of action of bradycardic agent, S 16257, on ionic currents of rabbit sinoatrial node cells. Br J Pharmacol 118:1051–1057

Borlak J, Thum T (2003) Hallmarks of ion channel gene expression in end-stage heart failure. FASEB J 17:1592–1608

Boyett MR, Honjo H, Kodama I (2000) The sinoatrial node, a heterogeneous pacemaker structure. Cardiovasc Res 47:658–687

Boyett MR, Dobrzynski H, Lancaster MK, Jones SA, Honjo H, Kodama I (2003) Sophisticated architecture is required for the sinoatrial node to perform its normal pacemaker function. J Cardiovasc Electrophysiol 14:104–106

Brown HF, DiFrancesco D (1980) Voltage-clamp investigations of membrane currents underlying pacemaker activity in rabbit sino-atrial node. J Physiol (Lond) 308:331–351

Bryant SM, Sears CE, Rigg L, Terrar DA, Casadei B (2001) Nitric oxide does not modulate the hyperpolarization-activated current, I_f, in ventricular myocytes from spontaneously hypertensive rats. Cardiovasc Res 51:51–58

Bucchi A, Baruscotti M, DiFrancesco D (2002) Current-dependent block of rabbit sino-atrial node I(f) channels by ivabradine. J Gen Physiol 120:1–13

Bucchi A, Baruscotti M, Robinson RB, DiFrancesco D (2003) I_f-dependent modulation of pacemaker rate mediated by cAMP in the presence of ryanodine in rabbit sino-atrial node cells. J Mol Cell Cardiol 35:905–913

Camelliti P, Green CR, LeGrice I, Kohl P (2004) Fibroblast network in rabbit sinoatrial node: structural and functional identification of homogeneous and heterogeneous cell coupling. Circ Res 94:828–835

Catterall WA (1992) Cellular and molecular biology of voltage-gated sodium channels. Physiol Rev 72:S15–S48

Cerbai E, Barbieri M, Mugelli A (1996) Occurrence and properties of the hyperpolarization-activated current I_f in ventricular myocytes from normotensive and hypertensive rats during aging. Circulation 94:1674–1681

Cerbai E, Sartiani L, DePaoli P, Pino R, Maccherini M, Bizzarri F, DiCiolla F, Davoli G, Sani G, Mugelli A (2001) The properties of the pacemaker current I_f in human ventricular myocytes are modulated by cardiac disease. J Mol Cell Cardiol 33:441–448

Chang F, Gao J, Tromba C, Cohen I, DiFrancesco D (1990) Acetylcholine reverses effects of beta-agonists on pacemaker current in canine cardiac Purkinje fibers but has no direct action. A difference between primary and secondary pacemakers. Circ Res 66:633–636

Chang F, Cohen IS, DiFrancesco D, Rosen MR, Tromba C (1991) Effects of protein kinase inhibitors on canine Purkinje fibre pacemaker depolarization and the pacemaker current i_f. J Physiol (Lond) 440:367–384

Chen J, Mitcheson JS, Lin M, Sanguinetti MC (2000) Functional roles of charged residues in the putative voltage sensor of the HCN2 pacemaker channel. J Biol Chem 275:36465–36471

Chen J, Mitcheson JS, Tristani-Firouzi M, Lin M, Sanguinetti MC (2001) The S4-S5 linker couples voltage sensing and activation of pacemaker channels. Proc Natl Acad Sci U S A 98:11277–11282

Cho HS, Takano M, Noma A (2003) The electrophysiological properties of spontaneously beating pacemaker cells isolated from mouse sinoatrial node. J Physiol (Lond) 550:169–180

Clapham DE (1998) Not so funny anymore: pacing channels are cloned. Neuron 21:5–7

Cohen CJ, Bean BP, Colatsky TJ, Tsien RW (1981) Tetrodotoxin block of sodium channels in rabbit Purkinje fibers. J Gen Physiol 78:383–411

Coraboeuf E, Carmeliet E (1982) Existence of two transient outward currents in sheep cardiac Purkinje fibers. Pflugers Arch 392:352–359

Decher N, Bundis F, Vajna R, Steinmeyer K (2003) KCNE2 modulates current amplitudes and activation kinetics of HCN4: influence of KCNE family members on HCN4 currents. Pflugers Arch 446:633–640

Decher N, Chen J, Sanguinetti MC (2004) Voltage-dependent gating of hyperpolarization-activated, cyclic nucleotide-gated pacemaker channels: molecular coupling between the S4-S5 and C-linkers. J Biol Chem 279:13859–13865

Denyer JC, Brown HF (1990) Rabbit sino-atrial node cells: isolation and electrophysiological properties. J Physiol (Lond) 428:405–424

DiFrancesco D (1981a) A new interpretation of the pace-maker current in calf Purkinje fibres. J Physiol (Lond) 314:359–376

DiFrancesco D (1981b) A study of the ionic nature of the pace-maker current in calf Purkinje fibres. J Physiol (Lond) 314:377–393

DiFrancesco D (1982) Block and activation of the pacemaker i_f channel in calf Purkinje fibres: effects of potasium, caesium and rubidium. J Physiol (Lond) 329:485–507

DiFrancesco D (1986) Characterization of single pacemaker channels in cardiac sino-atrial node cells. Nature 324:470–473

DiFrancesco D (1991) The contribution of the 'pacemaker' current (i_f) to generation of spontaneous activity in rabbit sino-atrial node myocytes. J Physiol (Lond) 434:23–40

DiFrancesco D (1993) Pacemaker mechanisms in cardiac tissue. Annu Rev Physiol 55:455–472

DiFrancesco D (1995) Cardiovascular controversies: the pacemaker current, i_f, plays an important role in regulating SA node pacemaker activity. Cardiovasc Res 30:307–308

DiFrancesco D, Ferroni A (1983) Delayed activation of the cardiac pacemaker current and its dependence on conditioning pre-hyperpolarizations. Pflugers Arch 396:265–267

DiFrancesco D, Tortora P (1991) Direct activation of cardiac pacemaker channels by intra-cellular cyclic AMP. Nature 351:145–147

DiFrancesco D, Tromba C (1988a) Inhibition of the hyperpolarization-activated current (i_f) induced by acetylcholine in rabbit sino-atrial node myocytes. J Physiol (Lond) 405:477–491

DiFrancesco D, Tromba C (1988b) Muscarinic control of the hyperpolarization-activated current (i_f) in rabbit sino-atrial node myocytes. J Physiol (Lond) 405:493–510

DiFrancesco D, Ducouret P, Robinson RB (1989) Muscarinic modulation of cardiac rate at low acetylcholine concentrations. Science 243:669–671

Er F, Larbig R, Ludwig A, Biel M, Hofmann F, Beuckelmann DJ, Hoppe UC (2003) Dominant-negative suppression of HCN channels markedly reduces the native pacemaker current I(f) and undermines spontaneous beating of neonatal cardiomyocytes. Circulation 107:485–489

Fernandez-Velasco M, Goren N, Benito G, Blanco-Rivero J, Bosca L, Delgado C (2003) Regional distribution of hyperpolarization-activated current (I_f) and hyperpolarization-activated cyclic nucleotide-gated channel mRNA expression in ventricular cells from control and hypertrophied rat hearts. J Physiol (Lond) 553:395–405

Gaudesius G, Miragoli M, Thomas SP, Rohr S (2003) Coupling of cardiac electrical activity over extended distances by fibroblasts of cardiac origin. Circ Res 93:421–428

Gintant GA, Datyner NB, Cohen IS (1984) Slow inactivation of a tetrodotoxin-sensitive current in canine cardiac Purkinje fibers. Biophys J 45:509–512

Gloss B, Trost S, Bluhm W, Swanson E, Clark R, Winkfein R, Janzen K, Giles W, Chassande O, Samarut J, Dillmann W (2001) Cardiac ion channel expression and contractile function in mice with deletion of thyroid hormone receptor alpha or beta. Endocrinology 142:544–550

Goldin AL, Barchi RL, Caldwell JH, Hofmann F, Howe JR, Hunter JC, Kallen RG, Mandel G, Meisler MH, Netter YB, Noda M, Tamkun MM, Waxman SG, Wood JN, Catterall WA (2000) Nomenclature of voltage-gated sodium channels. Neuron 28:365–368

Guo J, Ono K, Noma A (1995) A sustained inward current activated at the diastolic potential range in rabbit sino-atrial node cells. J Physiol (Lond) 483:1–13

Guo J, Mitsuiye T, Noma A (1997) The sustained inward current in sino-atrial node cells of guinea-pig heart. Pflugers Arch 433:390–396

Hagiwara N, Irisawa H, Kameyama M (1988) Contribution of two types of calcium currents to the pacemaker potentials of rabbit sino-atrial node cells. J Physiol (Lond) 395:233–253

Han W, Bao W, Wang Z, Nattel S (2002) Comparison of ion-channel subunit expression in canine cardiac Purkinje fibers and ventricular muscle. Circ Res 91:790–797

Hauswirth O, Noble D, Tsien RW (1968) Adrenaline: mechanism of action on the pacemaker potential in cardiac Purkinje fibers. Science 162:916–917

Herring N, Rigg L, Terrar DA, Paterson DJ (2001) NO-cGMP pathway increases the hyper-polarisation-activated current, I(f), and heart rate during adrenergic stimulation. Cardiovasc Res 52:446–453

Hiramatsu M, Furukawa T, Sawanobori T, Hiraoka M (2002) Ion channel remodeling in cardiac hypertrophy is prevented by blood pressure reduction without affecting heart weight increase in rats with abdominal aortic banding. J Cardiovasc Pharmacol 39:866–874

Hoffman BF, Cranefield PF (1976) Electrophysiology of the heart. Futura Publishing Co, Mount Kisco

Honjo H, Boyett MR, Kodama I, Toyama J (1996) Correlation between electrical activity and the size of rabbit sinoatrial node cells. J Physiol (Lond) 496:795–808

Honjo H, Boyett MR, Coppen SR, Takagishi Y, Opthof T, Severs NJ, Kodama I (2002) Heterogeneous expression of connexins in rabbit sinoatrial node cells: correlation between connexin isotype and cell size. Cardiovasc Res 53:89–96

Irisawa H, Brown HF, Giles W (1993) Cardiac pacemaking in the sinoatrial node. Physiol Rev 73:197–227

Ju YK, Allen DG (1998) Intracellular calcium and Na^+-Ca^{2+} exchange current in isolated toad pacemaker cells. J Physiol (Lond) 508:153–166

Kaupp UB, Seifert R (2001) Molecular diversity of pacemaker ion channels. Annu Rev Physiol 63:235–257

Kodama I, Nikmaram MR, Boyett MR, Suzuki R, Honjo H, Owen JM (1997) Regional differences in the role of the Ca^{2+} and Na^+ currents in pacemaker activity in the sinoatrial node. Am J Physiol 272:H2793–H2806

Kohl P (2003) Heterogeneous cell coupling in the heart: an electrophysiological role for fibroblasts. Circ Res 93:381–383

Lancaster MK, Jones SA, Harrison SM, Boyett MR (2004) Intracellular Ca^{2+} and pacemaking within the rabbit sinoatrial node: heterogeneity of role and control. J Physiol (Lond) 556:481–494

Lei M, Brown HF (1996) Two components of the delayed rectifier potassium current, I_K, in rabbit sino-atrial node cells. Exp Physiol 81:725–741

Lei M, Honjo H, Kodama I, Boyett MR (2001) Heterogeneous expression of the delayed-rectifier K^+ currents i(K,r) and i(K,s) in rabbit sinoatrial node cells. J Physiol (Lond) 535:703–714

Lei M, Jones SA, Liu J, Lancaster MK, Fung SSM, Dobrzynski H, Camelitti P, Maier S, Noble D, Boyett MR (2004) Requirement of neuronal- and cardiac-type sodium channels for murine sinoatrial node pacemaking. J Physiol (Lond) 559:835–848

Ludwig A, Zong X, Jeglitsch M, Hofmann F, Biel M (1998) A family of hyperpolarization-activated mammalian cation channels. Nature 393:587–591

Ludwig A, Zong X, Stieber J, Hullin R, Hofmann F, Biel M (1999) Two pacemaker channels from human heart with profoundly different activation kinetics. EMBO J 18:2323–2329

Ludwig A, Budde T, Stieber J, Moosmang S, Wahl C, Holthoff K, Langebartels A, Wotjak C, Munsch T, Zong X, Feil S, Feil R, Lancel M, Chien KR, Konnerth A, Pape HC, Biel M, Hofmann F (2003) Absence epilepsy and sinus dysrhythmia in mice lacking the pacemaker channel HCN2. EMBO J 22:216–224

Maier SK, Westenbroek RE, Yamanushi TT, Dobrzynski H, Boyett MR, Catterall WA, Scheuer T (2003) An unexpected requirement for brain-type sodium channels for control of heart rate in the mouse sinoatrial node. Proc Natl Acad Sci U S A 100:3507–3512

Mangoni ME, Couette B, Bourinet E, Platzer J, Reimer D, Striessnig J, Nargeot J (2003) Functional role of L-type Cav1.3 Ca^{2+} channels in cardiac pacemaker activity. Proc Natl Acad Sci U S A 100:5543–5548

Mannikko R, Elinder F, Larsson HP (2002) Voltage-sensing mechanism is conserved among ion channels gated by opposite voltages. Nature 419:837–841

Marx SO, Kurokawa J, Reiken S, Motoike H, D'Armiento J, Marks AR, Kass RS (2002) Requirement of a macromolecular signaling complex for beta adrenergic receptor modulation of the KCNQ1-KCNE1 potassium channel. Science 295:496–499

McAllister RE, Noble D, Tsien RW (1975) Reconstruction of the electrical activity of cardiac Purkinje fibres. J Physiol (Lond) 251:1–59

Miake J, Marban E, Nuss HB (2002) Gene therapy: biological pacemaker created by gene transfer. Nature 419:132–133

Mitsuiye T, Guo J, Noma A (1999) Nicardipine-sensitive Na^+-mediated single channel currents in guinea-pig sinoatrial node pacemaker cells. J Physiol (Lond) 521:69–79

Mitsuiye T, Shinagawa Y, Noma A (2000) Sustained inward current during pacemaker depolarization in mammalian sinoatrial node cells. Circ Res 87:88–91

Mobley BA, Page E (1972) The surface area of sheep cardiac Purkinje fibres. J Physiol (Lond) 220:547–563

Moosmang S, Stieber J, Zong X, Biel M, Hofmann F, Ludwig A (2001) Cellular expression and functional characterization of four hyperpolarization-activated pacemaker channels in cardiac and neuronal tissues. Eur J Biochem 268:1646–1652

Noble D, Tsien RW (1968) The kinetics and rectifier properties of the slow potassium current in cardiac Purkinje fibres. J Physiol (Lond) 195:185–214

Pachucki J, Burmeister LA, Larsen PR (1999) Thyroid hormone regulates hyperpolarization-activated cyclic nucleotide-gated channel (HCN2) mRNA in the rat heart. Circ Res 85:498–503

Perez-Reyes E (2003) Molecular physiology of low-voltage-activated t-type calcium channels. Physiol Rev 83:117–161

Pinto JM, Sosunov EA, Gainullin RZ, Rosen MR, Boyden PA (1999) Effects of mibefradil, a T-type calcium current antagonist, on electrophysiology of Purkinje fibers that survived in the infarcted canine heart. J Cardiovasc Electrophysiol 10:1224–1235

Platzer J, Engel J, Schrott-Fischer A, Stephan K, Bova S, Chen H, Zheng H, Striessnig J (2000) Congenital deafness and sinoatrial node dysfunction in mice lacking class D L-type Ca^{2+} channels. Cell 102:89–97

Plotnikov AN, Sosunov EA, Qu J, Shlapakova IN, Anyukhovsky EP, Liu L, Janse MJ, Brink PR, Cohen IS, Robinson RB, Danilo PJ, Rosen MR (2004) Biological pacemaker implanted in canine left bundle branch provides ventricular escape rhythms that have physiologically acceptable rates. Circulation 109:506–512

Potapova I, Plotnikov A, Lu Z, Danilo P Jr, Valiunas V, Qu J, Doronin S, Zuckerman J, Shlapakova IN, Gao J, Pan Z, Herron AJ, Robinson RB, Brink PR, Rosen MR, Cohen IS (2004) Human mesenchymal stem cells as a gene delivery system to create cardiac pacemakers. Circ Res 94:952–959

Pourrier M, Zicha S, Ehrlich J, Han W, Nattel S (2003) Canine ventricular KCNE2 expression resides predominantly in Purkinje fibers. Circ Res 93:189–191

Protas L, Robinson RB (2000) Mibefradil, an I(Ca,T) blocker, effectively blocks I(Ca,L) in rabbit sinus node cells. Eur J Pharmacol 401:27–30

Protas L, DiFrancesco D, Robinson RB (2001) L-type, but not T-type calcium current changes during post-natal development in rabbit sino-atrial node. Am J Physiol 281:H1252–H1259

Qu J, Barbuti A, Protas L, Santoro B, Cohen IS, Robinson RB (2001) HCN2 over-expression in newborn and adult ventricular myocytes: distinct effects on gating and excitability. Circ Res 89:e8–e14

Qu J, Plotnikov AN, Danilo PJ, Shlapakova I, Cohen IS, Robinson RB, Rosen MR (2003) Expression and function of a biological pacemaker in canine heart. Circulation 107:1106–1109

Rigg L, Heath BM, Cui Y, Terrar DA (2000) Localisation and functional significance of ryanodine receptors during beta-adrenoceptor stimulation in the guinea-pig sino-atrial node. Cardiovasc Res 48:254–264

Robinson RB, Siegelbaum SA (2003) Hyperpolarization-activated cation currents: from molecules to physiological function. Annu Rev Physiol 65:453–480

Rosen MR, Brink PR, Cohen IS, Robinson RB (2004) Genes, stem cells and biological pacemakers. Cardiovasc Res 64:12–23

Santoro B, Tibbs GR (1999) The HCN gene family: molecular basis of the hyperpolarization-activated pacemaker channels. Ann N Y Acad Sci 868:741–764

Santoro B, Grant SGN, Bartsch D, Kandel ER (1997) Interactive cloning with the SH3 domain of N-src identifies a new brain specific ion channel protein, with homology to Eag and cyclic nucleotide-gated channels. Proc Natl Acad Sci USA 94:14815–14820

Santoro B, Liu DT, Yao H, Bartsch D, Kandel ER, Siegelbaum SA, Tibbs GR (1998) Identification of a gene encoding a hyperpolarization-activated pacemaker channel of brain. Cell 93:1–20

Satoh H (1995) Role of T-type Ca^{2+} channel inhibitors in the pacemaker depolarization in rabbit sino-atrial nodal cells. Gen Pharmacol 26:581–587

Schulze-Bahr E, Neu A, Friederich P, Kaupp UB, Breithardt G, Pongs O, Isbrandt D (2003) Pacemaker channel dysfunction in a patient with sinus node disease. J Clin Invest 111:1537–1545

Shah AK, Cohen IS, Datyner NB (1987) Background K^+ current in isolated canine cardiac Purkinje myocytes. Biophys J 52:519–526

Shi W, Wymore R, Yu H, Wu J, Wymore RT, Pan Z, Robinson RB, Dixon JE, McKinnon D, Cohen IS (1999) Distribution and prevalence of hyperpolarization-activated cation channel (HCN) mRNA expression in cardiac tissues. Circ Res 85:e1–e6

Shi W, Yu H, Wu J, Zuckerman J, Wymore R, Dixon J, Robinson RB, McKinnon D, Cohen IS (2000) The distribution and prevalence of HCN isoforms in the canine heart and their relation to the voltage dependence of I_f. Biophys J 78:353A (abstr)

Shinagawa Y, Satoh H, Noma A (2000) The sustained inward current and inward rectifier K^+ current in pacemaker cells dissociated from rat sinoatrial node. J Physiol (Lond) 523:593–605

Stieber J, Herrmann S, Feil S, Loster J, Feil R, Biel M, Hofmann F, Ludwig A (2003) The hyperpolarization-activated channel HCN4 is required for the generation of pacemaker action potentials in the embryonic heart. Proc Natl Acad Sci U S A 100:15235–15240

Thollon C, Cambarrat C, Vian J, Prost JF, Peglion JL, Vilaine JP (1994) Electrophysiological effects of S 16257, a novel sino-atrial node modulator, on rabbit and guinea-pig cardiac preparations: comparison with UL-FS 49. Br J Pharmacol 112:37–42

Tseng G-N, Boyden PA (1989) Multiple types of Ca^{2+} currents in single canine Purkinje cells. Circ Res 65:1735–1750

Ueda K, Nakamura K, Hayashi T, Inagaki N, Takahashi M, Arimura T, Morita H, Higashiue-sato Y, Hirano Y, Yasunami M, Takishita S, Yamashina A, Ohe T, Sunamori M, Hiraoka M, Kimura A (2004) Functional characterization of a trafficking-defective HCN4 mutation, D553N, associated with cardiac arrhythmia. J Biol Chem 279:27194–27198

Vaca L, Stieber J, Zong X, Ludwig A, Hofmann F, Biel M (2000) Mutations in the S4 domain of a pacemaker channel alter its voltage dependence. FEBS Lett 479:35–40

van Bogaert PP, Pittoors F (2003) Use-dependent blockade of cardiac pacemaker current (I_f) by cilobradine and zatebradine. Eur J Pharmacol 478:161–171

Varro A, Balati B, Iost N, Takacs J, Virag L, Lathrop DA, Csaba L, Talosi L, Papp JG (2000) The role of the delayed rectifier component I_{Ks} in dog ventricular muscle and Purkinje fibre repolarization. J Physiol (Lond) 523:67–81

Vassalle M (1995) Cardiovascular controversies: The pacemaker current, i_f, does not play an important role in regulating SA node pacemaker activity. Cardiovasc Res 30:309–310

Vassalle M, Yu H, Cohen IS (1995) The pacemaker current in cardiac Purkinje myocytes. J Gen Physiol 106:559–578

Vemana S, Pandey S, Larsson HP (2004) S4 Movement in a Mammalian HCN Channel. J Gen Physiol 123:21–32

Verheijck EE, van Ginneken AC, Wilders R, Bouman LN (1999) Contribution of L-type Ca_{2+f} current to electrical activity in sinoatrial nodal myocytes of rabbits. Am J Physiol 276:H1064–H1077

Vinogradova TM, Bogdanov KY, Lakatta EG (2002) beta-Adrenergic stimulation modulates ryanodine receptor Ca(2+) release during diastolic depolarization to accelerate pacemaker activity in rabbit sinoatrial nodal cells. Circ Res 90:73–79

Wainger BJ, DeGennaro M, Santoro B, Siegelbaum SA, Tibbs GR (2001) Molecular mechanism of cAMP modulation of HCN pacemaker channels. Nature 411:805–810

Wang J, Chen S, Siegelbaum SA (2001) Regulation of hyperpolarization-activated HCN channel gating and cAMP modulation due to interactions of COOH terminus and core transmembrane regions. J Gen Physiol 118:237–250

Wu JY, Cohen IS (1997) Tyrosine kinase inhibition reduces i(f) in rabbit sinoatrial node myocytes. Pflugers Arch 434:509–514

Wu JY, Cohen IS, Gaudette G, Krukenkamp I, Zuckerman J, Yu H (1999) Is I_f the only pacemaker current in mammalian atrial myocytes? Biophys J 76:A306 (abstr)

Wu JY, Yu H, Cohen IS (2000) Epidermal growth factor increases I_f in rabbit SA node cells by activating a tyrosine kinase. Biochim Biophys Acta 1463:15–19

Yamagihara K, Irisawa H (1980) Inward current activated during hyperpolarization in the rabbit sinoatrial node cell. Pflugers Arch 385:11–19

Yu H, Chang F, Cohen IS (1993a) Pacemaker current exists in ventricular myocytes. Circ Res 72:232–236

Yu H, Chang F, Cohen IS (1993b) Phosphatase inhibition by calyculin A increases i_f in canine Purkinje fibers and myocytes. Pflugers Arch 422:614–616

Yu H, Chang F, Cohen IS (1995) Pacemaker current i_f in adult cardiac ventricular myocytes. J Physiol (Lond) 485:469–483

Yu H, Wu J, Potapova I, Wymore RT, Holmes B, Zuckerman J, Pan Z, Wang H, Shi W, Robinson RB, El-Maghrabi R, Benjamin W, Dixon J, McKinnon D, Cohen IS, Wymore R (2001) MinK-related protein 1: A β subunit for the HCN ion channel subunit family enhances expression and speeds activation. Circ Res 88:e84–e87

Yu HG, Lu Z, Pan Z, Cohen IS (2004) Tyrosine kinase inhibition differentially regulates heterologously expressed HCN channels. Pflugers Arch 447:392–400

Zagotta WN, Olivier NB, Black KD, Young EC, Olson R, Gouaux E (2003) Structural basis for modulation and agonist specificity of HCN pacemaker channels. Nature 425:200–205

Zaza A, Robinson RB, DiFrancesco D (1996) Basal responses of the L-type Ca^{2+} and hyperpolarization-activated currents to autonomic agonists in the rabbit sino-atrial node. J Physiol (Lond) 491:347–355

Zhang H, Holden AV, Kodama I, Honjo H, Lei M, Varghese T, Boyett MR (2000) Mathematical models of action potentials in the periphery and center of the rabbit sinoatrial node. Am J Physiol Heart Circ Physiol 279:H397–H421

Zhang Z, Xu Y, Song H, Rodriguez J, Tuteja D, Namkung Y, Shin HS, Chiamvimonvat N (2002) Functional roles of Cav1.3 (α1D) calcium channel in sinoatrial nodes. Insight gained using gene-targeted null mutant mice. Circ Res 90:981–987

HEP (2006) 171:73–97

Proarrhythmia

D. M. Roden (✉) · M. E. Anderson

Division of Clinical Pharmacology, Vanderbilt University School of Medicine,
532 Medical Research Building I, Nashville TN, 37232, USA
dan.roden@vanderbilt.edu

Abstract The concept that antiarrhythmic drugs can exacerbate the cardiac rhythm disturbance being treated, or generate entirely new clinical arrhythmia syndromes, is not new. Abnormal cardiac rhythms due to digitalis or quinidine have been recognized for decades. This phenomenon, termed "proarrhythmia," was generally viewed as a clinical curiosity, since it was thought to be rare and unpredictable. However, the past 20 years have seen the recognition that proarrhythmia is more common than previously appreciated in certain populations, and can in fact lead to substantially increased mortality during long-term antiarrhythmic therapy. These findings, in turn, have moved proarrhythmia from a clinical curiosity to the centerpiece of antiarrhythmic drug pharmacology in at least two important respects. *First*, clinicians now select antiarrhythmic drug therapy in a particular patient not simply to maximize efficacy, but very frequently to minimize the likelihood of proarrhythmia. *Second*, avoiding proarrhythmia has become a key element of contemporary new antiarrhythmic drug development. Further, recognition of the magnitude of the problem has led to important advances in understanding basic mechanisms. While the phenomenon of proarrhythmia remains unpredictable in an individual patient, it can no longer be viewed as "idiosyncratic." Rather, gradations of risk can be assigned based on the current understanding of mechanisms, and these will doubtless improve with ongoing research at the genetic, molecular, cellular, whole heart, and clinical levels.

Keywords Proarrhythmia · Antiarrhythmic drugs · Ion channels · Pharmacogenetics

1
General Introduction

A key step in understanding proarrhythmia has been the description of specific syndromes, each with its distinctive clinical presentations and underlying mechanisms, and these are described herein. In addition, certain features are common.

The clinical presentations of proarrhythmia vary from an incidental finding of increased arrhythmia frequency in an asymptomatic patient to severe symptoms such as syncope or death. While management varies by specific syndrome (and putative mechanism), certain considerations are common. The first is that proarrhythmia should be considered in the diagnosis whenever a patient presents with new or worsening arrhythmias: that is, proarrhythmia must be recognized. The second is that any factor that exacerbates the clinical syndrome should also be recognized and treated or removed. Thus, for example, intercurrent abnormalities such as hypokalemia or hypoxemia may exacerbate many types of proarrhythmia. Another common mechanism increasing the likelihood of a proarrhythmic response to drug therapy is pharmacokinetic interactions that elevate plasma drug concentrations and hence increase adverse effects. A third common feature of proarrhythmia is that multiple individual risk factors can often be identified in affected patients. While this observation makes prediction of risk in an individual, or in a population, somewhat difficult, it has also provided the impetus for some interesting newer work examining the role of genetic variants in modulating proarrhythmia risk. Exploration of the hypothesis that proarrhythmia risk includes a genetic component has provided a very useful starting point for examining genetic modulation of other forms of adverse drug reactions. In addition, identification of DNA variants that modulate proarrhythmia risks may also provide a window into understanding genetic modulation of common arrhythmia presentations.

Table 1 lists recognized proarrhythmia syndromes. The phenomenon of increased mortality during long-term antiarrhythmic therapy is listed as a separate entry. This outcome has been reported with both sodium channel-blocking agents and QT prolonging agents, and it is, naturally, felt that these outcomes

Table 1 Proarrhythmia syndromes

Culprit drug(s)	Clinical manifestations	Likely mechanisms
Digitalis, including herbal remedies containing digitalis (foxglove tea, toad venom)	Cardiac: Sinus bradycardia or exit block; AV nodal block; atrial tachycardia, bi-directional ventricular tachycardia; virtually any other arrhythmia can occur Non-cardiac: nausea; visual disturbances; cognitive dysfunction	Intracellular calcium overload leading to enhanced I_{ti} and delayed afterdepolarizations
QT interval-prolonging drugs:	QT prolongation and distortion; torsades de pointes	Heterogeneity of action potential prolongation, early afterdepolarizations, unstable intramural reentry (see text)
Antiarrhythmics: disopyramide, dofetilide, ibutilide, procainamide, quinidine, sotalol Non-antiarrhythmics (rarer)[a]		
Sodium channel-blocking drugs:	Exacerbated VT:	Reentry due to:
Antiarrhythmics: disopyramide, flecainide, procainamide, propafenone, quinidine	Increased frequency of VT in a patient with reentrant VT	Slowed conduction, especially within established or potential reentrant circuits and/or
Other: tricyclic antidepressants, cocaine	New VT in a patient susceptible to VT (e.g., with a myocardial scar)	Enhanced heterogeneity of repolarization, especially in the right ventricular outflow tract
	Difficulty cardioverting VT; Incessant VT	
	VT that becomes poorly tolerated hemodynamically (even if rate is slower)	
	Atrial flutter with 1:1 AV conduction	
	Increased pacing or defibrillating thresholds	
Sudden death coincident with drug administration: 5-fluorouracil, ephedra, anti-migraine agents (triptans), cocaine		Unknown. ? coronary spasm
Increased mortality during placebo-controlled trials: Flecainide, moricizine, and other sodium channel blockers d-Sotalol		Not established; likely related to torsades de pointes or unstable reentry (see text)

AV, atrioventricular; I_{ti}, transient inward current; VT, ventricular tachycardia [a]Many drugs have been implicated; one list and the strength of evidence linking drugs to QT prolongation can be found at www.torsades.org

reflect an extreme manifestation of proarrhythmia, i.e., proarrhythmia that results in death [Cardiac Arrhythmia Suppression Trial (CAST) Investigators 1989; The Cardiac Arrhythmia Suppression Trial II Investigators 1991; Waldo et al. 1995]. There seems little doubt that this scenario does occur, since such deaths are occasionally witnessed and patients can be resuscitated. However, it remains possible that therapy with these drugs increases mortality by mechanisms that have yet to be described.

High drug concentrations occur in three distinct clinical settings: overdose, dysfunction of the major organs of elimination, and drug interactions. The greatest risk is with drugs that undergo elimination by a single pathway, such as metabolism by a specific hepatic cytochrome P450 (CYP superfamily member) or by renal excretion. This represents a "high-risk" pharmacokinetic scenario, since dysfunction of the single elimination pathway (by disease, genetic factors, or concomitant drug therapy) can then result in extraordinary increases in plasma drug concentration due to the absence of alternate pathways of elimination. Such high-risk pharmacokinetics is a common mechanism whereby drug interactions result in clinically important adverse effects; Table 2 lists examples that increase the risk of proarrhythmia.

Thus, the general management of all forms of proarrhythmia includes withdrawal of any potentially offending agents and correction of other exacerbating clinical conditions. In addition, specific therapies have been proposed for some proarrhythmia syndromes, and these are discussed further in Sects. 4 and 5. None has been formally tested in a double-blind, randomized, placebo-controlled trial, and because of the sporadic nature of proarrhythmia and the potentially serious consequences of withholding treatment, never will be. Nevertheless, particularly when such therapies are based on a clear understanding of underlying pathophysiology, they can be highly effective compared to historical controls.

2
Digitalis Intoxication

2.1
Clinical Features

Digitalis glycosides have been used for the therapy of congestive heart failure and for cardiac arrhythmias [notably atrial fibrillation (AF) with rapid ventricular responses] for centuries (Willius and Keys 1942). Excess digitalis not only causes arrhythmias but a variety of extra-cardiac symptoms including confusion, visual abnormalities, and nausea. It has been speculated that the striking luminescence characteristic of Van Gogh's paintings late in his life actually is a symptom of digitalis intoxication; one reason for this speculation is that his portraits of his physician and friend Dr. Gachet showed him holding branches of the foxglove plant from which digitalis is derived (Lee 1981).

Table 2 Drug interactions increasing proarrhythmia risk

Drug	Interacting drug	Effect
Increased concentration of arrhythmogenic drug		
Digoxin	Some antibiotics	Elimination of gut flora that metabolize digoxin (Lindenbaum et al. 1981), or P-glycoprotein inhibition
Digoxin	Amiodarone	Increased digoxin concentration and toxicity
	Quinidine	
	Verapamil	
	Cyclosporine	
	Itraconazole	
	Erythromycin	
Cisapride[a]	Ketoconazole	Increased drug levels
Terfenadine, astemizole[a]	Itraconazole	
	Erythromycin	
	Clarithromycin	
	Some Ca^{2+} channel blockers	
	Some HIV protease inhibitors (especially ritonavir)	
Propafenone	Quinidine (even ultra-low dose)	Increased β-blockade
	Fluoxetine	
	Some tricyclic antidepressants	
Flecainide	Quinidine (even ultra-low dose)	Increased adverse effects (usually only if renal dysfunction also present)
	Fluoxetine	
	Some tricyclic antidepressants	
Dofetilide	Verapamil	Increased plasma concentration
Decreased concentration of antiarrhythmic drug		
Digoxin	Antacids	Decreased digoxin effect due to decreased absorption
	Rifampin	Increased P-glycoprotein activity
Quinidine, mexiletine	Rifampin, barbiturates	Induced drug metabolism

Table 2 (continued)

Drug	Interacting drug	Effect
Synergistic pharmacologic activity causing arrhythmias		
QT-prolonging antiarrhythmics (see Table 1)	Diuretics	Increased torsades de pointes risk due to diuretic-induced hypokalemia
β-Blockers		Bradycardia when used in combination
Digoxin		Bradycardia when used in combination
Verapamil		Bradycardia when used in combination
Diltiazem		Bradycardia when used in combination
Clonidine		Bradycardia when used in combination
PDE5 inhibitors (sildenafil, vardenafil, and others)	Nitrates	Increased and persistent vasodilation; risk of myocardial ischemia

[a]No longer available, or availability highly restricted

The cardiovascular manifestations of digitalis intoxication reflect inhibition of sodium-potassium ATPase, ultimately resulting in intracellular calcium overload, as well as an "indirect" vagotonic action (Smith 1988). With very severe intoxication, ATPase inhibition can result in profound hyperkalemia. These mechanisms account for the common arrhythmias seen with digitalis intoxication: abnormal automaticity in the form of isolated ectopic beats or sustained automatic tachyarrhythmias [arising in the atrioventricular (AV) junction or in the ventricles] as well as sinus bradycardia and AV nodal block. Clinical situations that exacerbate these toxicities include hypokalemia and hypothyroidism.

The most widely used preparation of digitalis is digoxin, which is excreted unchanged primarily through the kidneys. In renal dysfunction, therefore, the risk of digitalis toxicity rises if doses are not appropriately adjusted downward. Monitoring plasma digoxin concentrations has been a useful adjunct to reduce the incidence of toxicity. Plasma concentrations exceeding 2 ng/ml increase the risk of digitalis intoxication, and severe cardiovascular manifestations are common with concentrations above 5 ng/ml. The diagnosis of digitalis toxicity is usually one of clinical suspicion in a patient with typical arrhythmias, extra-cardiac symptoms (notably nausea), and elevated serum digoxin concentrations. Suicidal digitalis overdose can produce cardiac inexcitability due to hyperkalemia, which can be extreme (>10 mEq/l) and difficult to manage.

It was recognized in the late 1970s that administration of quinidine to a patient receiving chronic digoxin therapy doubles serum digoxin concentrations and leads to toxicity (Leahey et al. 1978). The mechanism remained obscure until the recognition that digoxin is a substrate for the drug efflux transporter P-glycoprotein (Tanigawara et al. 1992) encoded by the gene *MDR1*, normally expressed in the kidney and biliary tract (where it promotes digoxin efflux), on the luminal aspect of enterocytes (where it limits digoxin bioavailability), and on the endothelial surface of the capillaries of the blood–brain barrier (where it serves to limit CNS drug penetration). Clinical studies (Angelin et al. 1987; De Lannoy et al. 1992; Su and Huang 1996), as well as studies in mice in which *MDR1* has been disrupted (Fromm et al. 1999), support the idea that quinidine doubles serum digoxin concentration by inhibiting P-glycoprotein, and thereby reducing renal and biliary excretion as well as increasing drug bioavailability. Increased levels of the drug in the CNS may also contribute, particularly to the vagotonic (bradycardic) and "non-cardiac" effects. Similarly, a range of structurally and mechanistically unrelated drugs produce effects similar to those of quinidine; these include amiodarone, itraconazole, erythromycin, cyclosporine, and verapamil. The common mechanism appears to be P-glycoprotein inhibition, and these clinically important interactions represent examples of "high-risk" pharmacokinetics.

2.2
Mechanisms

The major target for digitalis glycosides is the sodium–potassium ATPase. Inhibition of this electrogenic pump leads to intracellular sodium overload, with resultant increased activity of the sodium–calcium exchanger, ultimately resulting in intracellular calcium overload. This long-hypothesized requirement for the sodium–calcium exchanger has been verified in heart cells from sodium–calcium exchanger knock-out mice, which fail to develop calcium overload even after exposure to very high levels of digitalis glycosides (Reuter et al. 2002). Intracellular calcium overload is exacerbated by stimulation at fast rates, and action potentials recorded from digitalis-intoxicated preparations show spontaneous depolarizations, termed delayed afterdepolarizations (DADs), following episodes of rapid pacing. DAD amplitude is determined by calcium release from intracellular calcium components, or stores, and is generally increased with a longer duration or a more rapid rate of antecedent pacing. DADs that reach threshold may generate single or sustained DAD-dependent action potentials. Presumably, this is the mechanism that underlies isolated or sustained ectopic activity in digitalis-intoxicated patients (Antman and Smith 1986). The nature of the inward current (I_{ti}) that underlies a DAD has not been established for all models and experimental systems, but the sodium–calcium exchange current is a leading candidate (Schlotthauer and Bers 2000).

2.3
Treatment

In mild forms of toxicity (few serious arrhythmias; serum concentrations <4–5 ng/ml), monitoring cardiac rhythm while the drug is eliminated may be sufficient. Occasionally, temporary pacing may be required. When arrhythmias are sufficiently severe as to warrant therapy, the treatment of choice is anti-digoxin antibody. In the largest clinical series reported to date, response occurred rapidly, within 4 h, and the treatment was remarkably effective: over half of the patients who presented with a cardiac arrest actually survived hospitalization (Antman et al. 1990). The rapid removal of active digitalis circulation by the Fab antibody can result in an increased ventricular rate during AF and exacerbation of heart failure, as well as hypokalemia, as the glycoside is rapidly bound to the antibody. Because the drug is still present in the circulation (albeit bound), serum digoxin concentrations cannot be interpreted, and thus serum digoxin measurement is not indicated after the antibody has been administered. Older approaches to therapy with antiarrhythmic drugs such as lidocaine or phenytoin have been supplanted by specific anti-digoxin antibody therapy.

2.4
Genetics

Polymorphisms have been described in the human MDR1 gene, and one of these, C3435T, has been associated with the variability in digoxin concentrations (Hoffmeyer et al. 2000; Kim et al. 2001). While this polymorphism is located in the coding region of the gene, it is synonymous (i.e., there is no predicted amino acid change). It seems likely the functional effects described here reflect the fact that this polymorphism itself modulates P-glycoprotein expression or it is in linkage disequilibrium with a polymorphism that modulates P-glycoprotein expression.

3
Drug-Induced Torsades de Pointes

3.1
Clinical Features

Quinidine was introduced into clinical drug therapy in the early 1920s (Wenckebach 1923), and syncope following the initiation of the drug was recognized in occasional patients shortly thereafter. The advent of online electrocardiographic monitoring in the 1960s established that quinidine syncope was caused by what we now recognize as torsades de pointes (Selzer and Wray 1964). Interestingly, the actual term was coined to describe the arrhythmia in a different context, an elderly woman with heart block and recurring episodes of syncope due to torsades de pointes (Dessertenne 1966). The initial descriptions of torsades de pointes actually did not highlight the QT interval prolongation of antecedent sinus beats that is now recognized as an important component of the syndrome. In typical drug-induced cases, a stereotypical series of cycle length changes ("short-long-short"; Fig. 1) is almost inevitably present (Kay et al. 1983; Roden et al. 1986).

Clinical studies have identified a series of risk factors for torsades de pointes listed in Table 3. These have provided an important starting point for "bedside to bench" research to address fundamental mechanisms, as described further below. In some cases, such as hypokalemia, these mechanisms are reasonably well understood. In other cases, such as female gender (Makkar et al. 1993) or a period of increased risk after conversion of AF to normal rhythm (Choy et al. 1999), they remain poorly understood. Similarly, the mechanisms whereby QT prolongation by amiodarone is associated with a much smaller risk of torsades de pointes than that by other drugs are not well understood (Lazzara 1989). A large clinical trial of a QT-prolonging antiarrhythmic, the non-β-blocking d-isomer of sotalol, showed higher mortality with drug compared to placebo (Waldo et al. 1995).

While antiarrhythmic drugs were the first recognized cause of drug-induced torsades de pointes, the syndrome has been increasingly recognized with "non-

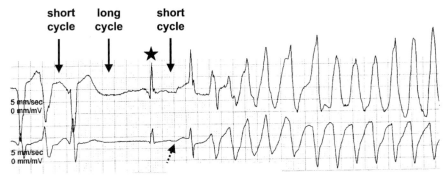

Fig. 1 Two-lead ECG recording during a typical episode of drug-induced torsades de pointes, in this case attributed to accumulation of the active metabolite N-acetyl procainamide (NAPA; plasma concentration 27 mg/ml) in a patient who developed renal failure while receiving procainamide. The stereotypical "short-long-short" series of cycle-length changes prior to the polymorphic tachycardia is indicated. Note that the second "short" cycle is actually the interrupted QT interval of the last supraventricular beat (shown by a *star*). The *broken arrow* indicates QTU deformity of this beat, most evident in the *lower tracing*

Table 3 Risk factors for drug-induced torsades de pointes

Factor	Reference(s)
Female gender	Makkar et al. 1993
Hypokalemia	Kay et al. 1983; Roden et al. 1986
Bradycardia	Kay et al. 1983; Roden et al. 1986
Recent conversion from atrial fibrillation	Houltz et al. 1998; Tan and Wilde 1998; Choy et al. 1999
Congestive heart failure	Torp-Pedersen et al. 1999
Digitalis therapy	Houltz et al. 1998
Subclinical congenital long QT syndrome	Donger et al. 1997; Napolitano et al. 1997, 2000; Yang et al. 2002
DNA polymorphisms	Abbott et al. 1999; Splawski et al. 2002; Sesti et al. 2000
High drug concentration (except quinidine)	Neuvonen et al. 1981; Woosley et al. 1993; Roden et al. 1986
Rapid rate of drug administration	Carlsson et al. 1993
Baseline QT prolongation	Houltz et al. 1998
Severe hypomagnesemia	Reddy et al. 1984

cardiovascular" therapies (Roden 2004a). Indeed, QT prolongation and torsades de pointes have been the single most common cause of withdrawal of marketed drugs in the past decade. The problem of torsades de pointes during treatment with "non-cardiovascular" drugs became particularly apparent in the early 1990s with the recognition of the problem with the antihistamine terfenadine (Monahan et al. 1990) and the gastric pro-kinetic drug cisapride (Bran et al. 1995). These agents represent another important example of "high-risk" pharmacokinetics, since they are both very potent QT-prolonging agents, but undergo very rapid (and indeed near-complete) pre-systemic biotransformation by the CYP3A enzyme system, and the resulting metabolites are devoid of QT-prolonging activity (Woosley et al. 1993). The risk of torsades de pointes with these agents appears almost exclusively confined to settings in which this protective presystemic clearance has been bypassed: patients receiving CYP3A inhibitors, such as erythromycin or ketoconazole, and those with advanced liver disease or overdose. In contrast to other drugs, torsades de pointes with quinidine occurs at low dosages and plasma concentrations, and investigation of the underlying mechanisms has been quite informative, as discussed in the following section.

One of the first tools used to study marked QT prolongation and torsades de pointes was intravenous administration of cesium, a relatively nonspecific potassium current blocker, in dogs (Brachmann et al. 1983). Interestingly, torsades de pointes has now been reported in patients receiving relatively large doses of cesium orally as "alternative therapy" for cancer (Pinter et al. 2002).

3.2
Mechanisms

Basic electrophysiologic considerations dictate that the QT prolongation characteristic of drug-induced torsades de pointes reflects prolongation of action potential durations in at least some ventricular cells. In turn, such action potential prolongation must reflect increased inward current or decreased outward current during the plateau of the action potential. Studies in congenital LQTS have elegantly confirmed these assumptions by demonstrating that disease-associated mutations in the cardiac sodium channel increase inward current, while those in the genes encoding the rapid or slow components of the delayed rectifier (I_{Kr} and I_{Ks}) reduce outward current (Keating and Sanguinetti 2001). Virtually all drugs that prolong the QT interval do so by blocking I_{Kr}. Compounds that enhance sodium current during the plateau also prolong action-potential duration (Kuhlkamp et al. 2003), but these are not clinically used. A number of compounds also block I_{Ks}, but "pure" I_{Ks} blockers have not undergone clinical trials.

Because the issue of drug-induced torsades de pointes has become an important consideration in risk–benefit evaluations by regulatory agencies such as the Food and Drug Administration, in vitro studies and animal models described below have been used to assess the potential for a new drug to cause torsades de pointes. Such screening often starts with description of the effects of a new drug on I_{Kr} and on action potentials recorded in cardiac tissues from non-rodent mammals (guinea pig, rabbit, dog) and may continue to animal models in which susceptibility to the arrhythmia can be more directly assessed.

3.2.1
Ionic Currents and Action Potential Prolongation

I_{Kr} is generated by expression of the human ether a-go-go-related gene (*HERG*, now termed *KCNH2*). The electrophysiological characteristics of heterologously expressed KCNH2 are very similar, but not identical to, I_{Kr} recorded from human cells; one commonly observed difference is in the rates of deactivation. A commonly invoked explanation for this discrepancy is that KCNH2 associates with other protein(s) in at least some (but perhaps not all) human myocytes to generate I_{Kr}. Candidate function-modifying proteins include members of the KCNE family (notably KCNE2) (Abbott et al. 1999), as well as the α-subunit that generates I_{Ks} (KCNQ1) (Ehrlich et al. 2004). Differences in post-translational modification may represent another mechanism. Site-directed mutagenesis and structural modeling studies have identified key features of the HERG/KCNH2 protein, absent in other potassium channels, that seem to underlie the fact that many structurally unrelated drugs inhibit I_{Kr} (Mitcheson et al. 2000). Drugs block the channel by accessing the pore region from the intracellular side of the channel. The HERG/KCNH2 protein, unlike other K^+ channels, lacks proline groups within S6, and the resultant lack of "kinking" of the S6 region is thought to facilitate access of even relatively bulky drugs to the pore region. In addition, the S6 also includes two aromatic residues, absent in other K^+ channels, oriented to face the pore, and these are thought to provide high-affinity drug binding sites with many drugs. Since the channel is a tetramer, there are actually eight such potential high-affinity sites within the pore, a feature also absent in other K^+ channels.

Screening new drugs for I_{Kr} block can be done using myocytes from a number of mammalian species (dog, rabbit, cat, guinea pig; but not adult mouse or rat), in cultured neonatal mouse cells (AT1 or HL1 cells), or by heterologous expression of HERG/KCNH2. The results obtained in such studies can generally define whether or not a drug is a potent blocker of the current, one important component of assessing the balance of potential risk versus anticipated benefit for a new drug.

When conditions mimicking those seen in torsades de pointes (hypokalemia, slow drive rates, QT-prolonging drug) are used in vitro, action potentials in cells of the conduction system (Purkinje fibers) markedly prolong and generate distinctive discontinuities and spontaneous upstrokes, arising from phase 3 of the action potential (Strauss et al. 1970; Dangman and Hoffman 1981; Roden and Hoffman 1985). These events are termed early afterdepolarizations (EADs), distinguishing them from DADs that arise after the action potential is fully repolarized. The ionic current that underlies the upstroke represented by EADs has not been fully defined. In some experiments, it is clear that reactivation of L-type calcium channels (enabled by the long phase 2 of prolonged action potentials) contributes (January and Riddle 1989). Indeed, in in vitro experiments L-type calcium channel blockers are highly effective in eliminating the triggered upstroke and reducing action potential prolongation (Nattel and Quantz 1988). Nevertheless, these agents have not been terribly effective at preventing torsades de pointes (although there is no randomized prospective trial). Other evidence points to a role for intercellular calcium overload and an I_{ti}-like mechanism, especially for EADs arising during phase 3 of the action potential (Wu et al. 1999). For example, although EADs are generally considered to be "bradycardia-dependent," they can also be elicited by rate acceleration, followed by a brief pause (Burashnikov and Antzelevitch 1998).

3.2.2
Action Potential Prolongation and Arrhythmogenesis

Action potential prolongation provides the initial electrophysiologic change that ultimately generates torsades de pointes. EADs are readily elicited in canine and rabbit cardiac Purkinje fibers but occur much less readily in ventricular muscle. Thus, an initial concept was that an EAD-triggered upstroke elicited in the conduction system propagated through the myocardium to generate torsades de pointes. Within the last 10 years, Charles Antzelevitch's laboratory has popularized a canine "wedge" preparation in which action potentials can be recorded from multiple layers of the myocardium (Belardinelli et al. 2003). The wedge preparation has defined the properties of a group of cells located in the mid-myocardium ("M cells") that respond to torsades de pointes-generating conditions in much the same way as Purkinje fibers, with marked action potential prolongation, and occasionally EADs. Further, the cell layers abutting the M cell layer (epicardium and endocardium) display much less dramatic changes in action potential duration and only rarely show EADs.

The electrocardiographic morphology of torsades de pointes, with a gradually "twisting" QRS axis, has been reproduced by pacing the right and left ventricles in isolated rabbit hearts at slightly different rates (D'Alnoncourt et al. 1982). This result likely reflects varying activation from the two pacemaker sites, and may or may not be relevant to the unusual morphology of torsades de pointes. Studies in the wedge preparation and using three-dimensional mapping techniques in dogs suggest that the unusual morphology arises from time-dependent functional arcs of block usually located at the M cell/epicardial boundary, that allow reentrant excitation across the thickness of the myocardium to occur, but with a slightly different activation sequence in each succeeding beat (El-Sherif et al. 1997; Akar et al. 2002). Thus, a contemporary view holds that physiologic transmural heterogeneities of action potential duration are exaggerated by torsades de pointes-generating conditions, and that this defines an important proximate substrate for the genesis of torsades de pointes. Whether the initiating beat is a triggered upstroke in the Purkinje network or elsewhere has not been fully defined. In the wedge preparation, torsades de pointes can be readily elicited by programmed electrical stimulation, but usually from the epicardium (Shimizu and Antzelevitch 1999), whereas programmed electrical stimulation in humans (from the endocardium) rarely elicits torsades de pointes. Interestingly, initiation of polymorphic ventricular tachycardia (VT) has been reported in the setting of advanced heart disease and left ventricular epicardial pacing (Medina-Ravell et al. 2003).

Administration of a QT-prolonging drug is generally insufficient to elicit marked QT prolongation and torsades de pointes in experimental animals. Nevertheless, a number of animal models in which susceptibility to the arrhythmia can be assessed have been developed; these have the common characteristic that some intervention has been made to enhance susceptibility. A well-studied rabbit model involves pretreatment with methoxamine; the mechanism whereby this pretreatment enhances the likelihood that an I_{Kr} blocker will generate torsades de pointes is not completely understood (Carlsson et al. 1990). One possibility is that methoxamine blocks other repolarizing currents (notably the transient outward current) to thereby exaggerate the susceptibility of the repolarization process to I_{Kr} block. Methoxamine also engages the baroreflex, and the resultant heart rate slowing also promotes torsades de pointes. In the dog, destruction of the AV node to create complete heart block similarly sensitizes the heart to I_{Kr} blockers (Chezalviel-Guilbert et al. 1995; Vos et al. 1998). The effect is apparent immediately after creation of block, but over time the sensitivity of the animal to I_{Kr} blockers becomes exaggerated. Mechanistic studies suggest that increased heterogeneities of repolarization time (including between the right and left ventricles) as well as down-regulation of I_{Ks} contribute to this increased sensitivity (Volders et al. 1998).

3.2.3
Variability in Response to I_{Kr} Block

As already discussed, the extent of action potential prolongation by I_{Kr} block varies among cell types, strongly suggesting that the contribution of individual ionic currents to overall cardiac repolarization varies across tissues. Repolarization in the conduction system and the ventricle is a complex process, involving both waning inward calcium (and possibly sodium) currents and increasing outward current through multiple potassium channels, including those underlying I_{Kr} and I_{Ks}. The animal models that require some "sensitizer" to fully elicit the arrhythmogenic effects of I_{Kr} block are also consistent with variable contributions by multiple ionic currents to normal repolarization.

Variability in the extent of QT prolongation by I_{Kr} blockers in humans is also consistent with this notion. We have proposed that multiple mechanisms exist to maintain QT intervals in the normal range upon exposure to I_{Kr} block, and that susceptibility to torsades de pointes may therefore reflect subtle lesions in these protective mechanisms that become manifest as marked QT prolongation during challenge by an I_{Kr} blocker (Roden 1998). Such lesions often reduce other K^+ currents, as has been described in heart failure, in dogs with chronic heart block and in patients with sub-clinical congenital long QT syndrome (LQTS). In other situations, I_{Kr} itself may be unusually sensitive to block by drugs. For example, lowering extracellular potassium from 8 to 1 mM decreased the IC_{50} for I_{Kr} block by dofetilide ~40-fold and by quinidine ~10-fold (Yang and Roden 1996). This is entirely consistent with the clinical observation that hypokalemia potentiates torsades de pointes risk. In addition, simply lowering extracellular potassium reduces I_{Kr} amplitude, an effect opposite to that predicted by the Nernst equation (Yang and Roden 1996). Two explanations have been proposed: that lowering extracellular K^+ either enhances the fast inactivation the channel undergoes upon depolarization (Yang et al. 1997), or enhances an I_{Kr} blocking property of normal extracellular sodium concentrations (Numaguchi et al. 2000), or both.

The lack of increasing risk with increasing plasma quinidine concentrations can be understood in this framework. Quinidine blocks I_{Kr} at extraordinarily low, generally "sub-therapeutic" concentrations, whereas at higher concentrations it blocks inward sodium current (I_{Na}), transient outward current (I_{to}), and I_{Ks}. Thus, the effect of quinidine depends on the relative contributions of these currents to repolarization in an individual patient, and the extent of block (itself a function of important patient-specific factors such as heart rate). Indeed, in vitro, low quinidine concentrations readily generate EADs that can be reversed by increasing the concentration of the drug, presumably reflecting sodium channel block (Belardinelli et al. 2003).

3.3
Genetics

Torsades de pointes occurs in drug-induced and congenital LQTS, and less commonly in other settings. In addition, there are interesting similarities between the congenital and drug-associated forms of the syndrome, e.g., female preponderance, and exaggeration by clinical risk factors such as hypokalemia and bradycardia. These parallels, and the relatively unpredictable nature of the drug-induced form, suggest the hypothesis that susceptibility to the drug-induced form may be, in part, genetically determined. Two distinct mechanisms have been described whereby DNA variants may modulate susceptibility to drug-induced torsades de pointes. The first is exposure of a subclinical ("forme fruste") variant of the congenital syndrome, due to a mutation in a congenital LQTS disease gene, and the second is identification of more common DNA variants, polymorphisms, that appear to increase susceptibility.

The cloning of the LQTS disease genes has enabled genotyping within affected kindreds and has been followed by the demonstration of incomplete penetrance, i.e., normal QT intervals in individuals who are nevertheless mutation carriers. A number of reports now identify drug challenge in such kindreds as a mechanism exposing the congenital syndrome (Donger et al. 1997; Napolitano et al. 2000; Yang et al. 2002). A common finding in such reports involves mutations in KCNQ1 (encoding the pore-forming protein for I_{Ks}), and these mutations, when studied, confer relatively minor functional defects in vitro. This is consistent with the fact that individuals may have normal or near normal QT intervals prior to drug challenge but nevertheless represent a situation of reduced repolarization reserve. Individuals with drug-induced torsades de pointes later discovered to have subclinical mutations in *KCNH2* and in *SCN5A* (encoding the cardiac sodium channel) have also been reported (Yang et al. 2002; Makita et al. 2002).

LQTS is relatively rare (with mutation carrier frequency of perhaps 1/3,000-1/1,000) (Roden 2004b). By contrast, the identification of common polymorphisms modulating the risk of drug-induced torsades de pointes might have much more widespread public health implications. One variant, S1103Y in *SCN5A*, confers a subtle gain of function genotype that appears to have little effect on baseline QT intervals, or on computed action potentials (Splawski et al. 2002). However, an association study showed a strikingly high incidence of the minor (tyrosine) allele in subjects with a range of arrhythmia syndromes, including drug-induced arrhythmias. Interestingly, this polymorphism is relatively common in African-American populations, but not detected in other groups. Q9E in KCNE2 (originally termed MiRP1) was initially identified as a mutation increasing susceptibility to drug-induced torsades de pointes by increasing sensitivity of I_{Kr} channels to drug block (Abbott et al. 1999). The proband was also an African-American subject, and subsequent analyses have demonstrated that this variant, like S1103Y, is common in African-Americans, but not detected in other populations. KCNE2 has been difficult to detect in mammalian heart, but a recent report suggesting that its expression is confined largely to Purkinje fibers may be especially relevant to the issue of long QT-related arrhythmia (Pourrier et al. 2003).

A *KCNE1* (*minK*) polymorphism resulting in D85N in the intracellular C-terminus of the protein has been associated with altered channel gating (Wei et al. 1999). In silico simulations indicate that this I_{Ks} gating defect has no effect on baseline action potential durations, but does enhance the likelihood of EADs with I_{Kr} block. Association studies have not yet linked this polymorphism to increased torsades de pointes susceptibility. There are a number of other common polymorphisms in the LQTS disease genes that may alter channel function, but none of these has yet been convincingly linked to the drug-induced arrhythmia genotype.

3.4
Treatment

Following recognition, withdrawal of offending agents, and correction of serum potassium to high normal values, intravenous magnesium is the treatment of choice for drug-induced torsades de pointes. The mechanism whereby this therapy appears effective has not been fully elucidated and may involve an effect of the drug on L-type calcium channels. Interestingly, magnesium does not generally shorten the QT interval, but does appear to reduce the incidence of episodes of torsades de pointes. This is consistent with an effect of the drug on EAD on triggered upstrokes, and thus may involve an effect on L-type calcium channels.

The almost inevitable presence of a pause just prior to an episode of drug-induced torsades de pointes provides the rationale for other therapies used if magnesium is ineffective: Cardiac pacing or isoproterenol both increase heart rate and abolish pauses. Isoproterenol may also augment I_{Ks} and reduce I_{Ks} block by drug (Yang et al. 2003). To the extent that this mechanism contributes to prolonged QT intervals by mixed blockers such as quinidine, this may be an additional beneficial effect.

4
Proarrhythmia Due to Sodium Channel Block

4.1
Clinical Features

The initial description of sodium channel blocker-related proarrhythmia probably came with the aggressive use of high doses of quinidine to pharmacologically convert AF decades ago. In occasional patients, this dosing tactic resulted in a wide-complex, relatively slow VT (Wetherbee et al. 1951). This approach to AF management was supplanted by electrical cardioversion, but similar VTs were noted with the introduction of the potent sodium channel blockers encainide and flecainide (Winkle et al. 1981; Oetgen et al. 1983). Multiple proarrhythmia syndromes, each attributable to sodium channel block, have now been described (Table 1). Patients with sustained monomorphic VT due to macro reentry related to remote myocardial infarction may experience an increase in frequency of VT episodes with sodium channel blockers. Occasionally, the tachycardia is slower, but nevertheless may be more hemodynamically significant and may be more difficult to cardiovert. Patients with a VT substrate (e.g., those with remote myocardial infarction) but who have not yet experienced this arrhythmia may present after initiation of therapy with sodium channel blockers.

In patients receiving sodium channel blockers (quinidine, propafenone, flecainide, amiodarone) for management of atrial fibrillation, a frequent outcome of drug therapy is the "regularization" of atrial activity to an atrial flutter-like rhythm. This arrhythmia appears to be, like typical atrial flutter, macroreentrant and frequently involving an isthmus in the lower right atrium. However, unlike typical atrial flutter in the drug-free patient, the rate is slower ($\sim$200 vs $\sim$300/min). As a consequence of this slowing of atrial rate, 1:1 atrioventricular conduction can occur. Moreover, since sodium channel block is use-dependent and exaggerated at fast rates, impulse propagation within the ventricles under these conditions may actually be slower than in sinus rhythm. As a result of all of these abnormalities, the patient may present with a regular rhythm, at $\sim$200/min, with wide QRS complexes. Not surprisingly, this arrhythmia is readily confused with VT (Crijns et al. 1988; Falk 1989).

Loss of sodium channel function is an important mechanism in the Brugada syndrome, and (in analogy to the congenital LQTS) exposure to sodium channel blockers may unmask subclinical Brugada syndrome. Another important effect of sodium channel block is decreased excitability, and this may be clinically manifest as an increase in energy requirement for cardiac pacing and defibrillation (Echt et al. 1989).

Cocaine and some tricyclic antidepressants also have sodium channel blocking and QT-prolonging I_{Kr}-blocking properties (Zhang et al. 2001; Nattel 1985). Arrhythmias during exposure to cocaine may reflect either of these properties or superimposed myocardial ischemia due to coronary vasospasm. Tricyclic antidepressant overdose is characterized by sinus tachycardia due to anticholinergic effects, wide QRS durations due to sodium channel block, and CNS toxicity. Torsades de pointes, while reported, is rare.

Sodium channel blockers increase mortality when used in patients with recent myocardial infarction. This result was best shown in the Cardiac Arrhythmia Suppression Trial (CAST) [Cardiac Arrhythmia Suppression Trial (CAST) Investigators 1989], but was also hinted at in earlier studies with mexiletine and disopyramide (Impact Research Group 1984; UK Rythmodan Multicentre Study Group 1984). Reanalysis of the CAST database strongly suggests that therapy with encainide or flecainide in patients susceptible to recurrent myocardial ischemia was an especially important combination in increasing risk for sudden death (Akiyama et al. 1991). It seems reasonable to hypothesize that the increased mortality in CAST arose from unstable VT due to conduction slowing, increased transmural heterogeneity of action potentials (similar to the Brugada syndrome), or both. However, as discussed above, it is also possible that other, as-yet-unidentified mechanisms contribute.

As with other forms of proarrhythmia, higher drug doses and concentrations are generally thought to increase risk, and patient-specific characteristics (notably the presence of diseased myocardium) are believed to modulate this risk. Propafenone is metabolized almost exclusively by CYP2D6, but the downstream metabolite, 5-hydroxy-propafenone, is also a sodium channel blocker. Therefore, in individuals with deficient CYP2D6 activity (either on a genetic basis or due to drug interactions), the parent drug accumulates, with somewhat more sodium channel block, as assessed by QRS prolongation. This is generally not clinically significant. The parent molecule does have β-blocking activity, whereas the metabolite does not, and so adverse effects due to β-blockade are more common with the deficient CYP2D6 activity (Lee et al. 1990). Flecainide is also a CYP2D6 substrate but is also excreted unchanged by the kidneys. Therefore, the CYP2D6 genotype generally has little effect on flecainide actions. Occasionally patients with defective CYP2D6 activity and renal dysfunction may experience very high flecainide concentration and toxicity (Evers et al. 1994).

4.2
Mechanisms

Sodium channel block is exaggerated by myocardial ischemia and rapid heart rates. The extent of this modulation, interpretable within the framework of the "modulated receptor hypothesis" (Hondeghem and Katzung 1984) and more recent molecular interpretations of drug-channel interactions (Balser 2001), varies among drugs of this class. In the intact heart, the result is conduction slowing, particularly in "fast response" tissues, such as the atrium or ventricle. This is manifest on the surface electrocardiogram even in normal individuals by P wave, PR interval, and QRS interval widening. In dogs with remote myocardial infarction, rendering them susceptible to reentrant VT, conduction slowing conferred by flecainide increased the duration of VT and the ease of inducibility of VT, analogous to the clinical situation (Coromilas et al. 1995). In addition, loss of sodium function by blocking drugs (or by genetic lesions) can result in marked abbreviation of action potentials in the epicardium with much less effect in the endocardium (Krishnan and Antzelevitch 1993; Lukas and Antzelevitch 1993). The result, increased heterogeneity in action potentials particularly prominent in the right ventricular wall, is thought to underlie the distinctive electrocardiogram in the Brugada syndrome and represents a second mechanism linking sodium channel blocking drugs to enhanced susceptibility to serious ventricular arrhythmias (Antzelevitch et al. 2003).

4.3
Genetics

Proarrhythmia due to sodium channel blockers is a risk in patients with sub-clinical Brugada syndrome. The extent to which such mutations, or more common polymorphisms, modulate the risk of sodium channel blocker-induced proarrhythmia in other situations, such as widespread drug exposure in CAST, is unknown.

4.4
Treatment

VT due to sodium channel block may be difficult to treat because it may be resistant to cardioversion and frequently recurs within several beats after cardioversion. Clinical anecdotes and animal studies have suggested that infusion of sodium bicarbonate or sodium chloride may be beneficial in some cases (Chouty et al. 1987; Bajaj et al. 1989).

Atrial flutter with rapid AV conduction is acutely managed by recognition and administration of AV nodal block agents, such as diltiazem and verapamil. Occasional patients undergo ablation of a key portion of the atrial flutter circuit and can then be maintained on drugs, free of AF (Huang et al. 1998).

5
Other Forms of Proarrhythmia

Many other drugs have been associated with sudden death, presumably due to arrhythmias. Whether these cases represent a variant on one of the well-recognized mechanisms described here, coronary vasospasm, or other as-yet-unrecognized mechanisms, is uncertain (Table 1).

6
Summary

Proarrhythmia has moved from a clinical curiosity to the centerpiece of antiarrhythmic drug selection and development of antiarrhythmic and other drugs. Elucidation of multiple syndromes of proarrhythmia and their underlying mechanisms has been important not only in identifying and reducing the problem, but also in understanding more general issues, including the role of genetics and other factors in the genesis of cardiac arrhythmias.

Acknowledgements Supported in part by grants from the United States Public Health Service (HL46681, HL49989, HL65962, HL62494, HL70250). Dr. Roden is the holder of the William Stokes Chair in Experimental Therapeutics, a gift from the Dai-ichi Corporation. Dr. Anderson is an Established Investigator of the American Heart Association.

References

Abbott GW, Sesti F, Splawski I, Buck ME, Lehmann MH, Timothy KW, Keating MT, Goldstein SAN (1999) MiRP1 forms IKr potassium channels with HERG and is associated with cardiac arrhythmia. Cell 97:175–187

Akar FG, Yan GX, Antzelevitch C, Rosenbaum DS (2002) Unique topographical distribution of M cells underlies reentrant mechanism of torsade de pointes in the long-QT syndrome. Circulation 105:1247–1253

Akiyama T, Pawitan Y, Greenberg H, Kuo CS, Reynolds-Haertle RA, The CAST Investigators (1991) Increased risk of death and cardiac arrest from encainide and flecainide in patients after non-Q-wave acute myocardial infarction in the Cardiac Arrhythmia Suppression Trial. Am J Cardiol 68:1551–1555

Angelin B, Arvidsson A, Dahlqvist R, Hedman A, Schenck-Gustafsson K (1987) Quinidine reduces biliary clearance of digoxin in man. Eur J Clin Invest 17:262–265

Antman EM, Smith TW (1986) Digitalis toxicity. Mod Concepts Cardiovasc Dis 55:26–30

Antman EM, Wenger TL, Butler VPJ, Haber E, Smith TW (1990) Treatment of 150 cases of life-threatening digitalis intoxication with digoxin-specific Fab antibody fragments. Final report of a multicenter study. Circulation 81:1744–1752

Antzelevitch C, Brugada P, Brugada J, Brugada R, Towbin JA, Nademanee K (2003) Brugada syndrome: 1992–2002: a historical perspective. J Am Coll Cardiol 41:1665–1671

Bajaj AK, Woosley RL, Roden DM (1989) Acute electrophysiologic effects of sodium administration in dogs treated with O-desmethyl encainide. Circulation 80:994–1002

Balser JR (2001) The cardiac sodium channel: gating function and molecular pharmacology. J Mol Cell Cardiol 33:599–613

Belardinelli L, Antzelevitch C, Vos MA (2003) Assessing predictors of drug-induced torsade de pointes. Trends Pharmacol Sci 24:619–625

Brachmann J, Scherlag BJ, Rosenshtraukh LV, Lazzara R (1983) Bradycardia-dependent triggered activity: relevance to drug-induced multiform ventricular tachycardia. Circulation 68:846–856

Bran S, Murray WA, Hirsch IB, Palmer JP (1995) Long QT syndrome during high-dose cisapride. Arch Intern Med 155:765–768

Burashnikov A, Antzelevitch C (1998) Acceleration-induced action potential prolongation and early afterdepolarizations. J Cardiovasc Electrophysiol 9:934–948

Cardiac Arrhythmia Suppression Trial (CAST) Investigators (1989) Increased mortality due to encainide or flecainide in a randomized trial of arrhythmia suppression after myocardial infarction. N Engl J Med 321:406–412

Cardiac Arrhythmia Suppression Trial II Investigators (1991) Effect of the antiarrhythmic agent moricizine on survival after myocardial infarction. N Engl J Med 327:227–233

Carlsson L, Almgren O, Duker G (1990) QTU-prolongation and torsades de pointes induced by putative class III antiarrhythmic agents in the rabbit: etiology and interventions. J Cardiovasc Pharmacol 16:276–285

Carlsson L, Abrahamsson C, Andersson B, Duker G, Schiller-Linhardt G (1993) Proarrhythmic effects of the class III agent almokalant: importance of infusion rate, QT dispersion, and early afterdepolarisations. Cardiovasc Res 27:2186–2193

Chezalviel-Guilbert F, Weissenburger J, Davy JM, Guhennec C, Poirier JM, Cheymol G (1995) Proarrhythmic effects of a quinidine analog in dogs with chronic A-V block. Fundam Clin Pharmacol 9:240–247

Chouty F, Funck-Brentano C, Landau JM, Lardoux H (1987) Efficacité de fortes doses de lactate molaire par voie veineuse lors des intoxications au flecainide. Presse Med 16:808–810

Choy AMJ, Darbar D, Dell'Orto S, Roden DM (1999) Increased sensitivity to QT prolonging drug therapy immediately after cardioversion to sinus rhythm. J Am Coll Cardiol 34:396–401

Coromilas J, Saltman AE, Waldecker B, Dillon SM, Wit AL (1995) Electrophysiological effects of flecainide on anisotropic conduction and reentry in infarcted canine hearts. Circulation 91:2245–2263

Crijns HJ, van Gelder IS, Lie KI (1988) Supraventricular tachycardia mimicking ventricular tachycardia during flecainide treatment. Am J Cardiol 62:1303–1306

D'Alnoncourt CN, Zierhut W, Luderitz B (1982) "Torsade de pointes" tachycardia: re-entry or focal activity? Br Heart J 48:213–216

Dangman KH, Hoffman BF (1981) In vivo and in vitro antiarrhythmic and arrhythmogenic effects of N-acetyl procainamide. J Pharmacol Exp Ther 217:851–862

de Lannoy IAM, Koren G, Klein J, Charuk J, Silverman M (1992) Cyclosporin and quinidine inhibition of renal digoxin excretion: Evidence for luminal secretion of digoxin. Am J Physiol 263:F613–F622

Dessertenne F (1966) La tachycardie ventriculaire à deux foyers opposés variables. Arch Mal Coeur Vaiss 59:263–272

Donger C, Denjoy I, Berthet M, Neyroud N, Cruaud C, Bennaceur M, Chivoret G, Schwartz K, Coumel P, Guicheney P (1997) KVLQT1 C-terminal missense mutation causes a forme fruste long-QT syndrome. Circulation 96:2778–2781

Echt DS, Black JN, Barbey JT, Coxe DR, Cato EL (1989) Evaluation of antiarrhythmic drugs on defibrillation energy requirements in dogs: Sodium channel block and action potential prolongation. Circulation 79:1106–1117

Ehrlich JR, Pourrier M, Weerapura M, Ethier N, Marmabachi AM, Hebert TE, Nattel S (2004) KvLQT1 modulates the distribution and biophysical properties of HERG. A novel alpha-subunit interaction between delayed rectifier currents. J Biol Chem 279:1233–1241

El-Sherif N, Chinushi M, Caref EB, Restivo M (1997) Electrophysiological mechanism of the characteristic electrocardiographic morphology of torsade de pointes tachyarrhythmias in the long-QT syndrome: detailed analysis of ventricular tridimensional activation patterns. Circulation 96:4392–4399

Evers J, Eichelbaum M, Kroemer HK (1994) Unpredictability of flecainide plasma concentrations in patients with renal failure: relation to side effects and sudden death? Ther Drug Monit 16:349–351

Falk RH (1989) Flecainide-induced ventricular tachycardia and fibrillation in patients treated for atrial fibrillation. Ann Intern Med 111:107–111

Fromm MF, Kim RB, Stein CM, Wilkinson GR, Roden DM (1999) Inhibition of P-glycoprotein-mediated drug transport: a unifying mechanism to explain the interaction between digoxin and quinidine. Circulation 99:552–557

Hoffmeyer S, Burk O, von Richter O, Arnold HP, Brockmoller J, Johne A, Cascorbi I, Gerloff T, Roots I, Eichelbaum M, Brinkmann U (2000) Functional polymorphisms of the human multidrug-resistance gene: multiple sequence variations and correlation of one allele with P-glycoprotein expression and activity in vivo. Proc Natl Acad Sci U S A 97:3473–3478

Hondeghem LM, Katzung BG (1984) Antiarrhythmic agents: the modulated receptor mechanism of action of sodium and calcium channel-blocking drugs. Annu Rev Pharmacol Toxicol 24:387–423

Houltz B, Darpo B, Edvardsson N, Blomstrom P, Brachmann J, Crijns HJGM, Jensen SM, Svernhage E, Vallin H, Swedberg K (1998) Electrocardiographic and clinical predictors of torsades de pointes induced by almokalant infusion in patients with chronic atrial fibrillation or flutter. A prospective study. Pacing Clin Electrophysiol 21:1044–1057

Huang DT, Monahan KM, Zimetbaum P, Papageorgiou P, Epstein LM, Josephson ME (1998) Hybrid pharmacologic and ablative therapy: a novel and effective approach for the management of atrial fibrillation. J Cardiovasc Electrophysiol 9:462–469

IMPACT Research Group (1984) International mexiletine and placebo antiarrhythmic coronary trial. I. Report on arrhythmia and other findings. J Am Coll Cardiol 4:1148–1163

January CT, Riddle JM (1989) Early afterdepolarizations: mechanism of induction and block: a role for L-type Ca^{2+} current. Circ Res 64:977–990

Kay GN, Plumb VJ, Arciniegas JG, Henthorn RW, Waldo AL (1983) Torsades de pointes: the long-short initiating sequence and other clinical features: observations in 32 patients. J Am Coll Cardiol 2:806–817

Keating MT, Sanguinetti MC (2001) Molecular and cellular mechanisms of cardiac arrhythmias. Cell 104:569–580

Kim RB, Leake BF, Choo EF, Dresser GK, Kubba SV, Schwarz UI, Taylor A, Xie HG, McKinsey J, Zhou S, Lan LB, Schuetz JD, Schuetz EG, Wilkinson GR (2001) Identification of functionally variant MDR1 alleles among European Americans and African Americans. Clin Pharmacol Ther 70:189–199

Krishnan SC, Antzelevitch C (1993) Flecainide-induced arrhythmia in canine ventricular epicardium. Phase 2 reentry? Circulation 87:562–572

Kuhlkamp V, Mewis C, Bosch R, Seipel L (2003) Delayed sodium channel inactivation mimics long QT syndrome 3. J Cardiovasc Pharmacol 42:113–117

Lazzara R (1989) Amiodarone and torsades de pointes. Ann Intern Med 111:549–551

Leahey EB Jr, Reiffel JA, Drusin RE, Heissenbuttel RH, Lovejoy WP, Bigger JT Jr (1978) Interaction between quinidine and digoxin. JAMA 240:533–534

Lee JT, Kroemer HK, Silberstein DJ, Funck-Brentano C, Lineberry MD, Wood AJ, Roden DM, Woosley RL (1990) The role of genetically determined polymorphic drug metabolism in the beta-blockade produced by propafenone. N Engl J Med 322:1764–1768

Lee TC (1981) Van Gogh's vision. Digitalis intoxication? JAMA 20:245:727–729

Lindenbaum J, Rund DG, Butler VP, Tse-Eng D, Saha JR (1981) Inactivation of digoxin by the gut flora: reversal by antibiotic therapy. N Engl J Med 305:789–794

Lukas A, Antzelevitch C (1993) Differences in the electrophysiological response of canine ventricular epicardium and endocardium to ischemia. Role of the transient outward current. Circulation 88:2903–2915

Makita N, Horie M, Nakamura T, Ai T, Sasaki K, Yokoi H, Sakurai M, Sakuma I, Otani H, Sawa H, Kitabatake A (2002) Drug-induced long-QT syndrome associated with a subclinical SCN5A mutation. Circulation 106:1269–1274

Makkar RR, Fromm BS, Steinman RT, Meissner MD, Lehmann MH (1993) Female gender as a risk factor for torsades de pointes associated with cardiovascular drugs. JAMA 270:2590–2597

Medina-Ravell VA, Lankipalli RS, Yan GX, Antzelevitch C, Medina-Malpica NA, Medina-Malpica OA, Droogan C, Kowey PR (2003) Effect of epicardial or biventricular pacing to prolong QT interval and increase transmural dispersion of repolarization: does resynchronization therapy pose a risk for patients predisposed to long QT or torsade de pointes? Circulation 107:740–746

Mitcheson JS, Chen J, Lin M, Culberson C, Sanguinetti MC (2000) A structural basis for drug-induced long QT syndrome. Proc Natl Acad Sci U S A 97:12329–12333

Monahan BP, Ferguson CL, Killeavy ES, Lloyd BK, Troy J, Cantilena LR Jr (1990) Torsades de pointes occurring in association with terfenadine use. JAMA 264:2788–2790

Napolitano C, Priori SG, Schwartz PJ, Cantu F, Paganini V, Matteo PS, de Fusco M, Pinnavaia A, Aquaro G, Casari G (1997) Identification of a long QT syndrome molecular defect in drug-induced torsades de pointes. Circulation 96:I-211 (abstr)

Napolitano C, Schwartz PJ, Brown AM, Ronchetti E, Bianchi L, Pinnavaia A, Acquaro G, Priori SG (2000) Evidence for a cardiac ion channel mutation underlying drug-induced QT prolongation and life-threatening arrhythmias. J Cardiovasc Electrophysiol 11:691–696

Nattel S (1985) Frequency-dependent effects of amitriptyline on ventricular conduction and cardiac rhythm in dogs. Circulation 72:898–906

Nattel S, Quantz MA (1988) Pharmacological response of quinidine induced early afterdepolarisations in canine cardiac Purkinje fibres: insights into underlying ionic mechanisms. Cardiovasc Res 22:808–817

Neuvonen PJ, Elonen E, Vuorenmaa T, Laakso M (1981) Prolonged Q-T interval and severe tachyarrhythmias, common features of sotalol intoxication. Eur J Clin Pharmacol 20:85–89

Numaguchi H, Johnson JP Jr, Petersen CI, Balser JR (2000) A sensitive mechanism for cation modulation of potassium current. Nat Neurosci 3:429–430

Oetgen WJ, Tibbits PA, Abt MEO, Goldstein RE (1983) Clinical and electrophysiologic assessment of oral flecainide acetate for recurrent ventricular tachycardia: evidence for exacerbation of electrical instability. Am J Cardiol 52:746–750

Pinter A, Dorian P, Newman D (2002) Cesium-induced torsades de pointes. N Engl J Med 346:383–384

Pourrier M, Zicha S, Ehrlich J, Han W, Nattel S (2003) Canine ventricular KCNE2 expression resides predominantly in Purkinje fibers. Circ Res 93:189–191

Reddy CVR, Kiok JP, Khan RG, El-Sherif N (1984) Repolarization alternans associated with alcoholism and hypomagnesemia. Am J Cardiol 53:390–391

Reuter H, Henderson SA, Han T, Matsuda T, Baba A, Ross RS, Goldhaber JI, Philipson KD (2002) Knockout mice for pharmacological screening: testing the specificity of Na+-Ca2+ exchange inhibitors. Circ Res 91:90–92

Roden DM (1998) Taking the idio out of idiosyncratic—predicting torsades de pointes. Pacing Clin Electrophysiol 21:1029–1034

Roden DM (2004a) Drug-induced prolongation of the QT Interval. N Engl J Med 350:1013–1022

Roden DM (2004b) Human genomics and its impact on arrhythmias. Trends Cardiovasc Med 14:112–116

Roden DM, Hoffman BF (1985) Action potential prolongation and induction of abnormal automaticity by low quinidine concentrations in canine Purkinje fibers. Relationship to potassium and cycle length. Circ Res 56:857–867

Roden DM, Woosley RL, Primm RK (1986) Incidence and clinical features of the quinidine-associated long QT syndrome: implications for patient care. Am Heart J 111:1088–1093

Schlotthauer K, Bers DM (2000) Sarcoplasmic reticulum Ca(2+) release causes myocyte depolarization. Underlying mechanism and threshold for triggered action potentials. Circ Res 87:774–780

Selzer A, Wray HW (1964) Quinidine syncope, paroxysmal ventricular fibrillations occurring during treatment of chronic atrial arrhythmias. Circulation 30:17–26

Sesti F, Abbott GW, Wei J, Murray KT, Saksena S, Schwartz PJ, Priori SG, Roden DM, George AL Jr, Goldstein SA (2000) A common polymorphism associated with antibiotic-induced cardiac arrhythmia. Proc Natl Acad Sci U S A 97:10613–10618

Shimizu W, Antzelevitch C (1999) Cellular basis for long QT, transmural dispersion of repolarization, and torsade de pointes in the long QT syndrome. J Electrocardiol 32 Suppl:177–184

Smith TW (1988) Digitalis: mechanisms of action and clinical use. N Engl J Med 318:358–365

Splawski I, Timothy KW, Tateyama M, Clancy CE, Malhotra A, Beggs AH, Cappuccio FP, Sagnella GA, Kass RS, Keating MT (2002) Variant of SCN5A sodium channel implicated in risk of cardiac arrhythmia. Science 297:1333–1336

Strauss HC, Bigger JT, Hoffman BF (1970) Electrophysiological and beta-receptor blocking effects of MJ 1999 on dog and rabbit cardiac tissue. Circ Res 26:661–678

Su SF, Huang JD (1996) Inhibition of the intestinal digoxin absorption and exsorption by quinidine. Drug Metab Dispos 24:142–147

Tan HL, Wilde AA (1998) T wave alternans after sotalol: evidence for increased sensitivity to sotalol after conversion from atrial fibrillation to sinus rhythm. Heart 80:303–306

Tanigawara Y, Okamura N, Hirai M, Yasuhara M, Ueda K, Kioka N, Komano T, Hori R (1992) Transport of digoxin by human P-glycoprotein expressed in a porcine kidney epithelial cell line (LLC-PK1). J Pharmacol Exp Ther 263:840–845

Torp-Pedersen C, Moller M, Bloch-Thomsen PE, Kober L, Sandoe E, Egstrup K, Agner E, Carlsen J, Videbaek J, Marchant B, Camm AJ (1999) Dofetilide in patients with congestive heart failure and left ventricular dysfunction. Danish Investigations of Arrhythmia and Mortality on Dofetilide Study Group. N Engl J Med 341:857–865

UK Rythmodan Multicentre Study Group (1984) Oral disopyramide after admission to hospital with suspected acute myocardial infarction. Postgrad Med J 60:98–107

Volders PG, Sipido KR, Vos MA, Kulcsar A, Verduyn SC, Wellens HJ (1998) Cellular basis of biventricular hypertrophy and arrhythmogenesis in dogs with chronic complete atrioventricular block and acquired torsade de pointes. Circulation 98:1136–1147

Vos MA, de Groot SH, Verduyn SC, van der Zande J, Leunissen HD, Cleutjens JP, van Bilsen M, Daemen MJ, Schreuder JJ, Allessie MA, Wellens HJ (1998) Enhanced susceptibility for acquired torsade de pointes arrhythmias in the dog with chronic, complete AV block is related to cardiac hypertrophy and electrical remodeling. Circulation 98:1125–1135

Waldo AL, Camm AJ, DeRuyter H, Friedman PL, MacNeil DJ, Pitt B, Pratt CM, Rodda BE, Schwartz PJ (1995) Survival with oral d-Sotalol in patients with left ventricular dysfunction after myocardial infarction: rationale, design, and methods (the SWORD trial). Am J Cardiol 75:1023–1027

Wei J, Yang IC, Tapper AR, Murray KT, Viswanathan P, Rudy Y, Bennett PB, Norris K, Balser JR, Roden DM, George AL (1999) KCNE1 polymorphism confers risk of drug-induced long QT syndrome by altering kinetic properties of IKs potassium channels. Circulation (suppl I):495 (abstr)

Wenckebach KF (1923) Cinchona derivates in the treatment of heart disorders. JAMA 81:472–474

Wetherbee DG, Holzman D, Brown MG (1951) Ventricular tachycardia following the administration of quinidine. Am Heart J 42:89–96

Willius FA, Keys TE (1942) A remarkably early reference to the use of cinchona in cardiac arrhythmias. Mayo Clinic Staff Meetings (May 13):294–297

Winkle RA, Mason JW, Griffin JC, Ross D (1981) Malignant ventricular tachy-arrhythmias associated with the use of encainide. Am Heart J 102:857–864

Woosley RL, Chen Y, Freiman JP, Gillis RA (1993) Mechanism of the cardiotoxic actions of terfenadine. JAMA 269:1532–1536

Wu Y, Roden DM, Anderson ME (1999) CaM kinase inhibition prevents development of the arrhythmogenic transient inward current. Circ Res 84:906–912

Yang P, Kanki H, Drolet B, Yang T, Wei J, Viswanathan PC, Hohnloser SH, Shimizu W, Schwartz PJ, Stanton MS, Murray KT, Norris K, George ALJ, Roden DM (2002) Allelic variants in Long QT disease genes in patients with drug-associated Torsades de Pointes. Circulation 105:1943–1948

Yang T, Roden DM (1996) Extracellular potassium modulation of drug block of IKr: implications for torsades de Pointes and reverse use-dependence. Circulation 93:407–411

Yang T, Snyders DJ, Roden DM (1997) Rapid inactivation determines the rectification and [K+]o dependence of the rapid component of the delayed rectifier K+ current in cardiac cells. Circ Res 80:782–789

Yang T, Kanki H, Roden DM (2003) Phosphorylation of the IKs channel complex inhibits drug block. Novel mechanism underlying variable antiarrhythmic drug actions. Circulation 108:132–134

Zhang S, Rajamani S, Chen Y, Gong Q, Rong Y, Zhou Z, Ruoho A, January CT (2001) Cocaine blocks HERG, but Not KvLQT1+minK, potassium channels. Mol Pharmacol 59:1069–1076

HEP (2006) 171:99–121

Cardiac Na+ Channels as Therapeutic Targets for Antiarrhythmic Agents

I. W. Glaaser[1] · C. E. Clancy[2] (✉)

[1]Department of Pharmacology, College of Physicians and Surgeons of Columbia University, 630 W. 168th St., New York NY, 10032, USA

[2]Department of Physiology and Biophysics, Institute for Computational Biomedicine, Weill Medical College of Cornell University, 1300 York Avenue, LC-501E, New York NY, 10021, USA
clc7003@med.cornell.edu

Abstract There are many factors that influence drug block of voltage-gated Na+ channels (VGSC). Pharmacological agents vary in conformation, charge, and affinity. Different drugs have variable affinities to VGSC isoforms, and drug efficacy is affected by implicit tissue properties such as resting potential, action potential morphology, and action potential frequency. The presence of polymorphisms and mutations in the drug target can also influence drug outcomes. While VGSCs have been therapeutic targets in the management of cardiac arrhythmias, their potential has been largely overshadowed by toxic side effects.

Nonetheless, many VGSC blockers exhibit inherent voltage- and use-dependent properties of channel block that have recently proven useful for the diagnosis and treatment of genetic arrhythmias that arise from defects in Na^+ channels and can underlie idiopathic clinical syndromes. These defective channels suggest themselves as prime targets of disease and perhaps even mutation specific pharmacological interventions.

Keywords Na+ channel blocker · Lidocaine · Flecainide · Local anesthetic · Mutation · Channelopathies · Polymorphism · Structural determinants · Antiarrhythmic · Proarrhythmic · VGSC · TTX · Tonic block · Use-dependent block · $Na_V 1.5$ · $Na_V 1.1$ · SCN5A · SCN1A · Pharmacokinetics · Pharmacodynamics · Structural determinants · Recovery from block · Singh-Vaughan Williams · Sicilian Gambit · CAST · CYP · Cytochrome enzymes · Long-QT Syndrome · Brugada Syndrome · Conduction disorders · Isoform specificity · Molecular determinants

1
Introduction—Sodium Channels

Voltage-gated sodium channels (VGSC) cause the rapid depolarization that marks the rising phase of action potentials in the majority of excitable cells. Thus far, eleven genes have been shown to encode different isoforms of the α-sububunit of the VGSC, many of which have been cloned and characterized in terms of kinetics and regional tissue expression (Goldin et al. 2000; Goldin 2001, 2002). The isoform differences in the voltage dependence of channel activation, inactivation, and recovery from inactivation result in unique conductance and rate dependence in specific cell and tissue types (Goldin 2001, 2002). In some tissues, the α-subunit has been shown to associate with accessory β-subunits, which act as modulators of channel function (Qu et al. 1995; Abriel et al. 2001).

In the myocardium VGSCs are required for initiation of the fast action potential upstroke that is required for cardiac excitation and conduction. Even within the same tissue or cell, multiple Na^+ channel isoforms may be expressed and confer variable cellular electrical properties. While channel spatial distribution has long been known as an implicit property of neurons, recent data suggest variable localization of ion channel isoforms within the myocardium, and even within the same ventricular myocyte (Maier et al. 2002, 2003; Malhotra et al. 2001; Cohen 1996). The Na^+ channel population within the intercalated disks in atrial and ventricular myocytes is composed primarily of tetrodotoxin (TTX)-insensitive $Na_V 1.5$ α-subunits, encoded by the gene SCN5A. While $Na_V 1.5$ is preferentially distributed near gap junctions and is the major player in initiating and sustaining cardiac conduction (Kucera et al. 2002), an isoform predominantly found in brain ($Na_V 1.1$) has been found to specifically localize within the transverse (T) tubules of the ventricular myocardium (Malhotra et al. 2001; Maier et al. 2002). The kinetic properties of $Na_V 1.1$ differ from the predominant cardiac isoform $Na_V 1.5$ in presumably important ways. Moreover, the $Na_V 1.1$ isoform displays profound TTX sensitivity (nanomolar range) compared to $Na_V1.5$ (millimolar range) (Malhotra et al. 2001; Maier et al. 2002).

In the sinoatrial node (SAN), a unique collection of ligand and voltage-gated channels are required for automaticity, an implicit cellular property that initiates cardiac excitation (Honjo et al. 1996; Kodama et al. 1997; Kodama et al. 1996). A number of studies have demonstrated that the SAN node is sensitive to the application of TTX, suggesting Na^+ current as a contributor to electrical activity in the SAN (Honjo et al. 1996; Kodama et al. 1997; Baruscotti et al. 1996; Muramatsu et al. 1996; Baruscotti et al. 1997; Baruscotti et al. 2001). In some species, $Na_V1.5$ has been identified using electrophysiological and pharmacological methods (TTX insensitive, $IC_{50} = \mu M$), while in others direct evidence using immunohistochemistry and low concentrations of TTX point to a central nervous system isoform $Na_V1.1$ (Kodama et al. 1997; Muramatsu et al. 1996; Baruscotti et al. 1997, 2001).

VGSC isoforms are functionally and structurally similar in that they are voltage-gated heteromultimeric protein complexes consisting of four heterologous domains, each containing six transmembrane spanning segments (Fig. 1). Positive residues are clustered in the S4 segments and constitute the voltage sensor (Stuhmer et al. 1989; Kontis et al. 1997). The intracellular linker between domains three and four, DIII/DIV, includes a hydrophobic isoleucine–phenylalanine–methionine (IFM) motif, which acts as a blocking inactivation particle and occludes the channel pore, resulting in channel inactivation subsequent to channel opening (West et al. 1992; Smith and Goldin 1997; Auld et al. 1990; Stuhmer et al. 1989). Recent studies also suggest a role for the C-terminus in channel inactivation in $Na_V1.1$ and $Na_V1.5$ (Cormier et al. 2002; Mantegazza et al. 2001). The S5 and S6 transmembrane segments of each domain constitute the putative channel pore and associated ion selectivity filter (Sun et al. 1997; Yamagishi et al. 2001).

All VGSCs make transitions between discrete conformational states via movement of charged portions of the channel within the lipid bilayer membrane (Ahern and Horn 2004). At negative membrane potentials, channels

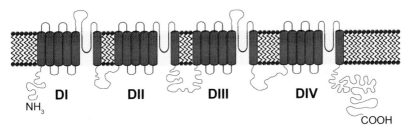

Fig. 1 Topological map of the cardiac voltage-gated sodium channel ($Na_V1.5$). Shown are the four heterologous domains (*DI–DIV*), each with six transmembrane spanning regions. The amino terminus and carboxy terminus (indicated NH_3 and *COOH*, respectively) are located in the intracellular membrane region

typically reside in closed and available resting states that represent a non-conducting conformation. Depolarization results in activation of the voltage sensors and channel opening, allowing for ion passage. Subsequent to channel activation, channels enter inactivated states that are non-conducting and refractory. Repolarization is required to alleviate inactivation with isoform-specific time and voltage dependence.

2
Antiarrhythmic Classification

The Singh–Vaughan Williams classification system is the most widely used and segregates antiarrhythmics into one of four classes based on their effects on the cardiac action potential (Vaughan Williams 1989). Antiarrhythmic drugs that cause sodium channel block fall into class I, and are further subdivided by kinetics of recovery from block (Harrison 1985). For example, several class Ib antiarrhythmic drugs commonly used therapeutically and in laboratory studies, lidocaine and mexiletine, are characterized by tonic and use-dependent block (UDB) and fast recovery from drug block (<1 s). Class Ia antiarrhythmics include procainamide and quinidine and have intermediate kinetics of recovery from drug block (1–10 s), while class Ic antiarrhythmics such as flecainide exhibit predominantly UDB and have slow kinetics of recovery from block (>10 s). This classification system has proved useful in its simplicity; however many drugs exhibit multiple electrophysiological actions and, as a result, fall into more than one class (Roden 1990). Moreover, drugs within the same class may result in vastly different clinical responses. In response to these shortcomings, the "Sicilian Gambit" proposed an alternate approach, whereby the arrhythmia is diagnosed and an attempt is made to identify the "vulnerable parameter", i.e., the electrophysiological component most susceptible to intervention that will terminate or suppress the arrhythmia with minimal toxicity (Task Force of the Working Group on Arrhythmias of the European Society of Cardiology 1991). While complex, the Sicilian Gambit approach provides a system for classifying drugs with multiple actions and identifying antiarrhythmic agents based on pathophysiological considerations.

3
Na⁺ Channel Blockers: Diagnosis and Treatment

Local anesthetic (LA) molecules such as lidocaine, mexiletine, and flecainide block Na^+ channels and have been used therapeutically to manage cardiac arrhythmias (Rosen and Wit 1983; Rosen et al. 1975; Wit and Rosen 1983). Despite the prospective therapeutic value of the inherent voltage- and use-dependent properties of channel block by these drugs in the treatment of tachyarrhythmias, their potential has been overshadowed by toxic side effects (Rosen and Wit 1987; Weissenburger et al. 1993).

There has been renewed interest in the study of voltage-gated Na^+ channels since the recent realization that genetic defects in Na^+ channels can underlie idiopathic clinical syndromes (Goldin 2001). Interestingly, all sodium channel-linked syndromes are characterized by episodic attacks and heterogeneous phenotypic manifestations (Lerche et al. 2001; Steinlein 2001). These defective channels suggest themselves as prime targets of disease and perhaps even mutation-specific pharmacological interventions (Carmeliet et al. 2001; Goldin 2001).

Na^+ channel blockade by flecainide is of particular interest as it had been shown to reduce QT prolongation in carriers of some Na^+ channel-linked long QT syndrome type 3 (LQT3) mutations, and to evoke ST-segment elevation, a hallmark of the Brugada syndrome (BrS), in patients with a predisposition to the disease (Brugada et al. 2000). Thus in the case of LQT3, flecainide has potential therapeutic application, whereas for BrS it has proved useful as diagnostic tool. However, in some cases, flecainide has been reported to provoke BrS symptoms (ST-segment elevation) in patients harboring LQT3 mutations (Priori et al. 2000). Furthermore, flecainide preferentially blocks some LQT3 or BrS-linked mutant Na^+ channels (Abriel et al. 2000; Grant et al. 2000; Liu et al. 2002; Viswanathan et al. 2001). Investigation of the drug interaction with these and other LQT3- and BrS-linked mutations may indicate the usefulness of flecainide in the detection and management of these disorders and determine whether or not it is reasonable to use this drug to identify potential disease-specific mutations.

Antiarrhythmic agents have effects in addition to channel blockade that may prove useful therapeutically. An LQTS-linked sodium channel mutation which resulted in reduced cell surface channel expression was shown to be partially rescued by mexiletine (Valdivia et al. 2002). This type of drug-induced rescue of channels had been previously demonstrated for loss of function K^+ channel mutations that are linked to arrhythmia (Zhou et al. 1999; Rajamani et al. 2002), but the study was the first such demonstration for Na^+ channel rescue. Drug rescue of channels has potential therapeutic value for loss of Na^+ channel function mutations that have been linked to the Brugada syndrome and conduction disorders (Valdivia et al. 2004).

4
Proarrhythmic Effects

A major concern for administration of currently used antiarrhythmic agents is that almost all can exhibit proarrhythmic effects and may exacerbate underlying arrhythmias (Roden 1990; Roden 2001). The mechanism varies between classes and between drugs within classes. However, extensive clinical studies examining agents that use sodium channel blockade as a mechanism to suppress cardiac arrhythmias have identified several potential proarrhythmic toxicities. Torsades de pointes is estimated to occur infrequently in patients exposed to sodium channel blockers, but has been seen in patients treated with quinidine, procainamide, and disopyramide. This reaction is difficult to predict, but can be exacerbated by other factors, including underlying heart disease (Fenichel et al. 2004).

Patients with histories of sustained ventricular tachyarrhythmia and patients recovering from myocardial infarction (MI) have also been found to exhibit proarrhythmic effects upon treatment with sodium channel blockade. In the latter case, the Cardiac Arrhythmia Suppression Trial (CAST) (Ruskin 1989) demonstrated a slight increase in mortality when post-MI patients were treated with flecainide or encainide. While these adverse cardiac effects resulting from the use of sodium channel blocking agents are more frequent in patients with additional contributing factors, they certainly must be considered in the administration of all antiarrhythmic agents.

5
Pharmacokinetics and Pharmacodynamics of Antiarrhythmic Agents

Antiarrhythmic agents vary widely in their clinical response. This disparity in efficacy may result from variability in drug absorption, distribution, metabolism, and elimination, collectively referred to as "pharmacokinetics." Pharmacokinetic variability can arise through differences in any of the component processes of drug absorption, distribution, metabolism, and elimination and is critical because variations in drug clearance can have proarrhythmic effects.

Drug metabolism is particularly important in pharmacokinetic variability among drugs. Many of the antiarrhythmic drugs are metabolized by the isoforms of the cytochrome P450 (CYP) enzymes. CYP enzymes are located primarily in the liver, although various isoforms are found in the intestines, kidneys, and lungs as well. The various CYP isoforms differ in their substrate specificities, and they can affect the plasma concentration of substrates through two mechanisms. In the first, genetic variants of CYP genes affect the efficacy of drug metabolism (Meyer et al. 1990). Among antiarrhythmic agents a polymorphism in the CYP isoform 2D6 (CYP2D6) that affects metabolism of the class III β-blocker propafenone is the only known example of this type of action, which is relatively rare (Lee et al. 1990). The second, more common effect, results from drug-induced inhibition or facilitation of the various CYP isoforms. In these cases, a drug is a substrate for a specific CYP isoform upon which a concurrently administered drug acts as an inhibitor or inducer. If the metabolic pathway is inhibited, drug can accumulate to toxic concentrations. Conversely, if the metabolic pathway is induced, the substrate drug may be rapidly eliminated, resulting in sub-therapeutic drug concentration (Roden 2000).

Differences in the biochemical and physiological actions of drugs and the mechanisms for these actions, termed "pharmacodynamics," may also affect clinical efficacy (Roden 1990; Roden 2000). Pharmacodynamic variability generally occurs as the result of two mechanisms. The first is variability within the entire biological environment within which the drug–receptor interaction occurs (Roden and George 2002). This can be as a result of genetic heterogeneity or due to changes in the environment as a result of disease states. A second mechanism is the occurrence of polymorphisms in the molecular target for drug action that affect function, as discussed in the next section.

6
Mutations and/or Polymorphisms May Increase Susceptibility to Drug-Induced Arrhythmias

Within the context of arrhythmia, pharmacogenomic considerations are important to determine the potential for genetic heterogeneity to directly affect drug targets and interfere with drug interactions. Mutations or polymorphisms may directly interfere with drug binding (Liu et al. 2002) or can result in a physiological substrate that increases predisposition to drug-induced arrhythmia (Splawski et al. 2002).

A recent study investigated the increased susceptibility to drug-induced arrhythmia in African-American carriers (4.6 million) of a common polymorphism (S1102 to Y1102) in $Na_V1.5$ (Splawski et al. 2002). The study used a combined experimental and theoretical investigation. Although the experimental data suggested that the polymorphism Y1102 had subtle effects on Na^+ channel function, the integrative model simulations revealed an increased susceptibility to arrhythmogenic-triggered activity in the presence of drug block (Splawski et al. 2002). Action potential simulations with cells containing S1102 or Y1102 channels showed that the subtle changes in gating did not alter action potentials (Fig. 2). However, in the presence of concentration-dependent block of the rapidly activating delayed rectifier potassium currents (I_{Kr}), a common side effect of many medications and hypokalemia, the computations predicted that Y1102 would induce action potentialprolongation and early afterdepolarizations (EADs) (Splawski et al. 2002). EADs are a cellular trigger for ventricular tachycardia. Thus, computational analyses indicated that Y1102 increased the likelihood of QT prolongation, EADs, and arrhythmia in response to drugs (or drugs coupled with hypokalemia) that inhibit cardiac repolarization. While most of these carriers will never have an arrhythmia because the effect of Y1102 is subtle, in combination with additional acquired risk factors— particularly common factors such as medications, hypokalemia, or structural heart disease—these individuals are at increased risk (Splawski et al. 2002).

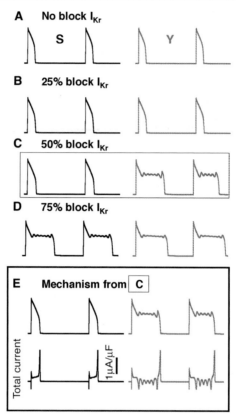

Fig. 2a–e SCN5A Y1102 increases arrhythmia susceptibility in the simulated presence of cardiac potassium channel blocking medications. Action potentials (19th and 20th after pacing from equilibrium conditions) for S1102 and Y1102 at cycle length = 2,000 ms are shown for a range of I_{Kr} block. I_{Kr} is frequently blocked as an unintended side effect of many medications. Under the conditions of no block and a 25% I_{Kr} block (**a** and **b**, respectively), both S1102- and Y1102-containing cells exhibit normal phenotypes. As I_{Kr} block is increased (50% block; **c**), the Y1102 variant demonstrates abnormal repolarization. **d** With 75% I_{Kr} block, both S1102 and Y1102 exhibit similar abnormal cellular phenotypes. The mechanism of this effect is illustrated in **e** by comparing action potentials in **c** with the underlying total cell current during the action potentials. Faster V_{max} (dV/dt) during the upstroke caused by Y1102 results in larger initial repolarizing current but not enough (due to drug block) to cause premature repolarization. This results in faster initial repolarization, which increases depolarizing current through sodium and L-type calcium channels. The net effect is prolongation of action potential duration, reactivation of calcium channels, early after depolarizations (EADs), and risk of arrhythmia. (From Splawski et al. 2002)

Genetic mutations or polymorphisms may affect drug binding by altering the length of time that a channel resides in a particular state. For example, the epilepsy-associated R1648H mutation in $Na_V1.1$ reduces the likelihood that a mutant channel will inactivate and increases the channel open probability

(Lossin et al. 2002). Hence, an agent that interacts with open channels will have increased efficacy, while one that interacts with inactivation states may have reduced efficacy. However, even this type of analysis may not predict actual drug–receptor interactions (Liu et al. 2002, 2003). The I1768V mutation increases the cardiac Na$^+$ channel isoform propensity for opening, suggesting that an open channel blocker would be more effective, but in fact the mutation is in close proximity to the drug-binding site, which may render open channel blockers non-therapeutic (Liu et al. 2002, 2003).

Recent findings revealed the differential properties of certain drugs on mutant and wild-type cardiac sodium channels. One such example is the preferential blockade by flecainide of persistent sodium current in the ΔKPQ sodium channel mutant (Nagatomo et al. 2000). It was also shown that some LQT-associated mutations were more sensitive to blockade by mexiletine, a drug with similar properties to lidocaine, than wild-type channels (Wang et al. 1997). In three mutations, ΔKPQ, N1325S, and R1644H, mexiletine displayed a higher potency for blocking late sodium current than peak sodium current (Wang et al. 1997).

One study showed that flecainide, but not lidocaine, showed a more potent interaction with a C-terminal D1790G LQT3 mutant than with wild-type channels and a correction of the disease phenotype (Abriel et al. 2001; Liu et al. 2002). The precise mechanism underlying these differences is unclear. Lidocaine has a pK_a of 7.6–8.0 and thus may be up to 50% neutral at physiologic pH. In contrast, flecainide has a pK_a of approximately 9.3, leaving less than 1% neutral at pH 7.4 (Strichartz et al. 1990; Schwarz et al. 1977; Hille 1977). Thus, one possibility underlying differences in the voltage-dependence of flecainide and lidocaine-induced modulation of cardiac Na$^+$ channels is restricted access to a common site that is caused by the ionized group of flecainide. Another possibility is that distinctive inactivation gating defects in the D1790G channel may underlie these selective pharmacologic effects. Indeed, recently it was shown mutations that promote inactivation (shift channel availability in the hyperpolarizing direction) enhance flecainide block. Interestingly, the data also showed that flecainide sensitivity is mutation, but not disease, specific (Liu et al. 2002).

These studies are important in the demonstration that effects of drugs segregate in a mutation-specific manner that is not correlated with disease phenotype, suggesting that some drugs may not be effective agents for diagnosing or treating genetically based disease. The nature of the interaction between pharmacologic agents and wild-type cardiac sodium channels has been extensively investigated. However, the new findings of drug action on mutant channels in long-QT and BrS have stimulated a renewed interest in a more detailed understanding of the molecular determinants of drug action with the specific aim of developing precise, disease-specific therapy for patients with inherited arrhythmias.

7
Modulated Receptor Hypothesis

The modulated receptor hypothesis (MRH) derives from the concept of conformational dependence of binding affinity of allosteric enzymes and was first proposed by Hille (1977) to describe the interaction of local anesthetic (LA) molecules with Na^+ channels. The idea is that the drug binding affinity is determined, and modulated by, the conformational state of the channel (closed, open, or inactivated). Moreover, once bound, a drug alters the gating kinetics of the channel.

8
Effect of Charge on Drug Binding: Tonic Versus Use-Dependent Block

LAs including lidocaine, procaine, and cocaine, exist in two forms at physiological pH (Hille 1977; Liu et al. 2003; Strichartz et al. 1990). The uncharged form accounts for approximately 50% of the drug, while the protonated charged form is in equal proportion. The uncharged base form is highly lipophilic and therefore easily crosses cell membranes and blocks Na^+ channels intracellularly. Quaternary ammonium (QA) compounds are positively charged permanently

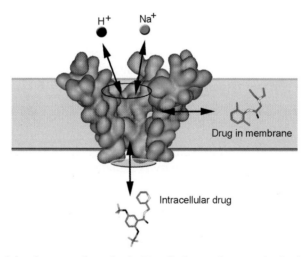

Fig. 3 The modulated receptor hypothesis. Two distinct pathways exist for drug block. The hydrophilic pathway (*vertical arrows*), is the likely path of a charged flecainide molecule, and requires channel opening for access to the drug receptor. Neutral drug such as lidocaine can reach the receptor through a hydrophobic "sideways movement" membrane pathway (*horizontal arrows*). Extracellular Na^+ ions (*gray circle*) and H^+ (*black circle*) can reach bound drug molecules through the selectivity filter shown as a *black ellipse*. The inactivation gate is shown as a transparent ellipse on the intracellular side of the pore. Figure adapted from Hille (1977)

and cannot cross cell membranes easily, but are effective Na$^+$ channel blockers when applied intracellularly. Flecainide is similar in structure to LAs, but is 99% charged at pH 7.4. Like flecainide, mexiletine has a pK_a of 9.3, and is therefore 99% charged at physiological pH (Liu et al. 2003).

Application of lidocaine or flecainide results in limited block of Na$^+$ channels at rest [tonic block (TB)] and likely results from neutral drug species interacting with the drug binding site via hydrophobic pathways through the cell membrane (Fig. 3; Liu et al. 2003). In other words, drug migration to the receptor occurs via "sideways" movement in the membrane, not by entry via the mouth of the channel pore (Hille 1977). Hence, neutral drug species are more effective tonic blockers, as they interact even when channels are inactivated by interaction of the intracellular linker between domains III and IV with residues within the channel pore. This inactivation process acts as a barrier to drug access via the hydrophilic pathway by preventing access of the drug to the receptor site within the channel pore (Fig. 3).

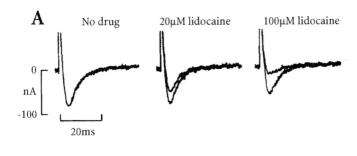

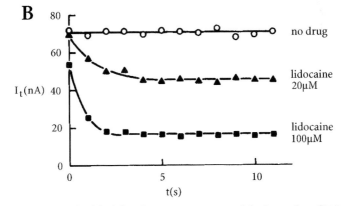

Fig. 4a,b Use-dependent block by lidocaine. I_{Na} was measured during trains of 500-ms pulses from −105 mV to −35 mV at 1.0 Hz. **a** The membrane currents were measured on the 1st and 12th pulses in (*from left to right*) 0, 20, and 100 µM lidocaine. **b** Peak sodium current amplitudes were measured for each of the pulses. The decrease in current magnitude has been fitted by an exponential curve, with $t = 1.3$ s in 20 µM lidocaine and $t = 0.7$ s in 100 µM lidocaine. (From Bean et al. 1983)

When channels are open, all Na^+ channel blockers have the opportunity to interact with the drug receptor via intracellular access to the pore. Subjecting channels to repetitive depolarizing voltage steps results in a profound build-up of channel block and as a result, accumulation of channel inhibition. This property is referred to as use-dependent block (UDB) and suggests that channel opening facilitates drug binding to the receptor, presumably by increasing the probability of drug access to the binding domain (Fig. 4; Ragsdale et al. 1994; Hille 1977; Liu et al. 2002). This idea is supported by the fact that mutations (like Y1795C, a naturally occurring gain-of-function LQT3 mutation) that act to increase the open time of the Na^+ channel exhibit increased *rate* of UDB

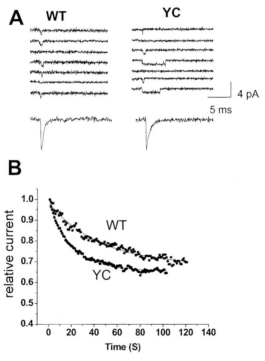

Fig. 5a,b Mutations that affect channel open times alter use-dependent block (UDB). Cell-attached patch recordings are shown for *WT* and Y1795C (*YC*) channels. Recordings were obtained in response to test pulses (−30 mV, 100 ms) applied at 2 Hz from −120 mV. **a** Current from consecutive single channel recordings is shown to emphasize the effects of inherited mutations on channel opening kinetics. Ensemble currents (constructed by averaging 500 consecutive sweeps) are shown for each construct below the individual sweeps. **b** Time course of the onset of UDB (1 Hz, 10 µM flecainide) during pulse trains applied to WT and YC channels. The data were normalized to the current amplitude of the first pulse in the train and fit with a single exponential function (A×exp-t/+base), the time constant for WT and YC were 45.29 s^{-1} and 20.09 s^{-1} ($p<0.01$ vs WT; $n=3$ cells per condition). (Adapted from Liu et al. 2002)

(Fig. 5; Liu et al. 2002). It should be noted that although UDB occurs more rapidly with longer channel openings, the *degree of block* (i.e., percentage of steady-state block) *is the same as observed in WT channels*. This suggests that although the drug can more easily access the receptor site, the affinity for the site is unchanged compared to WT. This is consistent with the notion that channel openings are required for UDB, but is not dependent on the open state to promote block. The repolarizing pulses between depolarizing steps do little to alleviate block, although unbinding does occur at sufficiently long hyperpolarized intervals. UDB has an implicit voltage dependence that exists in addition to the voltage dependence of activation gating. At increasingly depolarized potentials, much enhanced drug block is observed, despite the reduction in channel open times, which occurs due to fast voltage-dependent inactivation (Ragsdale et al. 1994). These are features of a positively charged drug that is expected to move within the electrical field of the membrane from inside the cell to access the drug binding site (Hille 1977).

9
Is It All Due to Charge?

Because the physical chemical properties of drugs are different, it is impossible to absolutely determine that drug access to the receptor and TB, UDB, and recovery from block profiles are fully attributable to differences in drug charge. For example, although the charge on flecainide is likely to restrict access of the drug to a receptor site, confer the voltage dependence of UDB, and account for recovery from block kinetics, a direct test has not been possible because of the differences in distribution between neutral and charged forms of each compound.

A recent study developed two custom-synthesized flecainide analogues, NU-FL and QX-FL, to investigate the role of charge in determining the profile of flecainide activity (Liu et al. 2003; Fig. 6). NU-FL has nearly identical hydrophobicity and very similar three-dimensional structure compared with flecainide, but has a very different pK_a. As measured by titration, NU-FL has an approximate pK_a value of 6.4 (Liu et al. 2003). Consequently, it should be nearly 90% neutral at physiological pH, thus more closely resembling the ionization profile of lidocaine. QX-FL shares a very similar three-dimensional structure with the parent compound flecainide, but is fully charged at physiological pH, and thus is well suited to discriminate between hydrophilic and hydrophobic access to its receptor (Liu et al. 2003).

The results indicated that, like lidocaine, the tertiary flecainide analog (NU-FL) interacts preferentially with inactivated channels without prerequisite channel openings (i.e., tonic block), while flecainide and QX-FL are ineffective in blocking channels that inactivate without first opening (Liu et al. 2003). Interestingly, slow recovery of channels from QX-FL block was impeded

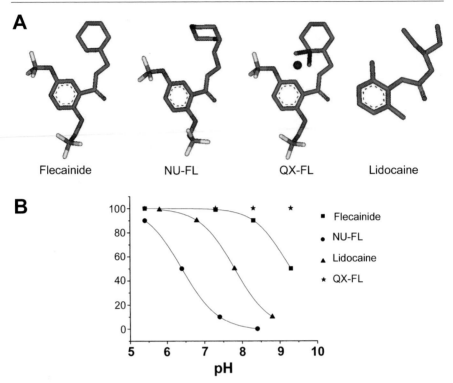

Fig. 6a,b Antiarrhythmic drug structure and drug charge as a function of pH. **a** Structural comparison of (from *left* to *right*) flecainide, and its novel analogs neutral flecainide (NU-FL), permanently charged flecainide (QX-FL), and the local anesthetic lidocaine. *White regions* represent nitrogen, *black regions* represent oxygen, *dark gray* elements are carbon, and *light gray* are fluorine. The *circle* in the QX-FL structure represents an iodine atom. **b** Plot of estimated concentrations of charged drugs as a function of pH. The pK_a values of each compound are 9.3 for flecainide, 6.4 for NU-FL, 7.8 for lidocaine. At relevant physiological pH values, flecainide is greater than 99% charged, QX-FL is fully ionized, lidocaine is approximately 50:50, and NU-FL is more than 90% neutral. (Adapted from Liu et al. 2003)

by outer pore block by tetrodotoxin, suggesting that the drug can diffuse away from channels via the outer pore. The data strongly suggest that it is the difference in degree of ionization (pK_a) between lidocaine and flecainide, rather than differences in their gross structural features, that determines distinction in block of cardiac Na^+ channels (Liu et al. 2003). The study also suggests that the two drugs share a common receptor, but, as outlined in the modulated receptor hypothesis, reach this receptor by distinct routes.

Differences in apparent UDB may also stem from differences between the kinetics of the recovery from block by neutral and charged drug forms (Liu et al. 2002, 2003). The disparity in the recovery kinetics is attributed to rapid unblock of neutral drug-bound channels and very slow unblock of charged drug-bound channels (Fig. 7). As proposed by Hille in the analysis of the pH

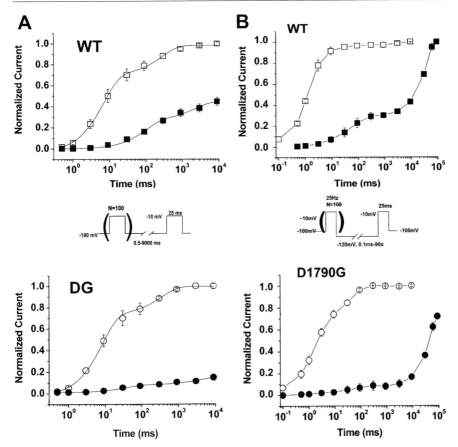

Fig. 7a,b Mutations and drug concentration affect the time course of recovery from drug block. Recovery from flecainide block of WT and D1790G. **a** UDB by 10 μM flecainide was induced by trains of 100 pulses (−10 mV, 25 ms, 25 Hz) from a −100-mV holding potential. Test pulses were then imposed after variable recovery intervals at −100 mV. Currents were normalized to steady-state current levels during slow pacing (once every 30 s) and plotted against recovery interval in the absence and presence of flecainide. *Open symbols* represent drug-free, and *filled symbols* drug-containing, conditions; $n = 3$–5 cells per condition. **b** Very slow recovery from 30 μM flecainide block of WT and D1790G channels. (Adapted from Liu et al. 2002)

dependence of UDB of Na$^+$ channels in muscle and nerve, during interpulse intervals, bound charged drug is trapped within the channel until the drug molecule is deprotonated. Neutral drug, which is less restricted, can dissociate from the channel via "sideways" movement through the membrane. At physiological pH, the fact that the recovery from block is faster for NU-FL than for flecainide may simply be due to the greater contribution (90%) of drug block by the neutral NU-FL component compared to charged component, while flecainide remains more than 99% charged (Liu et al. 2003). On the other hand, according the scheme described above, it is possible that deprotonation of NU-FL, which can occur when channels are closed and at rest, may occur faster than deprotonation of flecainide. Hence, that the differences in recovery kinetics occur not only because of the greater fraction of neutral NU-FL molecules at this pH, but also because the ionized-bound drug deprotonates faster than ionized-bound flecainide and leaves the vicinity of the receptor via a hydrophobic pathway (Liu et al. 2003). It would seem that UDB develops predominantly as a function of differences between the recovery kinetics of ionized and neutral drug molecules.

Neutral flecainide (NU-FL) preferentially interacts with inactivated channels and does not require channel openings to develop, a suggestion that predicts drug-dependent alteration of the voltage dependence of channel availability (Liu et al. 2003). Flecainide has little effect on channel availability, while lidocaine causes a well-documented negative shift in channel availability under the same voltage conditions. The tertiary flecainide analog NU-FL also shifts channel availability without conditioning pulses, similar to lidocaine but in contrast to flecainide (Liu et al. 2003). Thus, although nearly identical to flecainide in structure, NU-FL interacts with the inactivated state without mandatory channel openings similar to lidocaine, a drug with a significant neutral component at physiological pH (Liu et al. 2003). When all the data are taken together, it is likely that external flecainide diffuses into cells through rapid equilibrium via its neutral component, and, once inside, equilibrium is again established with more than 99% of intracellular drug ionized.

10
Molecular Determinants of Drug Binding

Much evidence suggests that antiarrhythmics bind in the pore of the channel on the intracellular side of the selectivity filter (Ragsdale et al. 1994, 1996). Mutagenesis experiments have revealed multiple sites that affect drug binding on the S6 segments of domains I, III, and IV, and that dramatic changes in drug affinity can result from mutations near to the putative drug receptor sites on DIVS6 (Fig. 8). For example, mutations of I409 and N418 in DIS6 moderately altered drug interaction affinity in the brain VGSC $Na_V1.2$ (Yarov-Yarovoy et al. 2002). Mutagenesis studies of DIIIS6 in $Na_V1.2$ suggest that L1465, N1466, and I1469 are involved in drug binding, since mutation of these residues reduced affinity of the LA etidocaine (Yarov-Yarovoy et al. 2001). Experiments using the rat skeletal muscle isoform found that residues corresponding to human $Na_V1.2$ L1465 (L1280) and S1276 modulated LA affinity as well as the affinity of the channel activator batrachotoxin (Wang et al. 2000b; Nau et al. 2003). Similar systematic mutagenesis of DIIS6 found no residues that had significant effects on drug binding (Yarov-Yarovoy et al. 2002). However, mutations of residues F1764 and Y1771 on DIVS6 in $Na_V1.2$ resulted in dramatic decreases in both

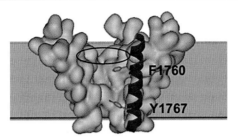

Fig. 8 Structural determinants of drug binding. Surface representation of the sodium channel with a *helix* representing DIVS6. Shown are the side chains for primary residues implicated in drug binding, F1760 and Y1767. The selectivity filter is indicated by a *black ellipse*

TB and UDB for lidocaine (Ragsdale et al. 1994, 1996). Subsequent studies in cardiac, skeletal muscle, and other brain sodium channel isoforms suggested these same residues to be important for drug interaction (Wright et al. 1998). Mutation of F1764 to alanine alone reduced the affinity of lidocaine for the inactivated state by almost 25-fold, although the UDB for flecainide was less dramatically affected by the single mutation compared to mutation of both F1764 and Y1771. Mutations of pore residues suggest that charged portions of drugs interact with the selectivity filter and mutations of pore residues, and residues responsible for TTX affinity affect drug access to, and egress from, the binding site (Sunami et al. 1997; Sunami et al. 2000; Sasaki et al. 2004).

It should be noted that different VGSC isoforms have different pharmacological and biophysical profiles, which would be expected to have diverse effects on drug binding. Also, several different antiarrhythmics, anticonvulsants, and LA agents were tested in the studies described above. Hence, the differences observed between drugs and isoforms may be attributable to any one of these variables. Finally, mutations may alter kinetic properties of channels that result in secondary effects on drug binding that are independent of the structural effect of the mutation.

11
Molecular and Biophysical Determinants of Isoform Specificity

There are many factors that contribute to efficacy of VGSC blockade. Drugs have variable affinity to different isoforms, and implicit tissue properties such as resting potential, action potential morphology, and action potential frequency affect in vivo drug responses. For example, antiarrhythmic agents are highly cardioselective and bind with higher affinity to cardiac sodium channel isoforms compared to brain and skeletal muscle. There is some debate as to the molecular mechanism of cardioselectivity: Does it result from intrinsically higher drug binding affinity (Wang et al. 1996), or as a secondary effect of isoform-specific kinetics (Wright et al. 1997), which may increase the probability of drug interaction with the binding site?

Two studies have identified amino acid differences between skeletal and cardiac isoforms that appear to be partial structural determinants of cardioselectivity. One study identifies a residue on the S4–S5 linker of DI that contains heterologous amino acids in rat heart (A252) and skeletal muscle (S251) isoforms (Kawagoe et al. 2002). Mutation of the rat skeletal muscle residue (S251) to alanine increased mexiletine affinity, although not nearly to the levels of wild-type rat heart, with respect to both tonic block and UDB. Another study found that mutation of rat skeletal muscle L1373, located on DIVS1S2 linker, to the glutamate found in the cardiac isoform shifted UDB by lidocaine toward that of the human cardiac isoform (Meisler et al. 2002). Interestingly, these residues are located on the opposite sides of the membrane, S251 located intracellularly and L1373 on the extracellular loop. In addition to the amino acid changes, the intrinsic affinity of the heart and skeletal muscle isoforms for LAs has been shown to be affected by its association with, or lack of association with β-subunits. The association with β-subunits shifts the midpoints of availability much more in the depolarizing direction for skeletal muscle isoform and modestly increases resting affinity for lidocaine, while the association with the β_1-subunit had the opposite effect on the cardiac isoform (Makielski et al. 1996, 1999).

12
Summary

Most antiarrhythmic agents were developed when there was relatively minimal information regarding the molecular and physicochemical basis of drug–receptor interactions. Since the advent of gene cloning, a wealth of information regarding these processes has been gathered. As our understanding of the basis for drug–receptor interactions becomes more complete, it will increasingly become possible to not only better understand the mechanism by which currently used antiarrhythmic agents exert their action, but to develop other more specific agents to suppress arrhythmias. Improvement in our understanding of drug–channel interactions sets the stage for a new era of "genetic medicine," where pharmacological agents can be developed to treat patients based on individual genotypic profile.

References

Abriel H, Wehrens XHT, Benhorin J, Kerem B, Kass RS (2000) Molecular pharmacology of the sodium channel mutation D1790G linked to the long-QT syndrome. Circulation 102:921–925
Abriel H, Cabo C, Wehrens XHT, Rivolta I, Motoike HK, Memmi M, Napolitano C, Priori SG, Kass RS (2001) Novel arrhythmogenic mechanism revealed by a Long-QT syndrome mutation in the cardiac Na+ channel. Circ Res 88:740–745

Ahern CA, Horn R (2004) Stirring up controversy with a voltage sensor paddle. Trends Neurosci 27:303–307

Auld VJ, Goldin AL, Krafte DS, Catterall WA, Lester HA, Davidson N, Dunn RJ (1990) A neutral amino-acid change in segment-IIS4 dramatically alters the gating properties of the voltage-dependent sodium-channel. Proc Natl Acad Sci USA 87:323–327

Baruscotti M, DiFrancesco D, Robinson RB (1996) A TTX-sensitive inward sodium current contributes to spontaneous activity in newborn rabbit sino-atrial node cells. J Physiol (Lond) 492:21–30

Baruscotti M, Westenbroek R, Catterall WA, DiFrancesco D, Robinson RB (1997) The newborn rabbit sino-atrial node expresses a neuronal type I-like Na+ channel. J Physiol (Lond) 498:641–648

Baruscotti M, DiFrancesco D, Robinson RB (2001) Single-channel properties of the sinoatrial node Na+ current in the newborn rabbit. Pflugers Arch 442:192–196

Brugada R, Brugada J, Antzelevitch C, Kirsch GE, Potenza D, Towbin JA, Brugada P (2000) Sodium channel blockers identify risk for sudden death in patients with ST-Segment elevation and right bundle branch block but structurally normal hearts. Circulation 101:510–515

Carmeliet E, Fozzard HA, Hiraoka M, Janse MJ, Ogawa S, Roden DM, Rosen MR, Rudy Y, Schwartz PJ, Matteo PS, Antzelevitch C, Boyden PA, Catterall WA, Fishman GI, George AL, Izumo S, Jalife J, January CT, Kleber AG, Marban E, Marks AR, Spooner PM, Waldo AL, Weiss JM, Zipes DLP (2001) New approaches to antiarrhythmic therapy, part I—emerging therapeutic applications of the cell biology of cardiac arrhythmias. Circulation 104:2865–2873

Cohen SA (1996) Immunocytochemical localization of rH1 sodium channel in adult rat heart atria and ventricle: presence in terminal intercalated disks. Circulation 94:3083–3086

Cormier JW, Rivolta I, Tateyama M, Yang AS, Kass RS (2002) Secondary structure of the human cardiac Na+ channel C terminus—Evidence for a role of helical structures in modulation of channel inactivation. J Biol Chem 277:9233–9241

Fenichel RR, Malik M, Antzelevitch C, Sanguinetti M, Roden DM, Priori SG, Ruskin JN, Lipicky RJ, Cantilena LR (2004) Drug-induced torsades de pointes and implications for drug development. J Cardiovasc Electrophysiol 15:475–495

Goldin AL (2001) Resurgence of sodium channel research. Annu Rev Physiol 63:871–894

Goldin AL (2002) Evolution of voltage-gated Na+ channels. J Exp Biol 205:575–584

Goldin AL, Barchi RL, Caldwell JH, Hofmann F, Howe JR, Hunter JC, Kallen RG, Mandel G, Meisler MH, Netter YB, Noda M, Tamkun MM, Waxman SG, Wood JN, Catterall WA (2000) Nomenclature of voltage-gated sodium channels. Neuron 28:365–368

Grant AO, Chandra R, Keller C, Carboni M, Starmer CF (2000) Block of wild-type and inactivation-deficient cardiac sodium channels IFM/QQQ stably expressed in mammalian cells. Biophys J 79:3019–3035

Harrison DC (1985) A rational scientific basis for subclassification of antiarrhythmic drugs. Trans Am Clin Climatol Assoc 97:43–52

Hille B (1977) Local-anesthetics—hydrophilic and hydrophobic pathways for drug-receptor reaction. J Gen Physiol 69:497–515

Honjo H, Boyett MR, Kodama I, Toyama J (1996) Correlation between electrical activity and the size of rabbit sino-atrial node cells. J Physiol (Lond) 496:795–808

Kambouris NG, Nuss HB, Johns DC, Marban E, Tomaselli GF, Balser JR (2000) A revised view of cardiac sodium channel "blockade" in the long-QT syndrome. J Clin Invest 105:1133–1140

Kawagoe H, Yamaoka K, Kinoshita E, Fujimoto Y, Maejima H, Yuki T, Seyama I (2002) Molecular basis for exaggerated sensitivity to mexiletine in the cardiac isoform of the fast Na channel. FEBS Lett 513:235–241

Kodama I, Boyett MR, Suzuki R, Honjo H, Toyama J (1996) Regional differences in the response of the isolated sino-atrial node of the rabbit to vagal stimulation. J Physiol (Lond) 495:785–801

Kodama I, Nikmaram MR, Boyett MR, Suzuki R, Honjo H, Owen JM (1997) Regional differences in the role of the Ca2+ and Na+ currents in pacemaker activity in the sinoatrial node. Am J Physiol Heart Circ Physiol 41:H2793–H2806

Kontis KJ, Rounaghi A, Goldin AL (1997) Sodium channel activation gating is affected by substitutions of voltage sensor positive charges in all four domains. J Gen Physiol 110:391–401

Kucera JP, Rohr S, Rudy Y (2002) Localization of sodium channels in intercalated disks modulates cardiac conduction. Circ Res 91:1176–1182

Lee JT, Kroemer HK, Silberstein DJ, Funck-Brentano C, Lineberry MD, Wood AJ, Roden DM, Woosley RL (1990) The role of genetically determined polymorphic drug metabolism in the beta-blockade produced by propafenone. N Engl J Med 322:1764–1768

Lerche H, Jurkat-Rott K, Lehmann-Horn F (2001) Ion channels and epilepsy. Am J Med Genet 106:146–159

Liu H, Atkins J, Kass R (2003) Common molecular determinants of flecainide and lidocaine block of heart Na(+) channels: evidence from experiments with neutral and quaternary flecainide analogues. J Gen Physiol 121:199–214

Liu HJ, Tateyama M, Clancy CE, Abriel H, Kass RS (2002) Channel openings are necessary but not sufficient for use-dependent block of cardiac Na+ channels by flecainide: evidence from the analysis of disease-linked mutations. J Gen Physiol 120:39–51

Lossin C, Wang DW, Rhodes TH, Vanoye CG, George AL (2002) Molecular basis of an inherited epilepsy. Neuron 34:877–884

Maier SKG, Westenbroek RE, Schenkman KA, Feigl EO, Scheuer T, Catterall WA (2002) An unexpected role for brain-type sodium channels in coupling of cell surface depolarization to contraction in the heart. Proc Natl Acad Sci USA 99:4073–4078

Makielski JC, Limberis JT, Chang SY, Fan Z, Kyle JW (1996) Coexpression of beta 1 with cardiac sodium channel alpha subunits in oocytes decreases lidocaine block. Mol Pharmacol 49:30–39

Makielski JC, Limberis J, Fan Z, Kyle JW (1999) Intrinsic lidocaine affinity for Na channels expressed in Xenopus oocytes depends on alpha (hH1 vs. rSkM1) and beta 1 subunits. Cardiovasc Res 42:503–509

Malhotra JD, Chen CL, Rivolta I, Abriel H, Malhotra R, Mattei LN, Brosius FC, Kass RS, Isom LL (2001) Characterization of sodium channel alpha- and beta-subunits in rat and mouse cardiac myocytes. Circulation 103:1303–1310

Mantegazza M, Yu FH, Catterall WA, Scheuer T (2001) Role of the C-terminal domain in inactivation of brain and cardiac sodium channels. Proc Natl Acad Sci USA 98:15348–15353

Meisler MH, Kearney JA, Sprunger LK, MacDonald BT, Buchner DA, Escayg A (2002) Mutations of voltage-gated sodium channels in movement disorders and epilepsy. Novartis Found Symp 241:72–86

Meyer UA, Zanger UM, Skoda RC, Grant D, Blum M (1990) Genetic polymorphisms of drug metabolism. Prog Liver Dis 9:307–323

Muramatsu H, Zou AR, Berkowitz GA, Nathan RD (1996) Characterization of a TTX-sensitive Na+ current in pacemaker cells isolated from rabbit sinoatrial node. Am J Physiol Heart Circ Physiol 39:H2108–H2119

Nagatomo T, January CT, Makielski JC (2000) Preferential block of late sodium current in the LQT3 DeltaKPQ mutant by the class I(C) antiarrhythmic flecainide. Mol Pharmacol 57:101–107

Nau C, Wang SY, Wang GK (2003) Point mutations at L1280 in Nav1.4 channel D3-S6 modulate binding affinity and stereoselectivity of bupivacaine enantiomers. Mol Pharmacol 63:1398–1406

Ong BH, Tomaselli GF, Balser JR (2000) A structural rearrangement in the sodium channel pore linked to slow inactivation and use dependence. J Gen Physiol 116:653–661

Priori SG, Napolitano C, Schwartz PJ, Bloise R, Crotti L, Ronchetti E (2000) The thin border between long QT and Brugada syndromes: the role of flecainide challenge. Circulation 102:676

Qu Y, Isom LL, Westenbroek RE, Rogers JC, Tanada TN, McCormick KA, Scheuer T, Catterall WA (1995) Modulation of cardiac Na+ channel expression in Xenopus oocytes by beta 1 subunits. J Biol Chem 270:25696–25701

Ragsdale DS, Mcphee JC, Scheuer T, Catterall WA (1994) Molecular determinants of state-dependent block of Na+ channels by local-anesthetics. Science 265:1724–1728

Ragsdale DS, McPhee JC, Scheuer T, Catterall WA (1996) Common molecular determinants of local anesthetic, antiarrhythmic, and anticonvulsant block of voltage-gated Na+ channels. Proc Natl Acad Sci USA 93:9270–9275

Rajamani S, Anderson CL, Anson BD, January CT (2002) Pharmacological rescue of human K(+) channel long-QT2 mutations: human ether-a-go-go-related gene rescue without block. Circulation 105:2830–2835

Roden D (1990) Antiarrhythmic drugs. In: Hardman JG, et al (eds) Goodman and Gilman's the pharmacological basis of therapeutics. McGraw-Hill, New York, pp 839–974

Roden D (2001) Principles in pharmacogenomics. Epilepsia 42:44–48

Roden DM (2000) Antiarrhythmic drugs: from mechanisms to clinical practice. Heart 84:339–346

Roden DM, George AL (2002) The genetic basis of variability in drug responses. Nat Rev Drug Discov 1:37–44

Rosen MR, Wit AL (1983) Electropharmacology of anti-arrhythmic drugs. Am Heart J 106:829–839

Rosen MR, Wit AL (1987) Arrhythmogenic actions of antiarrhythmic drugs. Am J Cardiol 59:E10–E18

Rosen MR, Hoffman BF, Wit AL (1975) Electrophysiology and pharmacology of cardiac-arrhythmias. 5. Cardiac antiarrhythmic effects of lidocaine. Am Heart J 89:526–536

Ruskin JN (1989) The Cardiac Arrhythmia Suppression Trial (CAST). N Engl J Med 321:386–388

Sasaki K, Makita N, Sunami A, Sakurada H, Shirai N, Yokoi H, Kimura A, Tohse N, Hiraoka M, Kitabatake A (2004) Unexpected mexiletine responses of a mutant cardiac Na+ channel implicate the selectivity filter as a structural determinant of antiarrhythmic drug access. Mol Pharmacol 66:330–336

Schwarz W, Palade PT, Hille B (1977) Local-anesthetics—effect of Ph on use-dependent block of sodium channels in frog muscle. Biophys J 20:343–368

Smith MR, Goldin AL (1997) Interaction between the sodium channel inactivation linker and domain III S4-S5. Biophys J 73:1885–1895

Splawski I, Timothy KW, Tateyama M, Clancy CE, Malhotra A, Beggs AH, Cappuccio FP, Sagnella GA, Kass RS, Keating MT (2002) Variant of SCN5A sodium channel implicated in risk of cardiac arrhythmia. Science 297:1333–1336

Steinlein OK (2001) Genes and mutations in idiopathic epilepsy. Am J Med Genet 106:139–145

Strichartz GR, Sanchez V, Arthur GR, Chafetz R, Martin D (1990) Fundamental properties of local anesthetics. II. Measured octanol:buffer partition coefficients and pKa values of clinically used drugs. Anesth Analg 71:158–170

Stuhmer W, Conti F, Suzuki H, Wang XD, Noda M, Yahagi N, Kubo H, Numa S (1989) Structural parts involved in activation and inactivation of the sodium channel. Nature 339:597–603

Sun YM, Favre I, Schild L, Moczydlowski E (1997) On the structural basis for size-selective permeation of organic cations through the voltage-gated sodium channel—Effect of alanine mutations at the DEKA locus on selectivity, inhibition by Ca2+ and H+, and molecular sieving. J Gen Physiol 110:693–715

Sunami A, Glaaser IW, Fozzard HA (2000) A critical residue for isoform difference in tetrodotoxin affinity is a molecular determinant of the external access path for local anesthetics in the cardiac sodium channel. Proc Natl Acad Sci U S A 97:2326–2331

Sunami A, Dudley SC Jr, Fozzard HA (1997) Sodium channel selectivity filter regulates antiarrhythmic drug binding. Proc Natl Acad Sci U S A 94:14126–14131

Task Force of the Working Group on Arrhythmias of the European Society of Cardiology (1991) The Sicilian Gambit. A new approach to the classification of antiarrhythmic drugs based on their actions on arrhythmogenic mechanisms. Circulation 84:1831–1851

Valdivia CR, Ackerman MJ, Tester DJ, Wada T, McCormack J, Ye B, Makielski JC (2002) A novel SCN5A arrhythmia mutation, M1766L, with expression defect rescued by mexiletine. Cardiovasc Res 55:279–289

Valdivia CR, Tester DJ, Rok BA, Porter CB, Munger TM, Jahangir A, Makielski JC, Ackerman MJ (2004) A trafficking defective, Brugada syndrome-causing SCN5A mutation rescued by drugs. Cardiovasc Res 62:53–62

Vaughan Williams E (1989) Classification of antiarrhythmic action. In: Vaughan Williams EM (ed) Handbook of experimental pharmacology vol. 89. Springer-Verlag, Berlin, Heidelberg, New York, pp 45–62

Viswanathan PC, Bezzina CR, George AL, Roden DM, Wilde AAM, Balser JR (2001) Gating-dependent mechanisms for flecainide action in SCN5A-linked arrhythmia syndromes. Circulation 104:1200–1205

Wang DW, Nie L, George AL Jr, Bennett PB (1996) Distinct local anesthetic affinities in Na+ channel subtypes. Biophys J 70:1700–1708

Wang DW, Yazawa K, Makita N, George AL Jr, Bennett PB (1997) Pharmacological targeting of long QT mutant sodium channels. J Clin Invest 99:1714–1720

Wang DW, Makita N, Kitabatake A, Balser JR, George AL (2000a) Enhanced Na+ channel intermediate inactivation in Brugada syndrome. Circ Res 87:E37–E43

Wang SY, Nau C, Wang GK (2000b) Residues in Na(+) channel D3-S6 segment modulate both batrachotoxin and local anesthetic affinities. Biophys J 79:1379–1387

Weissenburger J, Davy JM, Chezalviel F (1993) Experimental models of torsades de pointes. Fundam Clin Pharmacol 7:29–38

West JW, Patton DE, Scheuer T, Wang Y, Goldin AL, Catterall WA (1992) A cluster of hydrophobic amino acid residues required for fast Na(+)-channel inactivation. Proc Natl Acad Sci USA 89:10910–10914

Wit AL, Rosen MR (1983) Pathophysiologic mechanisms of cardiac-arrhythmias. Am Heart J 106:798–811

Wright SN, Wang SY, Kallen RG, Wang GK (1997) Differences in steady-state inactivation between Na channel isoforms affect local anesthetic binding affinity. Biophys J 73:779–788

Wright SN, Wang SY, Wang GK (1998) Lysine point mutations in Na+ channel D4-S6 reduce inactivated channel block by local anesthetics. Mol Pharmacol 54:733–739

Yamagishi T, Li RA, Hsu K, Marban E, Tomaselli GF (2001) Molecular architecture of the voltage-dependent Na channel: functional evidence for at helices in the pore. J Gen Physiol 118:171–181

Yarov-Yarovoy V, Brown J, Sharp EM, Clare JJ, Scheuer T, Catterall WA (2001) Molecular determinants of voltage-dependent gating and binding of pore-blocking drugs in transmembrane segment IIIS6 of the Na+ channel alpha subunit. J Biol Chem 276:20–27

Yarov-Yarovoy V, McPhee JC, Idsvoog D, Pate C, Scheuer T, Catterall WA (2002) Role of amino acid residues in transmembrane segments IS6 and IIS6 of the Na+ channel alpha subunit in voltage-dependent gating and drug block. J Biol Chem 277:35393–35401

Zhou Z, Gong G, January CT (1999) Correction of a defective protein trafficking of a mutant HERG potassium channel in human long QT syndrome. J Biol Chem 274:31123–31126

HEP (2006) 171:123–157
© Springer-Verlag Berlin Heidelberg 2006

Structural Determinants of Potassium Channel Blockade and Drug-Induced Arrhythmias

X. H. T. Wehrens

Center for Molecular Cardiology, Dept. of Physiology and Cellular Biophysics,
College of Physicians and Surgeons of Columbia University, 630 West 168th Street,
P&S 9-401, New York NY, 10032, USA
xw80@columbia.edu

Abstract Cardiac K^+ channels play an important role in the regulation of the shape and duration of the action potential. They have been recognized as targets for the actions of neurotransmitters, hormones, and anti-arrhythmic drugs that prolong the action potential duration (APD) and increase refractoriness. However, pharmacological therapy, often for the purpose of treating syndromes unrelated to cardiac disease, can also increase the vulnerability of some patients to life-threatening rhythm disturbances. This may be due to an underlying propensity stemming from inherited mutations or polymorphisms, or structural abnormalities that provide a substrate allowing for the initiation of arrhythmic triggers. A number of pharmacological agents that have proved useful in the treatment of allergic reactions, gastrointestinal disorders, and psychotic disorders, among others, have been shown to reduce repolarizing K^+ currents and prolong the Q-T interval on the electrocardiogram. Understanding the structural determinants of K^+ channel blockade might provide new insights into the mechanism and rate-dependent effects of drugs on cellular physiology. Drug-induced disruption of cellular repolarization underlies electrocardiographic abnormalities that are diagnostic indicators of arrhythmia susceptibility.

Keywords Potassium channel · Arrhythmias · Long QT syndrome · Delayed rectifier · Repolarization

1
Introduction

Abnormalities of cardiac rhythm are a major cause of morbidity and mortality in the Western world. More than 300,000 Americans suffer sudden cardiac death (SCD) due to arrhythmias each year, and many more require therapy for symptomatic arrhythmias (Zipes and Wellens 1998). Normal cardiac rhythm is governed by ordered propagation of excitatory stimuli resulting in rapid depolarization and slow repolarization of the myocardium, generating action potentials (AP) in individual myocytes (Roden et al. 2002). Abnormalities of impulse generation, propagation, or the duration of action potentials may underlie cardiac arrhythmias.

One proarrhythmic condition that has received particular attention is the congenital and drug-induced long QT syndrome (LQTS). In this case, inherited mutations or drugs prolong the duration of the action potential (APD) of ventricular myocytes, which can be observed as a prolongation of the Q-T interval of the surface electrocardiogram (ECG). APD prolongation is most frequently caused by a decrease in repolarizing potassium currents, in particular the delayed rectified K^+ current, I_K, which has both rapidly (I_{Kr}) and slowly (I_{Ks}) activating components (Sanguinetti and Jurkiewicz 1990b). Moreover, virtually every case of drug-induced QT prolongation can be traced to blockade of the I_{Kr} current (Roden et al. 2002; Wehrens et al. 2002).

In the past decade, molecular cloning experiments have defined the genes whose expression generates specific proteins, including pore-forming ion channels, responsible for individual ion currents in cardiac myocytes (Keating and Sanguinetti 2001). It has become apparent that the generation of potassium currents requires coordinated function of not only α- and β-subunits, but also multiple other gene products that determine intracellular functions such as trafficking, phosphorylation and dephosphorylation, assembly, and targeting to specific subcellular domains (Gutman et al. 2003). In this chapter, the structure–activity relationship of potassium channels in the heart will be reviewed. The focus will be on delayed rectifier K^+ channels, which have been linked to abnormalities in AP duration and drug-induced cardiac arrhythmias.

2
Ion Currents and the Cardiac Action Potential

Depolarization of the plasma membrane induces opening of voltage-gated Na^+ channels, allowing a large, rapid Na^+ influx, producing the typical rapid phase 0 depolarization (Fig. 1). In some myocytes, a rapid phase 1 repolarization then ensues, because of activation of transient outward K^+ channels. Phase 2, the long plateau phase of the action potential, reflects a delicate balance between inward L-type Ca^{2+} channels and outward current through delayed rectifier K^+ channels (Kass 1997). Repolarization occurs during phase 3 when outward movement of K^+ through delayed rectifier K^+ channels dominates over the inactivated Ca^{2+} channels, and the membrane potential returns to resting voltages.

The duration and configuration of the action potential vary in different regions of the heart (e.g., atrium versus ventricle) as well as in specific areas within those regions. Epicardial cells in the ventricle demonstrate a prominent

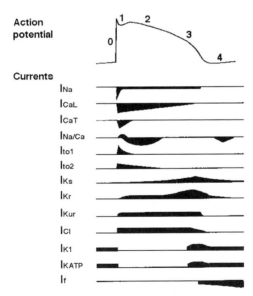

Fig. 1 Relationship between the ventricular action potential and individual ionic currents in ventricular myocytes. Schematic indication of the time course of depolarizing inward currents (*downward*) and repolarizing outward current (*upward*) in relation to the action potential. The amplitudes of the currents are not to scale

phase 1 notch, which is less prominent in the endocardium (Antzelevitch and Fish 2001). Purkinje and midmyocardial (M) cells display action potentials that are much longer than those in the epicardium and endocardium. These heterogeneities in action potential duration reflect variations in expression and/or function of the ion channels that constitute cardiac ion currents (Volders et al. 1999a). Increased heterogeneity, by ischemic heart disease, inherited ion channel mutations, or drug exposure, promotes reentrant excitation, a common mechanism for many ventricular arrhythmias.

3
Cardiac Delayed Rectifier Potassium Channels

Over 80 human K^+ channel-related genes have been cloned and characterized (Roden et al. 2002; Tamargo et al. 2004). Since many cDNAs encoding ion channels have been cloned from the mammalian heart, it has been difficult to assign cloned K^+ channel subunits to specific endogenous currents (Snyders 1999). Among the more than 12 cardiac K^+ currents, two types of voltage-gated K^+ channels play a dominant role in determining repolarization: the transient outward (I_{to}) and delayed rectifier (I_K) currents (Barry and Nerbonne 1996; Deal et al. 1996). Two components of the delayed rectifier I_K have been separated on the basis of the activation kinetics: a rapidly activating component (I_{Kr}) and slowly activating component (I_{Ks}) (Sanguinetti and Jurkiewicz 1990b).

3.1
The Molecular Basis of the I_{Kr} Current

3.1.1
Topology of the I_{Kr} Channel

The molecular basis of I_{Kr} was elucidated when *KCNH2* (*HERG*; human ether-a-go-go-related gene) was linked to a form of LQTS (LQT2) (Curran et al. 1995). The topology of KCNH2 channels is similar to many voltage-gated channels in that they are homo-tetramers of α-subunits containing six transmembrane spanning domains (S1–S6; Roden et al. 2002). A cluster of positive charges is localized in the S4 domain and acts as the putative activation voltage sensor (Warmke and Ganetzky 1994). A reentrant pore-loop between the fifth and sixth transmembrane helices contains a slightly modified version of the "signature sequence" of K^+-selective pores (Doyle et al. 1998; Heginbotham et al. 1994). Based on structure–function analyses of other voltage-gated K^+ channels, it is inferred that each functional KCNH2 channel is composed of four subunits surrounding a central aqueous pore, with the outer one-third lined by the pore-loops from the four subunits forming a narrow "selectivity filter," and the inner two-thirds lined by the carboxyl-halves of the S6 domains that form the inner vestibule of the pore (Doyle et al. 1998; Mitcheson et al. 2000a).

The channel encoded by *KCNH2* recapitulates indeed the major functional and pharmacological properties of I_{Kr}, including inward rectification, block by micromolar La^{3+}, and specific block by methanesulfonanilide antiarrhythmic agents such as E4031 and dofetilide (Sanguinetti et al. 1995; Snyders and Chaudhary 1996; Trudeau et al. 1995). There are, however, marked differences between native I_{Kr} and KCNH2-induced currents in heterologous expression systems in terms of gating (Abbott et al. 1999; Zhou et al. 1998), regulation by external K^+ (Abbott et al. 1999; McDonald et al. 1997; Shibasaki 1987), and sensitivity to antiarrhythmics (Sanguinetti et al. 1995). These data suggest the presence of a modulating subunit that co-assembles with KCNH2 in order to reconstitute native I_{Kr} currents.

3.1.1.1
Accessory Subunits of I_{Kr} Channels

A likely candidate is the minK-related protein 1 (*MiRP1*; *KCNE2*), which when co-expressed with *KCNH2* (*HERG*), results in currents similar to native I_{Kr} (Abbott et al. 1999). Expression of *KCNE2* in the heart has recently been shown at the protein level (Jiang et al. 2004). Furthermore, KCNE2 co-immunoprecipitates with KCNH2 when the two subunits are co-expressed in CHO cells, suggesting that they may co-assemble to form the I_{Kr} channel in the heart. Co-expression of *KCNE2* with *KCNH2* in *Xenopus* oocytes causes a +5–10 mV depolarizing shift in steady-state activation, accelerates the rate of deactivation, and decreases single channel conductance from 13 to 8 pS (Abbott et al. 1999). However, definitive biochemical evidence for a selective association between KCNH2 and KCNE2 in the human myocardium is currently lacking (Abbott and Goldstein 2001; Weerapura et al. 2002), and other factors may contribute to the functional differences between native I_{Kr} and HERG-induced currents in heterologous expression systems.

Other K^+ channel α-subunits may be able to modulate KCNH2 channel function. Treatment of a mouse atrial tumor cell line (AT-1) with anti-sense oligonucleotides against *KCNE1* (*minK*), thus suppressing *KCNE1* expression, reduced the I_{Kr} current amplitude (Yang et al. 1995). Additional evidence that KCNE1 can modulate I_{Kr} current amplitude and gating comes from the observation that the I_{Kr} amplitude is significantly smaller in homozygous *KCNE1* knockout mice (Kupershmidt et al. 1999). Finally, KCNH2 and KCNE1 can form a stable complex when co-expressed in HEK293 cells, and KCNH1 amplitude is augmented relative to cells expressing *KCNH2* alone (McDonald et al. 1997).

Recently, *MiRP2* (*KCNE3*) has been shown to suppress the expression of *KCNH2* in *Xenopus* oocytes, suggesting yet another β-subunit may modulate I_{Kr} channel function (Schroeder et al. 2000). It is possible that KCNE β-subunits (minK, MiRP1, MiRP2) can engage in interactions with α-subunits from more than one gene family, and their role in native K^+ channel function may be determined by relative expression levels, the subcellular distribution of different K^+ channel subunits, or both (Tseng 2001).

3.1.1.2
KCNH2 Splice Variants

Several N-terminal splice variants of the human and mouse *KCNH2* genes have been identified (Kupershmidt et al. 1998; Lees-Miller et al. 1997; London et al. 1997). The full-length *KNCH2* isoform has 396 amino acids in the N-terminus and is designated isoform 1a. The alternatively spliced isoform 1b has a much shorter and divergent N-terminus of 36 amino acids. Isoform 1b lacks the sequence of amino acids 2–16 and the Per-Arnt-Sim (PAS) motif, which are both important for the slow deactivation process of isoform 1a (Morais Cabral et al. 1998; Wang et al. 1998a, 2000b). Therefore, isoform 1b deactivates at a tenfold faster rate than isoform 1a. It has been suggested that both isoforms can form homomultimeric and heteromultimeric channels, and that the resulting deactivation rate will depend on the relative expression levels of these two isoforms (London et al. 1997). Studies by Nerbonne et al. have shown that isoform 1b is not expressed at the protein level in adult human, rat, and mouse hearts (Pond et al. 2000). However, more recent data have demonstrated expression of *KNCH2* isoform 1b in the mammalian heart using antibodies recognizing the different splice variants (Jones et al. 2004). These data suggest that isoforms 1a and 1b might indeed form heteromultimeric channels in cardiac myocytes.

A C-terminal splice variant of *KCNH2* has also been identified in the human heart (Kupershmidt et al. 1998). This isoform (KCNH2$_{USO}$) cannot form functional channels on its own since a critical C-terminal region is absent. However, when co-expressed with *KCNH2*, KCNH2$_{USO}$ suppresses the current amplitude and alters its gating kinetics (e.g., accelerates activation and shifts the voltage-dependence by -9 mV). The mRNA levels of KCNH2$_{USO}$ are twofold more abundant than those of KCNH2 in human heart (Kupershmidt et al. 1998). Although it has been suggested that KCNH2$_{USO}$ may play a role in determining the current amplitude and gating kinetics of I_{Kr} in cardiac myocytes, there has been no biochemical evidence so far supporting the expression of KCNH2$_{USO}$ in the human heart (Kupershmidt et al. 1998).

3.1.2
The Physiological Role of I_{Kr} in the Heart

The rapidly activating delayed rectifier current (I_{Kr}) can be distinguished from the slowly activating component (I_{Ks}) by its activation kinetics, as well at its sensitivity to block by class III antiarrhythmic drugs such as E-4031 (Follmer and Colatsky 1990) and dofetilide (Ficker et al. 1998; Jurkiewicz and Sanguinetti 1993; Snyders and Chaudhary 1996). An important feature of I_{Kr} is the inward rectification property that limits outward currents through the channel at positive voltages, which reduces the amount of inward current needed to maintain the action potential plateau phase. Detailed kinetic studies of I_{Kr} gating have revealed that the inward rectification is due to fast inactivation (Shibasaki 1987; Smith et al. 1996; Spector et al. 1996b). Unlike C-type inactivation in many other K^+ channels, I_{Kr} inactivation appears to be unique in that it possesses intrinsic voltage-dependence (Schonherr and Heinemann 1996; Smith et al. 1996).

I_{Kr} channels activate from closed to open states ($C \rightarrow O$) upon depolarization but pass very little outward current because they rapidly inactivate ($O \rightarrow I$). KCNH2 channels can also inactivate directly from closed states ($C \rightarrow I$; Kiehn et al. 1999). Inactivation from both pathways results in the accumulation of I_{Kr} channels in inactivated states during depolarization. Channels then reopen, or open for the first time, during repolarization as they recover from inactivation through the open state ($I \rightarrow O$). Deactivation of I_{Kr} ($I \rightarrow C$) is slow compared to other cardiac K^+ channels. These unique channel properties give rise to the I_{Kr} current during phase 3 repolarization of the cardiac action potential (Clancy and Rudy 2001; Kiehn et al. 1999).

3.1.3
Structural Basis of I_{Kr} Blockade

I_{Kr} is the primary target of highly specific and potent class III anti-arrhythmic drugs, methanesulfonanilides (dofetilide, E-4031, ibutilide, and MK-499; Spector et al. 1996a; Tamargo et al. 2004). I_{Kr} channels can also be blocked by myriad pharmacological agents with diverse chemical structures used for the treatment of both cardiac and non-cardiovascular disorders (Clancy et al. 2003). Recent studies have shed light on the molecular basis of the promiscuity of drug binding to the KCNH2 channel, and have provided further insight into the structure–function relationship of I_{Kr} channels.

The biophysical properties of KCNH2 blockade are consistent with a discrete state-dependent blocking mechanism (Kiehn et al. 1996a; Snyders and Chaudhary 1996). Most I_{Kr} blockers, including the methanesulfonanilides, gain access to the drug-binding site from the intracellular side of the membrane (Kiehn et al. 1996a; Kiehn et al. 1996b). Binding primarily occurs via the open state of the channel when the drugs can gain access to a high-affinity binding site located inside the channel vestibule (Tristani-Firouzi and Sanguinetti 2003). Once inside the pore, I_{Kr} blockers bind within the central cavity of the channel between the selectivity filter and the activation gate (see Fig. 2a). Unbinding of some methanesulfonalides (dofetilide, MK-499) is very slow and incomplete at negative voltages due to closure of the activation gate (deactivation) during repolarization, which traps the molecule within the cavity (Mitcheson et al. 2000b; Fig. 2b). If a drug is charged and appropriately sized, then block is nearly irreversible as long as the channels do not reopen even at negative potentials. This "drug-trapping" hypothesis was confirmed recently using a mutant KCNH2 channel (D540K) that opens in response to hyperpolarization. It was found that channel reopening at negative voltages allowed release of the drug MK-499 from the receptor (Mitcheson et al. 2000b; Sanguinetti and Xu 1999).

There are two structural features of the KCNH2 channels that contribute to their unique pharmacological properties: (1) The volume of the KCNH2 inner vestibule is larger that those of most other voltage-gated K^+ channels; and (2) two aromatic residues (Y652, F656), located in the S6 domain facing the channel vestibule, that form part of the contact points with inner mouth blockers are present (Fig. 2a).

The lack of the P-X-P sequence in the S6 domain of KCNH2 creates a large volume of the inner vestibule of the channel pore. Therefore, methanesulfonanilides (e.g., MK-499, with dimensions of 7×20 Å) can be trapped within the inner vestibule without affecting deactivation kinetics (Mitcheson et al. 2000b). Structurally, the larger inner vestibule can be explained by the lack of two proline residues that typically cause sharp bends in the S6 helices in all other voltage-gated K^+ channels (del Camino et al. 2000). Thus, the lack of such proline residues in KCNH2 makes the S6 domain more flexible and capable of forming a larger inner vestibule (Fig. 2a).

Recent studies have suggested that two aromatic residues in the S6 domain (Y652 and F656), which are unique to KCNH2 K^+ channels, may underlie the structural mechanism of preferential block of KCNH2 by a number of commonly prescribed drugs (Mitcheson et al. 2000a). Mutagenesis to alanine of both residues dramatically reduces the potency of channel block by a variety of KCNH2-blockers, including methanesulfonanilides, quinidine, cisapride, and terfenadine (Mitcheson et al. 2000a). The importance of residues Y652 and F656 was also demonstrated for the low-affinity ligand chloroquine, an antimalarial agent that appears to preferentially block open KCNH2 channels. Block of KCNH2 by chloroquine requires channel opening followed by interactions of the drug with the aromatic residues in the S6 domain that face the central cavity of the HERG channel pore (Sanchez-Chapula et al. 2002).

Homology modeling of the inner mouth structure of KCNH2, based on the crystal structure of KcsA (Doyle et al. 1998), suggests that the aromatic moieties of methanesulfonanilides and other drugs (e.g., cisapride, terfenadine) form electrostatic interactions with these two aromatic residues by π electron stacking (Mitcheson et al. 2000a). These two aromatic residues are unique to KCNH2, since the equivalent positions are occupied by isoleucines or valines in other voltage-gated K^+ channels. Thus, the features of the S6 domain in KCNH2 play a crucial role in determining the channel's unique pharmacological profile.

Mutations that result in loss of inactivation (S631A, G628C/S631C) reduce the affinity of methanesulfonanilides, while mutations that enhance inactivation (T432S, A443S, A453S) enhance drug block by dofetilide (Ficker et al. 2001; Tristani-Firouzi and Sanguinetti 2003). It has been hypothesized that the reduced affinity of noninactivating HERG mutant channels is not due to inac-

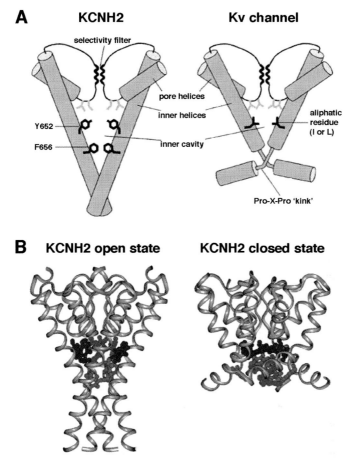

Fig. 2a,b Structural model of the drug-binding site in the KCNH2 channel. **a** The structures of two of the four subunits that form the pore and inner cavity of KCNH2 and Kv channels are shown. The inner helices and loops extending from the pore helices to the selectivity filter form the inner cavity and drug-binding site of HERG. Several structural features that help explain the nonspecific drug-binding properties of HERG are illustrated. The inner cavity of HERG is long, creating a relatively large space for trapping drugs and for channel–drug interactions. Aromatic residues (*black*) not found in Kv channels are critical sites for interaction for most compounds, but not for fluvoxamine. Other sites for drug interaction are polar residues (*gray*) located close to the selectivity filter. Kv channels have a Pro-X-Pro motif that is proposed to insert a 'kink' in the inner helices, resulting in a relatively small inner cavity. The inner cavity is lined by aliphatic rather than aromatic residues. Reproduced with permission from Mitcheson (2003). **b** Molecular model representing lowest score structures of propafenone docked into closed (*left*) and open-state (*right*) homology models (extracellular surface at *top*). Residue Y652 (*red*) and F656 (*yellow*) side-chains are displayed along with backbone ribbons (*gray*). Propafenone carbons are colored *green*. The model suggests that propafenone interacts with aromatic rings from Y652 and F656 when the channel is open. In the closed channel, drug trapping may occur via spatial restriction due to the ring of the four F656 side-chains. (Reproduced with permission from Witchel et al. 2004)

tivation per se but to inactivation gating-associated reorientation of residues Y652 and F656 in the S6 domain that mediate high-affinity drug binding (Chen et al. 2002a).

Recently, Milnes et al. (2003) showed that the selective serotonin reuptake inhibitor fluvoxamine exhibits KNCH2 channel blocking properties that are different from those previously described. The S6 domain mutation, Y652A and F656A, and the pore helix mutant S631A, only partially attenuated the block of the channel by fluvoxamine at concentrations causing profound inhibition of the wildtype KCNH2 channel (Milnes et al. 2003). This type of blockade is similar to that produced by canrenoic acid (CA), the main metabolite of the diuretic spironolactone, on KCNH2 channels expressed in CHO cells (Caballero et al. 2003). KCNH2 block by fluvoxamine and CA is far more rapid than that produced by methanesulfonanilides, suggesting that they cause either closed-state block or extremely rapidly developing open-state blockade. Thus, channel inactivation may not be a prerequisite for fluvoxamine- and CA-induced KCNH2 block.

3.1.4
Electrophysiological Consequences of I_{Kr} Block

Mutations in *KCNH2* are associated with LQTS, which is characterized by prolongation of the Q-T interval, ventricular arrhythmias, syncope, and sudden cardiac death (Wehrens et al. 2002). In cellular experiments, incorporation of mutant *KCNH2* subunits in the channel tetramer generally causes a reduction of I_{Kr} current (Kagan et al. 2000; Robertson 2000). Decreased repolarizing current through KCNH2 channels leads to prolongation of the ventricular action potential, which predisposes the heart to arrhythmogenic early afterdepolarizations (Clancy and Rudy 2001; Viswanathan et al. 1999).

Both the cellular effects of these congenital abnormalities and the resulting electrocardiographic abnormalities are analogous to those seen with pharmacological inhibition of KCNH2 channels by a variety of compounds. I_{Kr} blockers prolong the Q-T interval, and may cause torsade de pointes (TdP) arrhythmias that can degenerate into ventricular fibrillation and sudden cardiac death (Belardinelli et al. 2003). Moreover, reductions in I_{Kr} may result in increased dispersion of repolarization across the ventricular wall, which manifests on the ECG as widening of the T wave (Antzelevitch et al. 1996: Volders 1999). Thus, increased focal activity and reentry associated with an increased inhomogeneity of repolarization across the ventricular wall may lead to or predispose to the development of TdP arrhythmias (Antzelevitch and Fish 2001).

The prolongation of the APD produced by I_{Kr} blockers causes increased normal resting potentials and slow heart rates, while at depolarized potentials or during tachycardia this prolongation is much less marked or even absent (Tamargo 2000; Tamargo et al. 2004). Hence, I_{Kr} blockers are least effective when they are most needed. This reverse use-dependence limits anti-arrhythmic efficacy, while potentially maximizing the risk of TdP associated with bradycardia-dependent early afterdepolarizations. In guinea pigs, reverse use-dependence is attributed to a progressive I_{Ks} accumulation as the heart rate increases (due to the incomplete deactivation of this current), which shortens the APD and offsets the APD prolongation produced by the I_{Kr} blocker (Tseng 2001; see Sect. 3.2). Another explanation is that I_{Kr} block itself is reverse use-dependent, which might be attributed to the binding of KCNE2 to the channel complex (Abbott et al. 1999). An increased understanding of the mechanisms underlying KCNH2 block and APD prolongation may lead to the development of novel anti-arrhythmic drugs with safer use-dependence.

3.1.5
Modulation of I_{Kr} Channel Function

The amplitude and/or gating kinetics of the I_{Kr} current may be regulated by the autonomous nervous system. Pathological conditions of the heart may cause changes in local extracellular K^+ concentrations and acidosis, which may modulate I_{Kr}. Long-term disease of the heart can also alter I_{Kr} expression, which may contribute to abnormal cardiac electrical activity and arrhythmias.

3.1.5.1
Modulation by PKA and PKC

Pharmacological studies have revealed that activation of β-adrenergic receptors and elevation of intracellular cyclic AMP (cAMP) levels can regulate KCNH2 channels both through PKA-mediated effects and by direct interaction with the protein (Cui et al. 2000). PKA phosphorylation of KCNH2 reduces the current amplitude and induces a depolarizing shift in the voltage-dependent activation curve (Cui et al. 2000; Kiehn 2000). Sequence analysis has revealed that KCNH2 has four PKA phosphorylation sites and a cyclic nucleotide-binding domain (CNBD) in the C-terminus (Kiehn 2000). Mutation of all four PKA sites to alanines inhibits the shift in the voltage-dependence of activation (Cui et al. 2000).

KCNH2 channels are also regulated by PKC phosphorylation of the channel (Barros et al. 1998; Thomas et al. 2003). PKC-activator phorbol 12-myristate 13-acetate (PMA) causes a positive shift of activation and reduces I_{Kr} current. These effects, however, may be not specific to PKC phosphorylation of the KCNH2 subunit, since they are also observed after the PKC-dependent phosphorylation sites are altered by mutagenesis (Thomas et al. 2003). The α-adrenergic effects on I_{Kr} channel function may also be mediated by the endogenous phospholipid phosphatidylinositol 4,5-biphosphate (PIP2) generated following G protein-mediated activation of phospholipase C (PLC) (Bian et al. 2004).

The net effect of isoproterenol is to increase I_{Kr} current in guinea pig ventricular myocytes, an effect which could be inhibited by PKC inhibitor bisindolylmaleimide (Kiehn 2000). In rabbit sinoatrial cells, isoproterenol also increases I_{Kr} current, and this effect was inhibited by the PKA inhibitor H89 but not by bisindolylmaleimide (Lei et al. 2000). These results suggest that the regulation of I_{Kr} may be species- and tissue-specific and may also depend strongly on experimental conditions (Tamargo et al. 2004).

3.1.5.2
Modulation by Changes in Extracellular K^+ Concentrations

In contrast to most other K^+ currents, I_{Kr} amplitude increases upon elevation of extracellular K^+ concentrations $[K^+]_o$ despite a decrease in the driving force for outward current (Tristani-Firouzi and Sanguinetti 2003; Tseng 2001). A combination of mechanisms is thought to underlie this phenomenon. Elevation of $[K^+]_o$ reduces C-type inactivation by hindering the conformational changes in the outer mouth region necessary for the inactivation process (Baukrowitz and Yellen 1995; Wang et al. 1997; Yang et al. 1996). Since inactivation is the primary limiting factor for outward I_{Kr} currents at depolarized voltages, this is probably the most important mechanism by which elevating $[K^+]_o$ increases outward I_{Kr} current amplitudes (Tseng 2001). Secondly, elevating $[K^+]_o$ increases the single channel conductance of KCNH2 channels (Kiehn et al. 1996a; Zou et al. 1997). Finally, increasing $[K^+]_o$ could relieve channel blockade by extracellular Na^+ ions (Numaguchi et al. 2000). The latter two mechanisms will lead to an increase in both inward and outward K^+ currents through I_{Kr} channels. These mechanisms may explain why APD is reduced at higher $[K^+]_o$ and lengthened at lower $[K^+]_o$, and why QT prolongation may be more pronounced in patients with hypokalemia. On the other hand, modest elevations of $[K^+]_o$ using K^+ supplements and spironolactone in patients given I_{Kr} blockers or with LTQ2 significantly shortens the Q-T interval and may prevent TdP (Etheridge et al. 2003). Moreover, it is thought that the antiarrhythmic actions of I_{Kr} blockers can be reversed during ischemia, which is frequently accompanied by elevations of the $[K^+]_o$ in the narrow intercellular spaces and by catecholamine surges that occur with exercise or other activities associated with fast heart rates (Nattel 2000).

3.1.5.3
Extracellular Acidosis

Extracellular acidification (elevating $[H^+]_o$), for example during myocardial ischemia, induces a marked acceleration of I_{Kr} deactivation (Berube et al. 1999; Jiang et al. 1999; Vereecke and Carmeliet 2000). Since this occurs without significant changes in the current amplitude or the voltage-dependence of other gating transitions, the underlying mechanism probably does not involve pore blockade or screening of negative surface changes by protons. The effect of elevating $[H^+]_o$ on I_{Kr} deactivation is reduced when the N-terminal domain is removed (Jiang et al. 1999). This effect can also be prevented by pretreatment of the KCNH2 channel with extracellular diethylpyrocarbonate (DEPC), that can covalently modify the side chains of histidine and cysteine (Miles 1977). Together, these observations suggest that protonation of residues on the KCNH2 channel surface can induce allosteric changes in the channel, potentially destabilizing binding of the N-terminus to the activation gate at negative voltages, and accelerating channel deactivation.

3.2
The Molecular Basis of the I_{Ks} Current

3.2.1
Topology of the I_{Ks} Channel

In guinea pig cardiomyocytes, long depolarizing pulses in the presence of specific I_{Kr} blockers expose a large, slowly activating, outwardly rectifying K^+ current, I_{Ks} (Sanguinetti and Jurkiewicz 1990a,b). Initially, it was suggested that this current was conducted by KCNE1 channels, since expression of KCNE1 protein in *Xenopus* oocytes resulted in a current that resembled I_{Ks} (Varnum et al. 1993). The *KCNE1* gene was cloned from human cardiac tissue, and encodes a protein containing 129–130 amino acids consisting of a single transmembrane spanning domain (Folander et al. 1990; Murai et al. 1989). Other studies, however, suggested that KCNE1 activated both endogenous K^+ and Cl^- currents in the *Xenopus* oocytes (Attali et al. 1993).

Following linkage analysis of patients with LQT1, the K^+ channel gene *KCNQ1* (*KvLQT1*) was identified and cloned (Wang et al. 1996). The α-subunit of I_{Ks}, KCNQ1, shares topological homology with other voltage-gated K^+ channels in that its 676 amino acids consist of six transmembrane domains and a pore-forming region (Wang et al. 1996). However, the expressed KCNQ1 current displayed delayed rectifier characteristics unlike any previously identified current in the heart. Co-expression of *KCNQ1* with *KCNQ1*, however, fully recapitulated the kinetic features of the I_{Ks} current in cardiomyocytes (Barhanin et al. 1996; Sanguinetti et al. 1996b): slower activation and deactivation kinetics, a shift in the voltage-dependence of channel activation to more positive potentials, and an increase of the macroscopic current amplitude. Thus, the *KCNE1* (*minK*) controversy was ended by the demonstration that KCNE1 acts as a β-subunit that alters the intrinsic gating of KCNQ1.

The I_{Ks} current is believed to be generated by the co-assembly of four pore-forming KCNQ1 and two accessory KCNE1 subunits (Chen et al. 2003; Wang et al. 1998b), although the exact stoichiometry is not known. KCNE1 exhibits a single transmembrane spanning domain; the N-terminus is extracellular and the C-terminus intracellular. Controversial results have been reported regarding the sites of contact between the KCNQ1 and KCNE1 subunits. Although it has been suggested that KCNE1 forms part of the ion-conducting pore in the I_{Ks} channel complex (Tai and Goldstein 1998), it is difficult to reconcile the idea that KCNE1 could be part of the pore itself in a typical voltage-gated K^+ channel architecture of the selectivity filter (Doyle et al. 1998; Kurokawa et al. 2001b).

3.2.1.1
The I_{Ks} Channel Is a Macromolecular Signaling Complex

Marx et al. showed that the KCNQ1/KCNE1 channel forms a macromolecular signaling complex which allows for regulation of the I_{Ks} current by the sympathetic nervous system (Marx et al. 2002). A leucine zipper (LZ) motif in the C-terminus of KCNQ1 coordinates the binding of a targeting protein yotiao, which in turn binds to and recruits PKA and PP1 to the channel. Upon activation of the sympathetic nervous system, the signaling complex regulates PKA phosphorylation of Ser27 in the N-terminus of KCNQ1 (Kurokawa et al. 2003). Disruption of the LZ domain of KCNQ1 and the mutation S27A prevent cAMP-dependent upregulation of I_{Ks}, whereas in the absence of yotiao, KCNQ1/KCNE1 currents are not increased by intracellular cAMP (Marx et al. 2002; see Fig. 3).

   Artificial mutations such as substitution of Ala residues for the Leu in the second and third "d" positions within the KCNQ1 LZ motif abrogate its interaction with yotiao without disturbing the α-helical structure of the motif. Similarly, inherited mutations can disrupt LZs and uncouple signaling molecules from their substrates. The naturally occurring G589D mutation at an "e" position in the LZ motif of *KCNQ1* disrupts targeting of yotiao to KCNQ1. This mutation has been linked to LQT1 in Finnish families (Piippo et al. 2001). Moreover, the *KCNQ1*–G589D mutation disrupts the LZ motif in the C-terminus of KCNQ1, resulting in disruption of β-adrenergic-mediated regulation of the channel. Thus, the G589D mutation causes a defect in the regulation of the channel by preventing the assembly of the macromolecular complex that targets protein kinase A (PKA) and protein phosphatase 1 (PP1) to the C-terminus of the I_{Ks} channel. Interestingly, carriers of this mutation suffer from abnormal regulation of the Q-T interval during mental and physical stress (Paavonen et al. 2001), and are at risk for arrhythmia and SCD during exercise (Piippo et al. 2001).

3.2.1.2
KCNQ1 Splice Variants

The genomic structure of KCNQ1 reveals that at least six exons give rise to alternatively spliced mature isoforms. An alternatively spliced variant of KCNQ1 with a N-terminal deletion that produces a negative suppression of

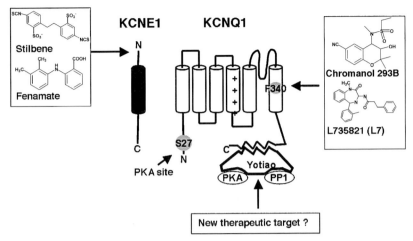

Fig. 3 Sympathetic regulation of I_{Ks} requires a macromolecular signaling complex. KCNQ1 and KCNE1 co-assemble to form the I_{Ks} channel. β-Adrenergic receptor stimulation results in activation of protein kinase A (PKA), which is recruited to the channel C-terminus in conjunction with protein phosphatase 1 (PP1) by yotiao [an A-kinase anchoring protein (AKAP) scaffolding protein]. PKA phosphorylation of serine 27 ensues and I_{Ks} amplitude is upregulated, allowing for rate-dependent adaptation of the action potential duration (APD). Both KCNQ1 and KCNE1 are targets for pharmacological agents. Stilbene and fenamate bind to the extracellular domain of KCNE1 and increase I_{Ks} (Abitbol et al. 1999). Chromanol 293B and L7 interact with the S6 segment of KCNQ1 and reduce I_{Ks} (Seebohm et al. 2003). (Adapted from Clancy et al. 2003)

KCNQ1 is preferentially expressed in M cells, which is consistent with the lower I_{Ks} density in this region (Mohammad-Panah et al. 1999; see Sect. 2). The native channel may represent a heterotetramer of KCNQ1 isoforms 1 and 2, together with KCNE1 (Demolombe et al. 1998). Transgenic mice overexpressing the spliced variant of *KCNQ1* present abnormalities of SA and AV node function, which suggests a role of KCNQ1 in normal automaticity (Demolombe et al. 2001).

3.2.2
Physiological Role of I_{Ks} in Cardiac Repolarization

The delayed rectifier K^+ current I_{Ks} is a current that activates slowly and does not inactivate. I_{Ks} is activated at potentials positive to -30 mV with a linear I–V relationship, reaching half-maximum activation at $+20$ mV (Kurokawa et al. 2001a; Sanguinetti and Jurkiewicz 1990b). Thus, I_{Ks} is a major contributor to repolarization of the cardiac action potential (Kass et al. 1996). Moreover, I_{Ks} is a dominant determinant of the physiological heart rate-dependent shortening of APD (Faber and Rudy 2000; Zeng et al. 1995). As heart rate increases, I_{Ks} channels have less time to deactivate, resulting in an accumulation of open channels and faster depolarization due to the build-up of instantaneous I_{Ks} repolarizing current at the AP onset (Faber and Rudy 2000). At slower rates, less repolarizing current exists during each action potential due to sufficient time between beats to allow for complete deactivation of I_{Ks} (Faber and Rudy 2000; Jurkiewicz and Sanguinetti 1993; Viswanathan et al. 1999). In canine left ventricle, I_{Ks} density is higher in epicardial and endocardial cells compared with the M cells (Liu and Antzelevitch 1995). The smaller I_{Ks} current in M cells may explain in part the steeper APD heart rate dependence and the greater tendency to display pronounced AP prolongation and afterdepolarization at slow heart rates or in response to QT-prolonging drugs (Liu and Antzelevitch 1995).

3.2.3
Structural Basis of I_{Ks} Block

The I_{Ks} current is resistant to methanesulfonanilides, but selectively blocked by the chromanol derivatives 293B and HMR 1556 (Busch et al. 1996; Gogelein et al. 2000) and by the benzodiazepine derivatives L-735,821 and L-768,673 (Selnick et al. 1997; Varro et al. 2000), whereas it is activated by L-364,272 (Salata et al. 1998). The open channel block produced by chromanols is enantioselective, (−)3R,4S-293B and (−)3R,4S-HMR 1556 being potent I_{Ks} blockers (Yang et al. 2000).

Investigation into the structural determinants of I_{Ks} block has only begun recently. Preliminary studies revealed a common site for binding of I_{Ks} blockers, including chromanol 293B and L-735,821 (L7), in the S6-domain (F340) of the KCNQ1 subunit (Fig. 3). Other putative interaction sites in the S6-domain (T312 and A344) and the pore-helix (I337) may lend specificity to pharmacological interactions (Seebohm et al. 2003). Interestingly, these binding sites are located near an aqueous crevice in KCNQ1 that is thought to be important for interactions with KCNE1 that allosterically affects pore geometry (Kurokawa et al. 2001b; Tapper and George 2000, 2001). Drug interaction sites for channel agonists stilbene and fenamate have also been elucidated on extracellular domains in KCNE1 (Abitbol et al. 1999; Fig. 3).

3.2.4
Electrophysiological Effects of I_{Ks} Block

Because I_{Ks} accumulates at faster stimulation rates due to slow deactivation, I_{Ks} blockers might be expected to be more useful in prolonging APD at fast rates (Jurkiewicz and Sanguinetti 1993; Viswanathan et al. 1999). Therefore, it has been proposed that I_{Ks} blockade may have less proarrhythmic potency compared to I_{Kr} blockers (Bosch et al. 1998). Because I_{Ks} activation occurs at 0 mV, which is more positive than the action potential plateau voltage in Purkinje fibers, I_{Ks} blockade is not expected to prolong the APD in these cells

(Tamargo et al. 2004). Conversely, in ventricular myocytes, the plateau voltage is more positive (+20 mV), allowing I_{Ks} to be substantially more activated, so that I_{Ks} block would be expected to prolong the APD markedly. The net result of both effects would be less drug-induced dispersion in repolarization and a reduced risk of arrhythmogenesis (Varro et al. 2000).

I_{Ks} blockers prolong the APD and suppress ventricular arrhythmias in animals with acute myocardial infarction and exercise superimposed on a healed myocardial infarction (MI) (Busch et al. 1996; Gogelein et al. 2000). This QT prolongation occurs in a dose-dependent manner, and can be accentuated by β-adrenergic stimulation (Shimizu and Antzelevitch 1998). In arterially perfused canine left ventricular wedge preparations, chromanol 293B prolongs the APD but does not induce TdP arrhythmias. However, in the presence of chromanol 293B, isoproterenol abbreviated the APD of epicardial and endocardial myocytes, but not in M cells, accentuating transmural dispersion of repolarization and inducing TdP (Shimizu and Antzelevitch 1998). These studies in canine preparations, however, may not be representative for humans, since canine repolarization appears to be less dependent upon I_{Ks} than other species (Mazhari et al. 2001; Stengl et al. 2003), and chromanol 293B was shown to markedly prolong human and guinea pig APD (Bosch et al. 1998). Furthermore, under normal conditions chromanol 293B and L-7 minimally prolong the APD regardless of pacing frequency in dog ventricular muscles and Purkinje fibers, probably because other K^+ currents may provide sufficient repolarizing reserve (Roden 1998). However, when the repolarizing reserve is decreased by QT-prolonging drugs (I_{Kr} or I_{K1} blockers), remodeling (hypertrophy, heart failure), or inherited disorders, I_{Ks} blockade can produce a marked prolongation of the ventricular APD, an enhanced dispersion of repolarization, and TdP arrhythmias (Shimizu and Antzelevitch 1998)

The presence of KCNE1 modulates the effects of I_{Ks} blockers and agonists (Busch et al. 1997; Wang et al. 2000a). KCNE1 is itself a distinct receptor for the I_{Ks} agonists stilbene and fenamate (Busch et al. 1997), which bind to an extracellular domain on KCNE1. Stilbene and fenamate and have been shown to be useful in reversing dominant-negative effects of some LQT5 C-terminal mutations and restoring I_{Ks} channel function (Abitbol et al. 1999). On the other hand, a 1,4-benzodiazepine compound, L364,373 was an effective agonistic on KCNQ1 currents only in the absence of KCNE1 (Salata et al. 1998). These types of studies illustrate the importance of accessory subunits in determining the pharmacological properties of I_{Ks}. Variable subunit expression may determine tissue selectivity or electrical heterogeneity of pharmacological action that could exacerbate dispersion of repolarization (Viswanathan et al. 1999). Finally, recent evidence suggests that PKA phosphorylation of the KCNQ1 subunit directly modulates drug access to a binding site on the channel (Yang et al. 2003).

3.2.5
Regulation of I_{Ks}

The I_{Ks} current is enhanced by β-adrenergic stimulation (Walsh and Kass 1988), α-adrenergic stimulation, PKC phosphorylation, or a rise in $[Ca^{2+}]_i$ (Tohse et al. 1987). Activation of β-adrenergic receptors increases PKA activity, which increases I_{Ks} current density and produces a rate-dependent shortening of the APD resulting from the slow deactivation of I_{Ks} (see Sect. 3.2.2). I_{Ks} amplitude is also directly mediated by β-adrenergic receptor (β-AR) stimulation through PKA phosphorylation of the channel macromolecular complex (Marx et al. 2002). PKA phosphorylation of I_{Ks} considerably increases current amplitude, by increasing the rate of channel activation (C→O transition) and reducing the rate of channel deactivation (O→C transition; Walsh and Kass 1991). Each of these outcomes acts to increase the channel open probability, leading to increased current amplitude and faster cardiac repolarization.

Lowering $[K^+]_o$ and $[Ca^{2+}]_o$ also increases I_{Ks} current (Tristani-Firouzi and Sanguinetti 2003). On the other hand, endothelin-1, a myocardial and endothelial peptide hormone, inhibits the I_{Ks} current, presumably through inhibition of adenylate cyclase via a PTX-sensitive G protein (Washizuka et al. 1997), and results in APD prolongation. Since both β-AR signaling and endothelin-A receptor signaling result in PKA phosphorylation, the molecular mechanisms of phosphorylation and dephosphorylation of I_{Ks} are of major interest as potential therapeutic targets (Fig. 3).

4
Potassium Channels Dysfunction in Cardiac Disease

4.1
Congenital Long QT Syndrome

The best-known evidence supporting the idea that potassium channel dysfunction can lead to SCD has come from the linkage of mutations in genes encoding cardiac K^+ channels to LQTS (Keating and Sanguinetti 2001). Mutations in at least five K^+ channels (i.e., KCNQ1, KCNH2, KCNE1, KCNE2, and KCNJ2) result in increased propensity to ventricular tachycardias and SCD (Wehrens et al. 2002). Most of the mutations identified in these K^+ channel α- and β-subunits are missense mutations, resulting in pathogenic single amino acid residue changes. The functional consequence of LQTS-linked K^+ channel mutations is a net reduction in outward K^+ current during the delicate plateau phase of the action potential, which disrupts the balance of inward and outward current leading to delayed repolarization. Prolongation of the APD manifests clinically as a prolongation of the Q-T interval on the electrocardiogram.

LQTS-associated mutations in KCNH2 have been shown to have heterogeneous cellular phenotypes. Pore mutations may result in a loss of function, sometimes due to trafficking defects (Petrecca et al. 1999), and may or may not co-assemble with wildtype subunits to exert dominant negative effects (Sanguinetti et al. 1996a). Other pore mutants give rise to altered kinetics leading to decreased repolarization current (Ficker et al. 1998; Smith et al. 1996). Nearby mutations in the S4–S5 linker have been shown to variably affect activation (Sanguinetti and Xu 1999). In either case, currents are typically reduced by 50% or more, leading to prolonged action potentials predisposing to arrhythmias.

Mutations in either *KCNQ1* or *KCNE1* can reduce I_{Ks} amplitude, resulting in abnormal cardiac phenotypes and the development of lethal arrhythmias (Splawski et al. 2000). In general, mutations in *KCNQ1* or *KCNE1* act to reduce I_{Ks} through dominant-negative effects (Chen et al. 1999; Chouabe et al. 1997, 2000; Roden et al. 1996; Russell et al. 1996; Wang et al. 1996; Wollnik et al. 1997), reduced responsiveness to β-AR signaling (Marx et al. 2002), or alterations in channel gating (Bianchi et al. 1999; Franqueza et al. 1999; Splawski et al. 1997). The latter effects typically manifest as either reduction in the rate of channel activation, such as R539W KCNQ1 (Chouabe et al. 2000), R555C KCNQ1 (Chouabe et al. 1997), or an increased rate of channel deactivation including S74L (Splawski et al. 1997), V47F, W87R (Bianchi et al. 1999), and W248R KCNQ1 (Franqueza et al. 1999). An LQTS-associated KCNQ1 C-terminal mutation, G589D, disrupts the leucine zipper motif and prevents cAMP-dependent regulation of I_{Ks} (Marx et al. 2002). The reduction of sensitivity to sympathetic activity likely prevents appropriate shortening of the action potential duration in response to increases in heart rate. Despite their distinct origins, congenital and drug-induced forms of ECG abnormalities related to alterations in I_{Ks} are remarkably similar. In either case, reduction in I_{Ks} results in prolongation of the Q-T interval on the ECG without an accompanying broadening of the T wave, as observed in other forms of LQTSs (Gima and Rudy 2002). Reduced I_{Ks} leads to loss of rate-dependent adaptation in APD, which is consistent with the clinical manifestation of arrhythmias associated with LQT1 and LQT5, which tend to occur due to sudden increases in heart rate.

4.2
Congenital Short QT Syndrome

Recent studies suggests that mutations in the same genes that cause delayed repolarization may results in a converse disorder, the "short QT syndrome" (SQTS) which is also believed to enhance SCD risk (Brugada et al. 2004). SQTS is a new clinical entity originally described as an inherited syndrome (Gussak et al. 2000). A missense mutation in *KCNH2* (N588K), linked to families with SQTS (Brugada et al. 2004), abolishes rectification of I_{Kr} and reduces the affinity of the channel for class III antiarrhythmic drugs. The net effect of the mutation is to increase the repolarizing currents active during the early phase of the AP, leading to abbreviation of the AP and thus shortening of the Q-T interval (Brugada et al. 2004). Recent data suggest that this disorder may be genetically heterogeneous, since a mutation in the *KCNQ1* gene was found in a patient with SQTS (Bellocq et al. 2004). Functional studies of the KCNQ1-V307L mutant linked to SQTS (alone or co-expressed with the wildtype channel, in the presence of KCNE1) revealed a pronounced shift of the half-activation potential and an acceleration of the activation kinetics, leading to a gain of function in I_{Ks} (Bellocq et al. 2004). Preliminary data suggest that quinidine may effectively prolong the Q-T interval and ventricular effective refractory period (ERP) in patients with SQTS, thereby preventing ventricular arrhythmias. This is particularly important because SQTS patients are at risk of sudden death from birth, and implantable cardioverter/defibrillator (ICD) implantation is not feasible in very young children (Gaita et al. 2004).

4.3
Polymorphisms in K⁺ Channels Predispose to Acquired Long QT Syndrome

In addition to rare mutations linked to congenital LQTS, common polymorphisms also exist in genes encoding cardiac K^+ channels. Common polymorphisms have been defined as nucleotide substitutions found in both control and patient populations, usually at a frequency of $\sim$1% or greater (Yang et al. 2002). When viewed in the context of pathological mutations, the presence of common non-synonymous single nucleotide polymorphisms (nSNPs) in apparently healthy populations suggests that they are well tolerated and likely to have wildtype-like physiology. However, the identification of common nSNPs in the KCNE2 K^+ channel β-subunit that alter channel physiology and drug sensitivity has challenged this point of view (Sesti et al. 2000). Indeed, these particular nSNPs have a functional phenotype *in vitro* and may mediate genetic susceptibility to fatal ventricular arrhythmias in the setting of acute myocardial infarction or exposure to QT-prolonging medications. Four nSNPs have been found within the *KCNH2* gene (Anson et al. 2004; Laitinen et al. 2000; Larsen et al. 2001; Yang et al. 2002). The most common nSNP identified to date, KCNH2-K897T, has been associated with altered channel biophysics and Q-T interval prolongation, although results vary between investigative groups (Bezzina et al. 2003; Laitinen et al. 2000; Paavonen et al. 2003; Scherer et al. 2002). In contrast to the *KCNE2* polymorphism T8A (Sesti et al. 2000), these *KCNH2* α-subunit polymorphisms do not convey increased sensitivity to drug block. Nevertheless, testing for ion channel polymorphisms could be used to reduce the risk of drug-induced arrhythmia and improve the risk stratification of common cardiac diseases that predispose to SCD.

4.4
Altered I_K Function in the Chronically Diseased Heart

Whereas inherited arrhythmogenic syndromes caused by K^+ channel mutations are rare disorders, changes in ion channel expression or function leading to prolongation of the APD are commonly observed in various disease states of the heart (Tomaselli and Marban 1999). Altered electrophysiological properties of diseased cardiomyocytes may provide a substrate for contractile dysfunction or fatal arrhythmias in patients with cardiac hypertrophy or heart failure (Tomaselli and Marban 1999; Wehrens and Marks 2003). It has also been established that repolarizing K^+ currents are reduced in human atrial and ventricular myocytes in a variety of pathological states (for more detailed review, see Tomaselli and Marban 1999). It is therefore important to consider these changes in K^+ channel function when designing therapeutic strategies for these pathological conditions of the heart.

4.4.1
Cardiac Hypertrophy

Cardiac hypertrophy secondary to hypertension is associated with a sixfold increase in the risk of SCD. It has been proposed that delayed ventricular repolarization due to electrical remodeling in the hypertrophied heart may predispose to acquired LQTS and TdP arrhythmias (Volders et al. 1999b). In a canine model of biventricular hypertrophy induced by chronic complete atrioventricular block, the I_{Ks} and I_{Kr} current densities were reduced in right ventricular myocytes (Volders et al. 1999b). However, I_{Kr} was not affected in myocytes from the left ventricular wall, indicating regional variation in I_{Kr} changes in the hypertrophied canine heart (Volders et al. 1999b). Studies using quantitative RT-PCR have demonstrated that the decrease in I_{Ks} current density is due to a downregulation of *KCNQ1* and *KCNE1* transcription. Similar reductions in current density of delayed rectifier currents have been observed in isolated myocytes from hypertrophied right and left ventricles of the cat and rabbit (Furukawa et al. 1994; Kleiman and Houser 1989; Tsuji et al. 2002).

4.4.2
Heart Failure

Usually, some degree of hypertrophy is present during the development of heart failure, often due to pressure or volume overload. Furthermore, the presence of compensatory hypertrophy in the non-infarcted myocardium in ischemic heart failure suggests similarities between electrophysiological changes in cardiac hypertrophy and failure (Nabauer and Kaab 1998). Prolongation of the action potential has been a consistent finding in animals with heart failure in a variety of experimental models and species. Depending on the species studied, different K^+ channels may be involved in similar phenotypic prolongation of the AP in heart failure (Nabauer and Kaab 1998; Tomaselli and Marban 1999).

Evidence for downregulation of cardiac potassium currents in heart failure has been derived from various animal models of heart failure (Pak et al. 1997; Rozanski et al. 1997) and from terminally failing human myocardium studied at the time of heart transplantation (Beuckelmann et al. 1993). There are, however, few studies on the delayed rectifier K^+ current in heart failure. Chen et al. (2002b) reported that it was hardly detectable in cardiomyopathic hamsters, and if detectable, it was small in both diseased and normal human myocytes. In a canine model of heart failure, I_{Ks} was found to be decreased, while I_{Kr} remained unchanged (Li et al. 2002). In a pacing-induced heart failure model of the rabbit, both I_{Kr} and I_{Ks} were reduced when measurements were made at physiological temperature (Tsuji et al. 2000). In addition to its potential contribution to primary ventricular tachyarrhythmias in heart failure, the decreased delayed rectifier currents in heart failure may sensitize patients to proarrhythmic effects of antiarrhythmic drugs. In fact, the presence of heart failure is known to be an important risk factor for drug-induced TdP (Lehmann et al. 1996).

Whereas additional studies are required to investigate the contribution of delayed rectifier currents to prolonged repolarization in heart failure, one of the most consistent changes in ionic currents in the failing heart is a significant reduction of the transient outward current (I_{to}) (Beuckelmann et al. 1993). Reduction of I_{to} is the most marked effect in myocytes from patients with severe heart failure and dogs with the pacing-induced heart failure model (Beuckelmann et al. 1993; Kaab et al. 1996). A remarkably good correlation has been found between the extent of reduction of I_{to} and reduction in mRNA transcripts encoding *KCND3* (Kv4.3) in human heart failure (Kaab et al. 1998). For a more detailed review about changes in I_{to} in heart failure, and other K^+ currents not discussed in this chapter, please see Janse (2004) and Nabauer and Kaab (1998).

5
Drug-Induced Ventricular Arrhythmias

Supraventricular tachyarrhythmias are often treated with class III anti-arrhythmic drugs (Vaughan Williams 1984). These K^+ channel blockers act by increasing the action potential duration and the effective refractory period in order to prevent premature re-excitation (Coumel et al. 1978). While these interventions can be useful in targeting tachyarrhythmias, they may predispose some patients to the development of other types of arrhythmia (Priori 2000). It has become apparent that drug-induced I_{Kr} block and QT prolongation are the likely molecular targets responsible for the cardiac toxicity of a wide range of pharmaceutical agents (Roden 2000; Sanguinetti and Jurkiewicz 1990b).

More than 50 commercially available agents (see *www.torsades.org*) or investigational drugs, often for the purpose of treating syndromes unrelated to cardiac disease, have been implicated with the drug-induced LQTS (Clancy et al. 2003). A number of these drugs have been withdrawn from the market in recent years (e.g., prenylamine, terodiline, and in some countries, terfenadine, astemizole, and cisapride) because their risk for triggering lethal arrhythmias was believed to outweigh therapeutic benefits (Walker et al. 1999). A number of histamine receptor-blocking drugs, including astemizole and terfenadine and more recently loratadine, have been shown to block I_{Kr} as an adverse side effect and prolong the Q-T interval of the electrocardiogram (Crumb 2000). Cisapride (Propulsid), a widely used gastrointestinal prokinetic agent in the treatment of gastroesophageal reflux disease and gastroparesis, also blocks KCNH2 K^+ channels and is associated with acquired LQTS and ventricular arrhythmias (Wysowski and Bacsanyi 1996). Cisapride produces a preferential prolongation of the APD of M cells, leading to the development of a large dispersion of APD between the M cell and epi/endocardium (Di Diego et al. 2003; Fig. 4). Changes in the morphology of the T wave were observed in more than 85% of patients treated for psychosis when the plasma concentration of the anti-psychotic drug thioridazine was greater than 1 μM (Axelsson and Aspenstrom 1982) due to blockade of I_{Kr} (IC$_{50}$, 1.25 μM) and I_{Ks} (IC$_{50}$, 14 μM). Since inadvertent side effects of drugs on cardiac K^+ channels are plentiful, the issue of Q-T interval prolongation has also become a major concern in the development of new pharmacological therapies (Shah 2004).

It is important to consider that in the majority of patients, drugs that block repolarizing currents may not produce an overt baseline Q-T interval prolongation, due to "repolarization reserve" (Roden 1998). However, a subclinical vulnerability stemming from genetic defects or polymorphisms, gender, hypokalemia, concurrent use of other medications, or structural heart abnormal-

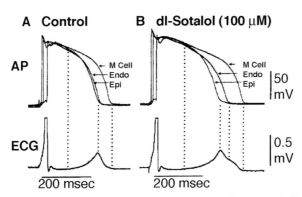

Fig. 4a,b Drug-induced prolongation of the Q-T interval and increased dispersion of repolarization. Each panel shows action potentials recorded from epicardial (Epi), M region (M), and endocardial (Endo) sites (*top*), and a transmural electrogram simulating an ECG (*bottom*). The traces were simultaneously recorded from an isolated arterially perfused canine wedge under control condition (**a**) and in the presence of the I_{Kr} blocker d,l-sotalol (100 mM, 30 min; **b**). Sotalol produced a preferential prolongation of the M cell action potential leading to the appearance of a long Q-T interval in the electrogram and the development of a large transmural dispersion of repolarization. (Reproduced with permission from Haverkamp et al. 2000)

ities may provide a substrate allowing for the initiation of arrhythmic triggers (De Ponti et al. 2002; Ebert et al. 1998). Many such arrhythmic events are heart rate-dependent and may be linked to sudden changes in heart rate due to exercise or auditory stimulation that may trigger life-threatening arrhythmias (Splawski et al. 2000). On the other hand, not all drugs that significantly prolong the Q-T interval are associated with arrhythmias. Amiodarone clearly prolongs the Q-T interval but rarely causes TdP arrhythmias (Zabel et al. 1997), although it may in the presence of polymorphisms in cardiac ion channels (Splawski et al. 2002). These findings have led to the belief that Q-T interval prolongation may not be an ideal predictor of proarrhythmia, and other parameters such as the Q-T interval dispersion, T wave vector loop, and T-U wave morphology analysis are currently being evaluated as screening tools in drug development (Anderson et al. 2002).

Recent experimental studies by Hondeghem et al. (2001a,b) have also suggested that prolongation of the APD is not inherently proarrhythmic. The cardiac electrophysiological effects of drugs known to block I_{Kr} were studied in rabbit Langendorff-perfused hearts. Beat-to-beat variability of APD, reverse frequency dependence of AP prolongation, and triangulation of AP repolarization were found to correlate with the induction of polymorphic VT. In contrast, agents that prolonged APD without instability (i.e., APD alternans) were antiarrhythmic. These data suggest that block of I_{Kr} may not be proarrhythmic per se, but that the specific mechanism of ion channel modulation and effects on other channels are critical.

6
Concluding Remarks

Cardiac K^+ channels play an important role in repolarization of the action potential, and have been recognized as potential therapeutic targets. The function and expression of K^+ channels differ widely in the different regions of the heart and are influenced by heart rate, neurohumoral state, cardiovascular diseases (cardiac hypertrophy, heart failure), and inherited disorders (short and long QT syndromes). Given the diversity of α- and β-subunits and splice variants that underlie the various K^+ channels in the heart, the precise role that each K^+ channel gene product plays in the regional heterogeneity of native currents or in the cellular pathophysiology in the human heart remains to be further investigated. The rational design of safer and more effective K^+ channel blockers, and attempts to prevent the proarrhythmic effects linked to the blockade of cardiac K^+ channels should be based on a better understanding of the molecular basis of the target channel, its cardiac distribution and function, and the type of drug–channel interaction.

References

Abbott GW, Goldstein SA (2001) Potassium channel subunits encoded by the KCNE gene family: physiology and pathophysiology of the MinK-related peptides (MiRPs). Mol Interv 1:95–107

Abbott GW, Sesti F, Splawski I, Buck ME, Lehmann MH, Timothy KW, Keating MT, Goldstein SA (1999) MiRP1 forms IKr potassium channels with HERG and is associated with cardiac arrhythmia. Cell 97:175–187

Abitbol I, Peretz A, Lerche C, Busch AE, Attali B (1999) Stilbenes and fenamates rescue the loss of I(KS) channel function induced by an LQT5 mutation and other IsK mutants. EMBO J 18:4137–4148

Anderson ME, Al-Khatib SM, Roden DM, Califf RM (2002) Cardiac repolarization: current knowledge, critical gaps, and new approaches to drug development and patient management. Am Heart J 144:769–781

Anson BD, Ackerman MJ, Tester DJ, Will ML, Delisle BP, Anderson CL, January CT (2004) Molecular and functional characterization of common polymorphisms in HERG (KCNH2) potassium channels. Am J Physiol Heart Circ Physiol 286:H2434–H2441

Antzelevitch C, Fish J (2001) Electrical heterogeneity within the ventricular wall. Basic Res Cardiol 96:517–527

Antzelevitch C, Sun ZQ, Zhang ZQ, Yan GX (1996) Cellular and ionic mechanisms underlying erythromycin-induced long QT intervals and torsade de pointes. J Am Coll Cardiol 28:1836–1848

Attali B, Guillemare E, Lesage F, Honore E, Romey G, Lazdunski M, Barhanin J (1993) The protein IsK is a dual activator of K^+ and Cl^- channels. Nature 365:850–852

Axelsson R, Aspenstrom G (1982) Electrocardiographic changes and serum concentrations in thioridazine-treated patients. J Clin Psychiatry 43:332–335

Barhanin J, Lesage F, Guillemare E, Fink M, Lazdunski M, Romey G (1996) K(V)LQT1 and lsK (minK) proteins associate to form the I(Ks) cardiac potassium current. Nature 384:78–80

Barros F, Gomez-Varela D, Viloria CG, Palomero T, Giraldez T, de la Pena P (1998) Modulation of human erg K+ channel gating by activation of a G protein-coupled receptor and protein kinase C. J Physiol 511:333–346

Barry DM, Nerbonne JM (1996) Myocardial potassium channels: electrophysiological and molecular diversity. Annu Rev Physiol 58:363–394

Baukrowitz T, Yellen G (1995) Modulation of K+ current by frequency and external [K+]: a tale of two inactivation mechanisms. Neuron 15:951–960

Belardinelli L, Antzelevitch C, Vos MA (2003) Assessing predictors of drug-induced torsade de pointes. Trends Pharmacol Sci 24:619–625

Bellocq C, van Ginneken AC, Bezzina CR, Alders M, Escande D, Mannens MM, Baro I, Wilde AA (2004) Mutation in the KCNQ1 gene leading to the short QT-interval syndrome. Circulation 109:2394–2397

Berube J, Chahine M, Daleau P (1999) Modulation of HERG potassium channel properties by external pH. Pflugers Arch 438:419–422

Beuckelmann DJ, Nabauer M, Erdmann E (1993) Alterations of K+ currents in isolated human ventricular myocytes from patients with terminal heart failure. Circ Res 73:379–385

Bezzina CR, Verkerk AO, Busjahn A, Jeron A, Erdmann J, Koopmann TT, Bhuiyan ZA, Wilders R, Mannens MM, Tan HL, Luft FC, Schunkert H, Wilde AA (2003) A common polymorphism in KCNH2 (HERG) hastens cardiac repolarization. Cardiovasc Res 59:27–36

Bian JS, Kagan A, McDonald TV (2004) Molecular analysis of phosphatidyl inositol 4,5-bisphosphate regulation of HERG/IKr. Am J Physiol Heart Circ Physiol 287:H2154–H2163

Bianchi L, Shen Z, Dennis AT, Priori SG, Napolitano C, Ronchetti E, Bryskin R, Schwartz PJ, Brown AM (1999) Cellular dysfunction of LQT5-minK mutants: abnormalities of IKs, IKr and trafficking in long QT syndrome. Hum Mol Genet 8:1499–1507

Bosch RF, Gaspo R, Busch AE, Lang HJ, Li GR, Nattel S (1998) Effects of the chromanol 293B, a selective blocker of the slow, component of the delayed rectifier K+ current, on repolarization in human and guinea pig ventricular myocytes. Cardiovasc Res 38:441–450

Brugada R, Hong K, Dumaine R, Cordeiro J, Gaita F, Borggrefe M, Menendez TM, Brugada J, Pollevick GD, Wolpert C, Burashnikov E, Matsuo K, Wu YS, Guerchicoff A, Bianchi F, Giustetto C, Schimpf R, Brugada P, Antzelevitch C (2004) Sudden death associated with short-QT syndrome linked to mutations in HERG. Circulation 109:30–35

Busch AE, Suessbrich H, Waldegger S, Sailer E, Greger R, Lang H, Lang F, Gibson KJ, Maylie JG (1996) Inhibition of IKs in guinea pig cardiac myocytes and guinea pig IsK channels by the chromanol 293B. Pflugers Arch 432:1094–1096

Busch AE, Busch GL, Ford E, Suessbrich H, Lang HJ, Greger R, Kunzelmann K, Attali B, Stuhmer W (1997) The role of the IsK protein in the specific pharmacological properties of the IKs channel complex. Br J Pharmacol 122:187–189

Caballero R, Moreno I, Gonzalez T, Arias C, Valenzuela C, Delpon E, Tamargo J (2003) Spironolactone and its main metabolite, canrenoic acid, block human ether-a-go-go-related gene channels. Circulation 107:889–895

Chen H, Kim LA, Rajan S, Xu S, Goldstein SA (2003) Charybdotoxin binding in the I(Ks) pore demonstrates two MinK subunits in each channel complex. Neuron 40:15–23

Chen J, Seebohm G, Sanguinetti MC (2002a) Position of aromatic residues in the S6 domain, not inactivation, dictates cisapride sensitivity of HERG and eag potassium channels. Proc Natl Acad Sci U S A 99:12461–12466

Chen Q, Zhang D, Gingell RL, Moss AJ, Napolitano C, Priori SG, Schwartz PJ, Kehoe E, Robinson JL, Schulze-Bahr E, Wang Q, Towbin JA (1999) Homozygous deletion in KVLQT1 associated with Jervell and Lange-Nielsen syndrome. Circulation 99:1344–1347

Chen X, Piacentino V 3rd, Furukawa S, Goldman B, Margulies KB, Houser SR (2002b) L-type Ca2+ channel density and regulation are altered in failing human ventricular myocytes and recover after support with mechanical assist devices. Circ Res 91:517–524

Chouabe C, Neyroud N, Guicheney P, Lazdunski M, Romey G, Barhanin J (1997) Properties of KvLQT1 K+ channel mutations in Romano-Ward and Jervell and Lange-Nielsen inherited cardiac arrhythmias. EMBO J 16:5472–5479

Chouabe C, Neyroud N, Richard P, Denjoy I, Hainque B, Romey G, Drici MD, Guicheney P, Barhanin J (2000) Novel mutations in KvLQT1 that affect Iks activation through interactions with Isk. Cardiovasc Res 45:971–980

Clancy CE, Rudy Y (2001) Cellular consequences of HERG mutations in the long QT syndrome: precursors to sudden cardiac death. Cardiovasc Res 50:301–313

Clancy CE, Kurokawa J, Tateyama M, Wehrens XH, Kass RS (2003) K+ channel structure-activity relationships and mechanisms of drug-induced QT prolongation. Annu Rev Pharmacol Toxicol 43:441–461

Coumel P, Krikler D, Rosen MR, Wellens HJ, Zipes DP (1978) Newer antiarrhythmic drugs. Pacing Clin Electrophysiol 1:521–528

Crumb WJ Jr (2000) Loratadine blockade of K(+) channels in human heart: comparison with terfenadine under physiological conditions. J Pharmacol Exp Ther 292:261–264

Cui J, Melman Y, Palma E, Fishman GI, McDonald TV (2000) Cyclic AMP regulates the HERG K(+) channel by dual pathways. Curr Biol 10:671–674

Curran ME, Splawski I, Timothy KW, Vincent GM, Green ED, Keating MT (1995) A molecular basis for cardiac arrhythmia: HERG mutations cause long QT syndrome. Cell 80:795–803

De Ponti F, Poluzzi E, Cavalli A, Recanatini M, Montanaro N (2002) Safety of non-antiarrhythmic drugs that prolong the QT interval or induce torsade de pointes: an overview. Drug Saf 25:263–286

Deal KK, England SK, Tamkun MM (1996) Molecular physiology of cardiac potassium channels. Physiol Rev 76:49–67

del Camino D, Holmgren M, Liu Y, Yellen G (2000) Blocker protection in the pore of a voltage-gated K+ channel and its structural implications. Nature 403:321–325

Demolombe S, Baro I, Pereon Y, Bliek J, Mohammad-Panah R, Pollard H, Morid S, Mannens M, Wilde A, Barhanin J, Charpentier F, Escande D (1998) A dominant negative isoform of the long QT syndrome 1 gene product. J Biol Chem 273:6837–6843

Demolombe S, Lande G, Charpentier F, van Roon MA, van den Hoff MJ, Toumaniantz G, Baro I, Guihard G, Le Berre N, Corbier A, de Bakker J, Opthof T, Wilde A, Moorman AF, Escande D (2001) Transgenic mice overexpressing human KvLQT1 dominant-negative isoform. Part I. Phenotypic characterisation. Cardiovasc Res 50:314–327

Di Diego JM, Belardinelli L, Antzelevitch C (2003) Cisapride-induced transmural dispersion of repolarization and torsade de pointes in the canine left ventricular wedge preparation during epicardial stimulation. Circulation 108:1027–1033

Doyle DA, Morais Cabral J, Pfuetzner RA, Kuo A, Gulbis JM, Cohen SL, Chait BT, MacKinnon R (1998) The structure of the potassium channel: molecular basis of K+ conduction and selectivity. Science 280:69–77

Ebert SN, Liu XK, Woosley RL (1998) Female gender as a risk factor for drug-induced cardiac arrhythmias: evaluation of clinical and experimental evidence. J Womens Health 7:547–557

Etheridge SP, Compton SJ, Tristani-Firouzi M, Mason JW (2003) A new oral therapy for long QT syndrome: long-term oral potassium improves repolarization in patients with HERG mutations. J Am Coll Cardiol 42:1777–1782

Faber GM, Rudy Y (2000) Action potential and contractility changes in [Na(+)](i) overloaded cardiac myocytes: a simulation study. Biophys J 78:2392–2404

Ficker E, Jarolimek W, Kiehn J, Baumann A, Brown AM (1998) Molecular determinants of dofetilide block of HERG K+ channels. Circ Res 82:386–395

Ficker E, Jarolimek W, Brown AM (2001) Molecular determinants of inactivation and dofetilide block in ether a-go-go (EAG) channels and EAG-related K(+) channels. Mol Pharmacol 60:1343–1348

Folander K, Smith JS, Antanavage J, Bennett C, Stein RB, Swanson R (1990) Cloning and expression of the delayed-rectifier IsK channel from neonatal rat heart and diethylstilbestrol-primed rat uterus. Proc Natl Acad Sci U S A 87:2975–2979

Follmer CH, Colatsky TJ (1990) Block of delayed rectifier potassium current, IK, by flecainide and E-4031 in cat ventricular myocytes. Circulation 82:289–293

Franqueza L, Lin M, Shen J, Splawski I, Keating MT, Sanguinetti MC (1999) Long QT syndrome-associated mutations in the S4-S5 linker of KvLQT1 potassium channels modify gating and interaction with minK subunits. J Biol Chem 274:21063–21070

Furukawa T, Myerburg RJ, Furukawa N, Kimura S, Bassett AL (1994) Metabolic inhibition of ICa, L and IK differs in feline left ventricular hypertrophy. Am J Physiol 266:H1121–H1131

Gaita F, Giustetto C, Bianchi F, Schimpf R, Haissaguerre M, Calo L, Brugada R, Antzelevitch C, Borggrefe M, Wolpert C (2004) Short QT syndrome: pharmacological treatment. J Am Coll Cardiol 43:1494–1499

Gima K, Rudy Y (2002) Ionic current basis of electrocardiographic waveforms: a model study. Circ Res 90:889–896

Gogelein H, Bruggemann A, Gerlach U, Brendel J, Busch AE (2000) Inhibition of IKs channels by HMR 1556. Naunyn Schmiedebergs Arch Pharmacol 362:480–488

Gussak I, Brugada P, Brugada J, Wright RS, Kopecky SL, Chaitman BR, Bjerregaard P (2000) Idiopathic short QT interval: a new clinical syndrome? Cardiology 94:99–102

Gutman GA, Chandy KG, Adelman JP, Aiyar J, Bayliss DA, Clapham DE, Covarriubias M, Desir GV, Furuichi K, Ganetzky B, Garcia ML, Grissmer S, Jan LY, Karschin A, Kim D, Kuperschmidt S, Kurachi Y, Lazdunski M, Lesage F, Lester HA, McKinnon D, Nichols CG, O'Kelly I, Robbins J, Robertson GA, Rudy B, Sanguinetti M, Seino S, Stuehmer W, Tamkun MM, Vandenberg CA, Wei A, Wulff H, Wymore RS (2003) International Union of Pharmacology. XLI. Compendium of voltage-gated ion channels: potassium channels. Pharmacol Rev 55:583–586

Haverkamp W, Breithardt G, Camm AJ, Janse MJ, Rosen MR, Antzelevitch C, Escande D, Franz M, Malik M, Moss A, Shah R (2000) The potential for QT prolongation and pro-arrhythmia by non-anti-arrhythmic drugs: clinical and regulatory implications. Report on a policy conference of the European Society of Cardiology. Cardiovasc Res 47:219–233

Heginbotham L, Lu Z, Abramson T, MacKinnon R (1994) Mutations in the K+ channel signature sequence. Biophys J 66:1061–1067

Hondeghem LM, Carlsson L, Duker G (2001a) Instability and triangulation of the action potential predict serious proarrhythmia, but action potential duration prolongation is antiarrhythmic. Circulation 103:2004–2013

Hondeghem LM, Dujardin K, De Clerck F (2001b) Phase 2 prolongation, in the absence of instability and triangulation, antagonizes class III proarrhythmia. Cardiovasc Res 50:345–353

Janse MJ (2004) Electrophysiological changes in heart failure and their relationship to arrhythmogenesis. Cardiovasc Res 61:208–217

Jiang M, Dun W, Tseng GN (1999) Mechanism for the effects of extracellular acidification on HERG-channel function. Am J Physiol 277:H1283–H1292

Jiang M, Zhang M, Tang DG, Clemo HF, Liu J, Holwitt D, Kasirajan V, Pond AL, Wettwer E, Tseng GN (2004) KCNE2 protein is expressed in ventricles of different species, and changes in its expression contribute to electrical remodeling in diseased hearts. Circulation 109:1783–1788

Jones EM, Roti Roti EC, Wang J, Delfosse SA, Robertson GA (2004) Cardiac IKr channels minimally comprise hERG1a and 1b subunits. J Biol Chem 279:44690–44694

Jurkiewicz NK, Sanguinetti MC (1993) Rate-dependent prolongation of cardiac action potentials by a methanesulfonanilide class III antiarrhythmic agent. Specific block of rapidly activating delayed rectifier K+ current by dofetilide. Circ Res 72:75–83

Kaab S, Nuss HB, Chiamvimonvat N, O'Rourke B, Pak PH, Kass DA, Marban E, Tomaselli GF (1996) Ionic mechanism of action potential prolongation in ventricular myocytes from dogs with pacing-induced heart failure. Circ Res 78:262–273

Kaab S, Dixon J, Duc J, Ashen D, Nabauer M, Beuckelmann DJ, Steinbeck G, McKinnon D, Tomaselli GF (1998) Molecular basis of transient outward potassium current downregulation in human heart failure: a decrease in Kv4.3 mRNA correlates with a reduction in current density. Circulation 98:1383–1393

Kagan A, Yu Z, Fishman GI, McDonald TV (2000) The dominant negative LQT2 mutation A561V reduces wild-type HERG expression. J Biol Chem 275:11241–11248

Kass RS (1997) Genetically induced reduction in small currents has major impact. Circulation 96:1720–1721

Kass RS, Davies MP, Freeman LC (1996) Functional differences between native and recombinant forms of IKs. In: Endoh H, Morad M, Scholz H, Iijima T (eds) Molecular and cellular mechanisms of cardiovascular regulation. Springer, New York, pp 33–46

Keating MT, Sanguinetti MC (2001) Molecular and cellular mechanisms of cardiac arrhythmias. Cell 104:569–580

Kiehn J (2000) Regulation of the cardiac repolarizing HERG potassium channel by protein kinase A. Trends Cardiovasc Med 10:205–209

Kiehn J, Lacerda AE, Wible B, Brown AM (1996a) Molecular physiology and pharmacology of HERG. Single-channel currents and block by dofetilide. Circulation 94:2572–2579

Kiehn J, Wible B, Lacerda AE, Brown AM (1996b) Mapping the block of a cloned human inward rectifier potassium channel by dofetilide. Mol Pharmacol 50:380–387

Kiehn J, Lacerda AE, Brown AM (1999) Pathways of HERG inactivation. Am J Physiol 277:H199–210

Kleiman RB, Houser SR (1989) Outward currents in normal and hypertrophied feline ventricular myocytes. Am J Physiol 256:H1450–H1461

Kupershmidt S, Snyders DJ, Raes A, Roden DM (1998) A K+ channel splice variant common in human heart lacks a C-terminal domain required for expression of rapidly activating delayed rectifier current. J Biol Chem 273:27231–27235

Kupershmidt S, Yang T, Anderson ME, Wessels A, Niswender KD, Magnuson MA, Roden DM (1999) Replacement by homologous recombination of the minK gene with lacZ reveals restriction of minK expression to the mouse cardiac conduction system. Circ Res 84:146–152

Kurokawa J, Abriel H, Kass RS (2001a) Molecular basis of the delayed rectifier current I(ks) in heart. J Mol Cell Cardiol 33:873–882

Kurokawa J, Motoike HK, Kass RS (2001b) TEA(+)-sensitive KCNQ1 constructs reveal pore-independent access to KCNE1 in assembled I(Ks) channels. J Gen Physiol 117:43–52

Kurokawa J, Chen L, Kass RS (2003) Requirement of subunit expression for cAMP-mediated regulation of a heart potassium channel. Proc Natl Acad Sci U S A 100:2122–2127

Laitinen P, Fodstad H, Piippo K, Swan H, Toivonen L, Viitasalo M, Kaprio J, Kontula K (2000) Survey of the coding region of the HERG gene in long QT syndrome reveals six novel mutations and an amino acid polymorphism with possible phenotypic effects. Hum Mutat 15:580–581

Larsen LA, Andersen PS, Kanters J, Svendsen IH, Jacobsen JR, Vuust J, Wettrell G, Tranebjaerg L, Bathen J, Christiansen M (2001) Screening for mutations and polymorphisms in the genes KCNH2 and KCNE2 encoding the cardiac HERG/MiRP1 ion channel: implications for acquired and congenital long Q-T syndrome. Clin Chem 47:1390–1395

Lees-Miller JP, Kondo C, Wang L, Duff HJ (1997) Electrophysiological characterization of an alternatively processed ERG K+ channel in mouse and human hearts. Circ Res 81:719–726

Lehmann MH, Hardy S, Archibald D, quart B, MacNeil DJ (1996) Sex difference in risk of torsade de pointes with d,l-sotalol. Circulation 94:2535–2541

Lei M, Brown HF, Terrar DA (2000) Modulation of delayed rectifier potassium current, iK, by isoprenaline in rabbit isolated pacemaker cells. Exp Physiol 85:27–35

Li GR, Lau CP, Ducharme A, Tardif JC, Nattel S (2002) Transmural action potential and ionic current remodeling in ventricles of failing canine hearts. Am J Physiol Heart Circ Physiol 283:H1031–H1041

Liu DW, Antzelevitch C (1995) Characteristics of the delayed rectifier current (IKr and IKs) in canine ventricular epicardial, midmyocardial, and endocardial myocytes. A weaker IKs contributes to the longer action potential of the M cell. Circ Res 76:351–365

London B, Trudeau MC, Newton KP, Beyer AK, Copeland NG, Gilbert DJ, Jenkins NA, Satler CA, Robertson GA (1997) Two isoforms of the mouse ether-a-go-go-related gene coassemble to form channels with properties similar to the rapidly activating component of the cardiac delayed rectifier K+ current. Circ Res 81:870–878

Marx SO, Kurokawa J, Reiken S, Motoike H, D'Armiento J, Marks AR, Kass RS (2002) Requirement of a macromolecular signaling complex for beta adrenergic receptor modulation of the KCNQ1-KCNE1 potassium channel. Science 295:496–499

Mazhari R, Greenstein JL, Winslow RL, Marban E, Nuss HB (2001) Molecular interactions between two long-QT syndrome gene products, HERG and KCNE2, rationalized by in vitro and in silico analysis. Circ Res 89:33–38

McDonald TV, Yu Z, Ming Z, Palma E, Meyers MB, Wang KW, Goldstein SA, Fishman GI (1997) A minK-HERG complex regulates the cardiac potassium current I(Kr). Nature 388:289–292

Miles EW (1977) Modification of histidyl residues in proteins by diethylpyrocarbonate. Methods Enzymol 47:431–442

Milnes JT, Crociani O, Arcangeli A, Hancox JC, Witchel HJ (2003) Blockade of HERG potassium currents by fluvoxamine: incomplete attenuation by S6 mutations at F656 or Y652. Br J Pharmacol 139:887–898

Mitcheson JS (2003) Drug binding to HERG channels: evidence for a 'non-aromatic' binding site for fluvoxamine. Br J Pharmacol 139:883–884

Mitcheson JS, Chen J, Lin M, Culberson C, Sanguinetti MC (2000a) A structural basis for drug-induced long QT syndrome. Proc Natl Acad Sci U S A 97:12329–12333

Mitcheson JS, Chen J, Sanguinetti MC (2000b) Trapping of a methanesulfonanilide by closure of the HERG potassium channel activation gate. J Gen Physiol 115:229–240

Mohammad-Panah R, Demolombe S, Neyroud N, Guicheney P, Kyndt F, van den Hoff M, Baro I, Escande D (1999) Mutations in a dominant-negative isoform correlate with phenotype in inherited cardiac arrhythmias. Am J Hum Genet 64:1015–1023

Morais Cabral JH, Lee A, Cohen SL, Chait BT, Li M, Mackinnon R (1998) Crystal structure and functional analysis of the HERG potassium channel N terminus: a eukaryotic PAS domain. Cell 95:649–655

Murai T, Kakizuka A, Takumi T, Ohkubo H, Nakanishi S (1989) Molecular cloning and sequence analysis of human genomic DNA encoding a novel membrane protein which exhibits a slowly activating potassium channel activity. Biochem Biophys Res Commun 161:176–181

Nabauer M, Kaab S (1998) Potassium channel down-regulation in heart failure. Cardiovasc Res 37:324–334

Nattel S (2000) Acquired delayed rectifier channelopathies: how heart disease and antiarrhythmic drugs mimic potentially-lethal congenital cardiac disorders. Cardiovasc Res 48:188–190

Numaguchi H, Johnson JP Jr, Petersen CI, Balser JR (2000) A sensitive mechanism for cation modulation of potassium current. Nat Neurosci 3:429–430

Paavonen KJ, Swan H, Piippo K, Hokkanen L, Laitinen P, Viitasalo M, Toivonen L, Kontula K (2001) Response of the QT interval to mental and physical stress in types LQT1 and LQT2 of the long QT syndrome. Heart 86:39–44

Paavonen KJ, Chapman H, Laitinen PJ, Fodstad H, Piippo K, Swan H, Toivonen L, Viitasalo M, Kontula K, Pasternack M (2003) Functional characterization of the common amino acid 897 polymorphism of the cardiac potassium channel KCNH2 (HERG). Cardiovasc Res 59:603–611

Pak PH, Nuss HB, Tunin RS, Kaab S, Tomaselli GF, Marban E, Kass DA (1997) Repolarization abnormalities, arrhythmia and sudden death in canine tachycardia-induced cardiomyopathy. J Am Coll Cardiol 30:576–584

Petrecca K, Atanasiu R, Akhavan A, Shrier A (1999) N-linked glycosylation sites determine HERG channel surface membrane expression. J Physiol 515:41–48

Piippo K, Swan H, Pasternack M, Chapman H, Paavonen K, Viitasalo M, Toivonen L, Kontula K (2001) A founder mutation of the potassium channel KCNQ1 in long QT syndrome: implications for estimation of disease prevalence and molecular diagnostics. J Am Coll Cardiol 37:562–568

Pond AL, Scheve BK, Benedict AT, Petrecca K, Van Wagoner DR, Shrier A, Nerbonne JM (2000) Expression of distinct ERG proteins in rat, mouse, and human heart. Relation to functional I(Kr) channels. J Biol Chem 275:5997–6006

Priori SG (2000) Long QT and Brugada syndromes: from genetics to clinical management. J Cardiovasc Electrophysiol 11:1174–1178

Robertson GA (2000) LQT2: amplitude reduction and loss of selectivity in the tail that wags the HERG channel. Circ Res 86:492–493

Roden DM (1998) Taking the "idio" out of "idiosyncratic": predicting torsades de pointes. Pacing Clin Electrophysiol 21:1029–1034

Roden DM (2000) Acquired long QT syndromes and the risk of proarrhythmia. J Cardiovasc Electrophysiol 11:938–940

Roden DM, Lazzara R, Rosen M, Schwartz PJ, Towbin J, Vincent GM (1996) Multiple mechanisms in the long-QT syndrome. Current knowledge, gaps, and future directions. The SADS Foundation Task Force on LQTS. Circulation 94:1996–2012

Roden DM, Balser JR, George AL Jr, Anderson ME (2002) Cardiac ion channels. Annu Rev Physiol 64:431–475

Rozanski GJ, Xu Z, Whitney RT, Murakami H, Zucker IH (1997) Electrophysiology of rabbit ventricular myocytes following sustained rapid ventricular pacing. J Mol Cell Cardiol 29:721–732

Russell MW, Dick M 2nd, Collins FS, Brody LC (1996) KVLQT1 mutations in three families with familial or sporadic long QT syndrome. Hum Mol Genet 5:1319–1324

Salata JJ, Jurkiewicz NK, Wang J, Evans BE, Orme HT, Sanguinetti MC (1998) A novel benzodiazepine that activates cardiac slow delayed rectifier K+ currents. Mol Pharmacol 54:220–230

Sanchez-Chapula JA, Navarro-Polanco RA, Culberson C, Chen J, Sanguinetti MC (2002) Molecular determinants of voltage-dependent human ether-a-go-go related gene (HERG) K+ channel block. J Biol Chem 277:23587–23595

Sanguinetti MC, Jurkiewicz NK (1990a) Lanthanum blocks a specific component of IK and screens membrane surface change in cardiac cells. Am J Physiol 259:H1881–H1889

Sanguinetti MC, Jurkiewicz NK (1990b) Two components of cardiac delayed rectifier K+ current. Differential sensitivity to block by class III antiarrhythmic agents. J Gen Physiol 96:195–215

Sanguinetti MC, Xu QP (1999) Mutations of the S4-S5 linker alter activation properties of HERG potassium channels expressed in Xenopus oocytes. J Physiol 514:667–675

Sanguinetti MC, Jiang C, Curran ME, Keating MT (1995) A mechanistic link between an inherited and an acquired cardiac arrhythmia: HERG encodes the IKr potassium channel. Cell 81:299–307

Sanguinetti MC, Curran ME, Spector PS, Keating MT (1996a) Spectrum of HERG K+-channel dysfunction in an inherited cardiac arrhythmia. Proc Natl Acad Sci U S A 93:2208–2212

Sanguinetti MC, Curran ME, Zou A, Shen J, Spector PS, Atkinson DL, Keating MT (1996b) Coassembly of K(V)LQT1 and minK (IsK) proteins to form cardiac I(Ks) potassium channel. Nature 384:80–83

Scherer CR, Lerche C, Decher N, Dennis AT, Maier P, Ficker E, Busch AE, Wollnik B, Steinmeyer K (2002) The antihistamine fexofenadine does not affect I(Kr) currents in a case report of drug-induced cardiac arrhythmia. Br J Pharmacol 137:892–900

Schonherr R, Heinemann SH (1996) Molecular determinants for activation and inactivation of HERG, a human inward rectifier potassium channel. J Physiol 493:635–642

Schroeder BC, Waldegger S, Fehr S, Bleich M, Warth R, Greger R, Jentsch TJ (2000) A constitutively open potassium channel formed by KCNQ1 and KCNE3. Nature 403:196–199

Seebohm G, Chen J, Strutz N, Culberson C, Lerche C, Sanguinetti MC (2003) Molecular determinants of KCNQ1 channel block by a benzodiazepine. Mol Pharmacol 64:70–77

Selnick HG, Liverton NJ, Baldwin JJ, Butcher JW, Claremon DA, Elliott JM, Freidinger RM, King SA, Libby BE, McIntyre CJ, Pribush DA, Remy DC, Smith GR, Tebben AJ, Jurkiewicz NK, Lynch JJ, Salata JJ, Sanguinetti MC, Siegl PK, Slaughter DE, Vyas K (1997) Class III antiarrhythmic activity in vivo by selective blockade of the slowly activating cardiac delayed rectifier potassium current IKs by (R)-2-(2,4-trifluoromethyl)-N-[2-oxo-5-phenyl-1-(2,2,2-trifluoroethyl)-2, 3-dihydro-1H-benzo[e][1,4]diazepin-3-yl]acetamide. J Med Chem 40:3865–3868

Sesti F, Tai KK, Goldstein SA (2000) MinK endows the I(Ks) potassium channel pore with sensitivity to internal tetraethylammonium. Biophys J 79:1369–1378

Shah RR (2004) Drug-induced QT interval prolongation: regulatory perspectives and drug development. Ann Med 36 Suppl 1:47–52

Shibasaki T (1987) Conductance and kinetics of delayed rectifier potassium channels in nodal cells of the rabbit heart. J Physiol 387:227–250

Shimizu W, Antzelevitch C (1998) Cellular basis for the ECG features of the LQT1 form of the long-QT syndrome: effects of beta-adrenergic agonists and antagonists and sodium channel blockers on transmural dispersion of repolarization and torsade de pointes. Circulation 98:2314–2322

Smith PL, Baukrowitz T, Yellen G (1996) The inward rectification mechanism of the HERG cardiac potassium channel. Nature 379:833–836

Snyders DJ (1999) Structure and function of cardiac potassium channels. Cardiovasc Res 42:377–390

Snyders DJ, Chaudhary A (1996) High affinity open channel block by dofetilide of HERG expressed in a human cell line. Mol Pharmacol 49:949–955

Spector PS, Curran ME, Keating MT, Sanguinetti MC (1996a) Class III antiarrhythmic drugs block HERG, a human cardiac delayed rectifier K+ channel. Open-channel block by methanesulfonanilides. Circ Res 78:499–503

Spector PS, Curran ME, Zou A, Keating MT, Sanguinetti MC (1996b) Fast inactivation causes rectification of the IKr channel. J Gen Physiol 107:611–619

Splawski I, Tristani-Firouzi M, Lehmann MH, Sanguinetti MC, Keating MT (1997) Mutations in the hminK gene cause long QT syndrome and suppress IKs function. Nat Genet 17:338–340

Splawski I, Shen J, Timothy KW, Lehmann MH, Priori S, Robinson JL, Moss AJ, Schwartz PJ, Towbin JA, Vincent GM, Keating MT (2000) Spectrum of mutations in long-QT syndrome genes. KVLQT1, HERG, SCN5A, KCNE1, and KCNE2. Circulation 102:1178–1185

Splawski I, Timothy KW, Tateyama M, Clancy CE, Malhotra A, Beggs AH, Cappuccio FP, Sagnella GA, Kass RS, Keating MT (2002) Variant of SCN5A sodium channel implicated in risk of cardiac arrhythmia. Science 297:1333–1336

Stengl M, Volders PG, Thomsen MB, Spatjens RL, Sipido KR, Vos MA (2003) Accumulation of slowly activating delayed rectifier potassium current (IKs) in canine ventricular myocytes. J Physiol 551:777–786

Tai KK, Goldstein SA (1998) The conduction pore of a cardiac potassium channel. Nature 391:605–608

Tamargo J (2000) Drug-induced torsade de pointes: from molecular biology to bedside. Jpn J Pharmacol 83:1–19

Tamargo J, Caballero R, Gomez R, Valenzuela C, Delpon E (2004) Pharmacology of cardiac potassium channels. Cardiovasc Res 62:9–33

Tapper AR, George AL Jr (2000) MinK subdomains that mediate modulation of and association with KvLQT1. J Gen Physiol 116:379–390

Tapper AR, George AL Jr (2001) Location and orientation of minK within the I(Ks) potassium channel complex. J Biol Chem 276:38249–38254

Thomas D, Zhang W, Wu K, Wimmer AB, Gut B, Wendt-Nordahl G, Kathofer S, Kreye VA, Katus HA, Schoels W, Kiehn J, Karle CA (2003) Regulation of HERG potassium channel activation by protein kinase C independent of direct phosphorylation of the channel protein. Cardiovasc Res 59:14–26

Tohse N, Kameyama M, Irisawa H (1987) Intracellular Ca2+ and protein kinase C modulate K+ current in guinea pig heart cells. Am J Physiol 253:H1321–H1324

Tomaselli GF, Marban E (1999) Electrophysiological remodeling in hypertrophy and heart failure. Cardiovasc Res 42:270–283

Tristani-Firouzi M, Sanguinetti MC (2003) Structural determinants and biophysical properties of HERG and KCNQ1 channel gating. J Mol Cell Cardiol 35:27–35

Trudeau MC, Warmke JW, Ganetzky B, Robertson GA (1995) HERG, a human inward rectifier in the voltage-gated potassium channel family. Science 269:92–95

Tseng GN (2001) I(Kr): the hERG channel. J Mol Cell Cardiol 33:835–849

Tsuji Y, Opthof T, Kamiya K, Yasui K, Liu W, Lu Z, Kodama I (2000) Pacing-induced heart failure causes a reduction of delayed rectifier potassium currents along with decreases in calcium and transient outward currents in rabbit ventricle. Cardiovasc Res 48:300–309

Tsuji Y, Opthof T, Yasui K, Inden Y, Takemura H, Niwa N, Lu Z, Lee JK, Honjo H, Kamiya K, Kodama I (2002) Ionic mechanisms of acquired QT prolongation and torsades de pointes in rabbits with chronic complete atrioventricular block. Circulation 106:2012–2018

Varnum MD, Busch AE, Bond CT, Maylie J, Adelman JP (1993) The min K channel underlies the cardiac potassium current IKs and mediates species-specific responses to protein kinase C. Proc Natl Acad Sci U S A 90:11528–11532

Varro A, Balati B, Iost N, Takacs J, Virag L, Lathrop DA, Csaba L, Talosi L, Papp JG (2000) The role of the delayed rectifier component IKs in dog ventricular muscle and Purkinje fibre repolarization. J Physiol 523:67–81

Vaughan Williams EM (1984) A classification of antiarrhythmic actions reassessed after a decade of new drugs. J Clin Pharmacol 24:129–147

Vereecke J, Carmeliet E (2000) The effect of external pH on the delayed rectifying K+ current in cardiac ventricular myocytes. Pflugers Arch 439:739–751

Viswanathan PC, Shaw RM, Rudy Y (1999) Effects of IKr and IKs heterogeneity on action potential duration and its rate dependence: a simulation study. Circulation 99:2466–2474

Volders PG, Sipido KR, Carmeliet E, Spatjens RL, Wellens HJ, Vos MA (1999a) Repolarizing K+ currents ITO1 and IKs are larger in right than left canine ventricular midmyocardium. Circulation 99:206–210

Volders PG, Sipido KR, Vos MA, Spatjens RL, Leunissen JD, Carmeliet E, Wellens HJ (1999b) Downregulation of delayed rectifier K(+) currents in dogs with chronic complete atrioventricular block and acquired torsades de pointes. Circulation 100:2455–2461

Walker AM, Szneke P, Weatherby LB, Dicker LW, Lanza LL, Loughlin JE, Yee CL, Dreyer NA (1999) The risk of serious cardiac arrhythmias among cisapride users in the United Kingdom and Canada. Am J Med 107:356–362

Walsh KB, Kass RS (1988) Regulation of a heart potassium channel by protein kinase A and C. Science 242:67–69

Walsh KB, Kass RS (1991) Distinct voltage-dependent regulation of a heart-delayed IK by protein kinases A and C. Am J Physiol 261:C1081–C1090

Wang HS, Brown BS, McKinnon D, Cohen IS (2000a) Molecular basis for differential sensitivity of KCNQ and I(Ks) channels to the cognitive enhancer XE991. Mol Pharmacol 57:1218–1223

Wang J, Trudeau MC, Zappia AM, Robertson GA (1998a) Regulation of deactivation by an amino terminal domain in human ether-a-go-go-related gene potassium channels. J Gen Physiol 112:637–647

Wang J, Myers CD, Robertson GA (2000b) Dynamic control of deactivation gating by a soluble amino-terminal domain in HERG K(+) channels. J Gen Physiol 115:749–758

Wang Q, Curran ME, Splawski I, Burn TC, Millholland JM, VanRaay TJ, Shen J, Timothy KW, Vincent GM, de Jager T, Schwartz PJ, Toubin JA, Moss AJ, Atkinson DL, Landes GM, Connors TD, Keating MT (1996) Positional cloning of a novel potassium channel gene: KVLQT1 mutations cause cardiac arrhythmias. Nat Genet 12:17–23

Wang S, Liu S, Morales MJ, Strauss HC, Rasmusson RL (1997) A quantitative analysis of the activation and inactivation kinetics of HERG expressed in Xenopus oocytes. J Physiol 502:45–60

Wang W, Xia J, Kass RS (1998b) MinK-KvLQT1 fusion proteins, evidence for multiple stoichiometries of the assembled IsK channel. J Biol Chem 273:34069–34074

Warmke JW, Ganetzky B (1994) A family of potassium channel genes related to eag in Drosophila and mammals. Proc Natl Acad Sci U S A 91:3438–3442

Washizuka T, Horie M, Watanuki M, Sasayama S (1997) Endothelin-1 inhibits the slow component of cardiac delayed rectifier K+ currents via a pertussis toxin-sensitive mechanism. Circ Res 81:211–218

Weerapura M, Nattel S, Chartier D, Caballero R, Hebert TE (2002) A comparison of currents carried by HERG, with and without coexpression of MiRP1, and the native rapid delayed rectifier current. Is MiRP1 the missing link? J Physiol 540:15–27

Wehrens XH, Marks AR (2003) Altered function and regulation of cardiac ryanodine receptors in cardiac disease. Trends Biochem Sci 28:671–678

Wehrens XH, Vos MA, Doevendans PA, Wellens HJ (2002) Novel insights in the congenital long QT syndrome. Ann Intern Med 137:981–992

Witchel HJ, Dempsey CE, Sessions RB, Perry M, Milnes JT, Hancox JC, Mitcheson JS (2004) The low potency, voltage-dependent HERG blocker propafenone—molecular determinants and drug trapping. Mol Pharmacol 66:1201–1212

Wollnik B, Schroeder BC, Kubisch C, Esperer HD, Wieacker P, Jentsch TJ (1997) Pathophysiological mechanisms of dominant and recessive KVLQT1 K+ channel mutations found in inherited cardiac arrhythmias. Hum Mol Genet 6:1943–1949

Wysowski DK, Bacsanyi J (1996) Cisapride and fatal arrhythmia. N Engl J Med 335:290–291

Yang IC, Scherz MW, Bahinski A, Bennett PB, Murray KT (2000) Stereoselective interactions of the enantiomers of chromanol 293B with human voltage-gated potassium channels. J Pharmacol Exp Ther 294:955–962

Yang P, Kanki H, Drolet B, Yang T, Wei J, Viswanathan PC, Hohnloser SH, Shimizu W, Schwartz PJ, Stanton M, Murray KT, Norris K, George AL Jr, Roden DM (2002) Allelic variants in long-QT disease genes in patients with drug-associated torsades de pointes. Circulation 105:1943–1948

Yang T, Kupershmidt S, Roden DM (1995) Anti-minK antisense decreases the amplitude of the rapidly activating cardiac delayed rectifier K+ current. Circ Res 77:1246–1253

Yang T, Kanki H, Roden DM (2003) Phosphorylation of the IKs channel complex inhibits drug block: novel mechanism underlying variable antiarrhythmic drug actions. Circulation 108:132–134

Yang ZK, Boyett MR, Janvier NC, McMorn SO, Shui Z, Karim F (1996) Regional differences in the negative inotropic effect of acetylcholine within the canine ventricle. J Physiol 492:789–806

Zabel M, Hohnloser SH, Behrens S, Woosley RL, Franz MR (1997) Differential effects of D-sotalol, quinidine, and amiodarone on dispersion of ventricular repolarization in the isolated rabbit heart. J Cardiovasc Electrophysiol 8:1239–1245

Zeng J, Laurita KR, Rosenbaum DS, Rudy Y (1995) Two components of the delayed rectifier K+ current in ventricular myocytes of the guinea pig type. Theoretical formulation and their role in repolarization. Circ Res 77:140–152

Zhou Z, Gong Q, Ye B, Fan Z, Makielski JC, Robertson GA, January CT (1998) Properties of HERG channels stably expressed in HEK 293 cells studied at physiological temperature. Biophys J 74:230–241

Zipes DP, Wellens HJ (1998) Sudden cardiac death. Circulation 98:2334–2351

Zou A, Curran ME, Keating MT, Sanguinetti MC (1997) Single HERG delayed rectifier K+ channels expressed in Xenopus oocytes. Am J Physiol 272:H1309–H1314

HEP (2006) 171:159–199
© Springer-Verlag Berlin Heidelberg 2006

Sodium Calcium Exchange as a Target for Antiarrhythmic Therapy

K. R. Sipido[1] (✉) · A. Varro[2] · D. Eisner[3]

[1]Lab. of Experimental Cardiology, KUL, Campus Gasthuisberg O/N 7th floor,
Herestraat 49, B-3000 Leuven, Belgium
Karin.Sipido@med.kuleuven.ac.be

[2]Department of Pharmacology and Pharmacotherapy,
Albert Szent-Gyorgyi Medical Center, University of Szeged, Hungary

[3]Unit of Cardiac Physiology, University of Manchester, Manchester UK

Abstract In search of better antiarrhythmic therapy, targeting the Na/Ca exchanger is an option to be explored. The rationale is that increased activity of the Na/Ca exchanger has been implicated in arrhythmogenesis in a number of conditions. The evidence is strong for triggered arrhythmias related to Ca^{2+} overload, due to increased Na^+ load or during adrenergic stimulation; the Na/Ca exchanger may be important in triggered arrhythmias in heart failure and in atrial fibrillation. There is also evidence for a less direct role of the Na/Ca exchanger in contributing to remodelling processes. In this chapter, we review this evidence and discuss the consequences of inhibition of Na/Ca exchange in the perspective of its physiological role in Ca^{2+} homeostasis. We summarize the current data on the use of available blockers of Na/Ca exchange and propose a framework for further study and development of such drugs. Very selective agents have great potential as tools for further study of the role the Na/Ca exchanger plays in arrhythmogenesis. For therapy, they may have their specific indications, but they carry the risk of increasing Ca^{2+} load of the cell. Agents with a broader action that includes Ca^{2+} channel block may have advantages in other conditions, e.g. with Ca^{2+} overload. Additional actions such as block of K^+ channels, which may be unwanted in e.g. heart failure, may be used to advantage as well.

Keywords Heart failure · Sodium/calcium exchange · Sarcoplasmic reticulum · Afterdepolarizations · Arrhythmias

1
Introduction

Sudden, presumed to be arrhythmic, death is a major cause of mortality (Zipes and Wellens 1998). It occurs in a variety of cardiac disease, from congenital ion channel mutations without structural heart disease to the complex setting of ischaemic cardiomyopathy. In this wide variety, several mechanisms can initiate and maintain atrial and ventricular arrhythmias. The goal has been to identify and tailor therapy toward the specific mechanisms involved (Members of the Sicilian Gambit 2001). For this purpose, agents that block or modulate specific ion channels have been developed. A particular example is drugs that block with high affinity a subset of K^+ channels in the atria (Nattel et al. 1999; Varro et al. 2004). The goal is to prolong the atrial action potential without affecting the ventricular action potential, and prolong the atrial refractory period to prevent atrial fibrillation. In the setting of ischaemic cardiomyopathy and heart failure, mortality due to arrhythmias is high and the search for efficient antiarrhythmic drugs has been particularly frustrating. Current practice advocates implantation of an implantable cardioverter defibrillator (ICD), an efficient approach but expensive, which also impacts on quality of life (Epstein 2004). This invasive approach developed in the face of disappointments with currently available medical therapy. Class I antiarrhythmics have been associated with increased mortality, and the negative outcome of the CAST studies has had a profound impact on further studies (The CAST Investigators 1989; The CAST II Investigators 1992; Myerburg et al. 1998). In heart failure, pure K^+ channel blockers prolonging the action potential are unlikely to be an option, given that the action potential is already prolonged and repolarization disturbed. Indeed, d-sotalol was associated with a higher mortality in the SWORD study (Waldo et al. 1996). The multi-action drug amiodarone has no negative effects but appears to be less efficient than an ICD (Bokhari et al. 2004). The most efficient medical therapy associated with a reduction of sudden death seems to be β-blockade, but the larger studies were not set up to test specifically for the antiarrhythmic effect (Kendall 2000).

In search of better antiarrhythmic therapy, targeting the Na/Ca exchanger is an option to be explored. The rationale is that in heart failure, in particular in ischaemic cardiomyopathy, triggered arrhythmias are most common and increased activity of the Na/Ca exchanger is causally involved (Pogwizd 2003). This approach is clearly distinct from earlier ion channel blockers as it would target directly what is thought to be the culprit in arrhythmia initiation.

The Na/Ca exchanger is an ion transporter, exchanging three Na^+ for one Ca^{2+} ion, and the generated ionic current can be inward or outward as the electrochemical driving force changes during the cardiac cycle. The Na/Ca exchange (NCX) current will thus alternately have a repolarizing as well as depolarizing effect, and its contribution to the action potential profile is complex; it also contributes prominently to abnormal depolarizations occurring after the action potential. Reducing or increasing NCX not only influences electrical activity but also directly affects Ca^{2+} handling and therefore contractility. This is another particular property that sets the Na/Ca exchanger apart as a target for antiarrhythmic therapy and it may even be the prime reason for choosing it. These dual effects on electrical activity and Ca^{2+} handling should always be considered. In this chapter, we will review the properties of the Na/Ca exchanger and consequent effects on electrical activity, the expected effects of NCX blockers and the current experience with such agents.

2
The Na/Ca Exchanger, Major Regulator of Ca^{2+} Balance in the Cardiac Cell

The Na/Ca exchanger is a Ca^{2+} and Na^+ transport protein found in most tissues, but it is particularly abundant in cardiac muscle where the dominant isoform expressed is NCX1. The molecular properties have been explored in detail, following the cloning of the cardiac exchanger (reviewed in Blaustein and Lederer 1999). An important characteristic of the transporter is its stoichiometry. Initially established at 3 Na^+:1 Ca^{2+} (Ehara et al. 1989), it was more recently reported to be closer to 4 Na$^+$:1 Ca^{2+} and variable (Fujioka et al. 2000). Kang et al. could subsequently demonstrate that in addition to the major 3 Na$^+$:1 Ca^{2+} mode, Na$^+$-only and Na$^+$/Ca^{2+} co-transport modes exist which can explain the earlier discrepancy, setting the overall stoichiometry at 3.2 Na$^+$:1 Ca^{2+} (Kang and Hilgemann 2004). Given this asymmetrical charge transport, the Na/Ca exchanger is electrogenic, with, for most conditions, one charge moved for one exchange cycle. The driving force for this ionic current, and thus for Ca^{2+} and Na$^+$ transport, is the electrochemical gradient, the difference between the membrane potential, E_m, and the reversal potential, E_{NCX}, determined by the concentrations of Ca^{2+} and Na$^+$. The effect of changes in ion concentrations is illustrated in Fig. 1. Panel a illustrates the situation of a resting, unstimulated

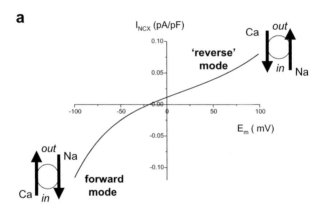

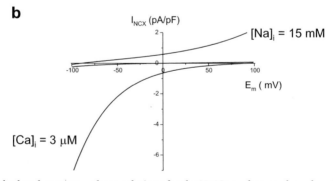

Fig. 1a,b Calculated current–voltage relations for the Na/Ca exchanger, based on the relation $I_{NCX}=k([Na^+]_i^3 \ [Ca^{2+}]_o e^{(rE_mF/RT)}-[Na^+]_o^3 \ [Ca^{2+}]_i \ e^{((1-r)E_mF/RT)})$. **a** For a cell at rest, with $[Ca^{2+}]_i$ 50 nM, $[Na^+]_i$ 5 mM, and $[Na^+]_o$ 130 mM, $[Ca^{2+}]_o$ 1.8 mM. Inward current corresponds to Ca^{2+} removal or forward mode, outward current to Ca^{2+} influx or reverse mode. **b** During stimulation $[Na^+]_i$ rises to 15 mM and $[Ca^{2+}]_i$ in diastole to 100 nM and Na/Ca exchange current is illustrated by the *upper curve*. With Ca^{2+} release from the sarcoplasmic reticulum, $[Ca^{2+}]_i$ rises to a few micromolar corresponding to the lower curve

cell. With an increase in $[Na^+]_i$ to 15 nM, the curve of the cell at rest would be as indicated in panel b. Typically during a single cardiac cycle, E_m changes quickly during the action potential, but rapid changes in $[Ca^{2+}]_i$ will also affect E_{NCX}, resulting in the lower curve of Fig. 1b. The predicted NCX Ca^{2+} flux during an action potential is an initial Ca^{2+} influx with outward current, due to the strong depolarization at initially low $[Ca^{2+}]_i$, followed by Ca^{2+} efflux and inward current with the increase in $[Ca^{2+}]_i$ that shifts the reversal potential to more positive values. This dual transport by the exchanger is incorporated in most models of the cardiac action potential, illustrated by an example of a simulation in the Oxsoft Heart model (Janvier and Boyett 1996; Fig. 2). Although this sequence of events is generally accepted, there is uncertainty regarding the magnitude and duration of the Ca^{2+} influx/outward current and the resultant effect on the action potential time course, as discussed in Sect. 3.

For the purpose of the current discussion we review in more detail some relevant features of NCX regulation, namely its role in Ca^{2+} removal and regulation by $[Ca^{2+}]_i$, its regulation by $[Na^+]_i$ and we examine the alternative pathways for Ca^{2+} removal if NCX were inhibited.

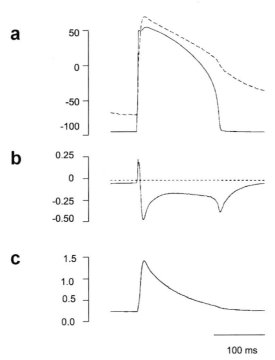

Fig. 2a–c Theoretical modelling of the Na/Ca exchange current during the action potential in a guinea-pig ventricular myocyte. **a** Superimposed on the action potential, the changes in reversal potential, E_{NCX} (*dashed line*), which are related to the increase in $[Ca^{2+}]_i$, illustrated in **c**. **b** The calculated Na/Ca exchange current. (Reproduced from Janvier et al. 1996)

2.1
Ca²⁺ Removal and Regulation by [Ca²⁺]ᵢ

The process of excitation-contraction coupling is schematically illustrated in Fig. 3 (reviewed in Bers 2002; Trafford and Eisner 2002). During each cardiac cycle, a certain amount of Ca^{2+} enters the cell through voltage-activated Ca^{2+} channels, with a small additional amount entering through the Na/Ca exchanger, depending on the $[Na^+]_i$. This Ca^{2+} acts as a trigger to activate the Ca^{2+} channel in the sarcoplasmic reticulum (SR), the ryanodine receptor (RyR), and more Ca^{2+} is released from the SR, this being the major source for Ca^{2+} to activate the myofilaments. As Ca^{2+} channels inactivate and RyRs close, Ca^{2+} is removed from the cytosol by re-uptake into the SR by the ATP-driven

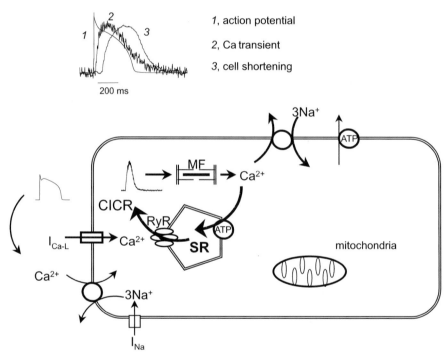

Fig. 3 Schematic of excitation–contraction coupling. During the action potential Ca^{2+} enters the cell through voltage-activated Ca^{2+} channels, with a small additional amount entering through the Na/Ca exchanger, depending on the $[Na^+]_i$. This Ca^{2+} acts as trigger to activate the Ca^{2+} channel in the sarcoplasmic reticulum (SR), the ryanodine receptor (*RyR*), and more Ca^{2+} is released from the SR (*CICR*), this being the major source for Ca^{2+} to activate the myofilaments. As Ca^{2+} channels inactivate and RyRs close, Ca^{2+} is removed from the cytosol by re-uptake into the SR by the ATP-driven Ca^{2+} pump (SERCA), and by efflux through the Na/Ca exchanger. Some Ca^{2+} is removed by an ATP-driven Ca^{2+} pump in the sarcolemma, PMCA. The mitochondria can also participate in the Ca^{2+} flux (see text and Fig. 5)

Ca^{2+} pump (SERCA), and by efflux through the Na/Ca exchanger. To maintain a steady state, the same amount of Ca^{2+} that entered via Ca^{2+} channels and via reverse NCX has to be removed from the myocyte within the same cardiac cycle.

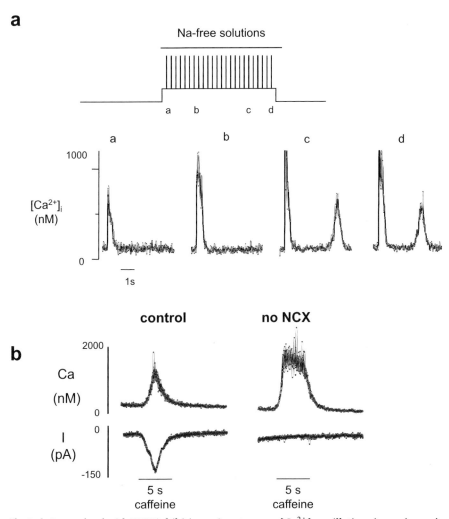

Fig. 4a,b Ca overload with NCX inhibition. **a** Spontaneous $[Ca^{2+}]_i$ oscillations in a guinea-pig ventricular myocyte when repeatedly stimulated after removing Na from the solution (both intracellular and extracellular solution, whole-cell patch clamp). The *top panel* illustrates the experimental protocol. **b** *Left panel*: Caffeine-induced release of Ca^{2+} from the SR induces an inward Na/Ca exchange current. *Right panel*: This current is absent when a similar caffeine application is done in the absence of extracellular Na, and the decline of $[Ca^{2+}]_i$ is virtually abolished on this time scale, until removal of caffeine. (Reproduced from Sipido et al. 1995)

Most evidence indicates that the Na/Ca exchanger is the major extrusion pathway for maintaining this beat-to-beat balance (Bridge et al. 1990; Trafford et al. 2002). Experimentally, inhibition of Ca^{2+} removal by NCX (by removing Na^+ from the extracellular fluid) leads to rapid Ca^{2+} accumulation and spontaneous Ca^{2+} release (Fig. 4a) unless Ca^{2+} influx is severely reduced. Sudden increases in $[Ca^{2+}]_i$, as could occur during Ca^{2+} overload (see Sect. 3.2.1), will induce a rapid shift in E_{NCX}, inducing an inward current and Ca^{2+} removal. The importance of NCX in Ca^{2+} removal can be demonstrated experimentally by the time course of $[Ca^{2+}]_i$ decline following a caffeine-induced Ca^{2+} release from the SR (Fig. 4b). In control conditions, $[Ca^{2+}]_i$ declines rapidly accompanied by an inward current, whereas when NCX is blocked, the decline slows down several-fold. Note that in this particular example in a guinea-pig ventricular myocyte, there is no inward current in the absence of NCX. The previous examples illustrate the immediate 'activation' of the NCX current following an increase in $[Ca^{2+}]_i$. There are two important additional aspects to this regulation. The first issue is that transport by the Na/Ca exchanger is dictated by $[Ca^{2+}]$ near the Ca^{2+} binding sites. This concentration may deviate substantially from what is measured experimentally by cytosolic $[Ca^{2+}]_i$ indicators, in particular with the large fluxes during Ca^{2+} release from the SR. The restricted space where such deviations from the global cytosolic concentrations can occur has been named the 'fuzzy' space (Lederer et al. 1990). This fuzzy space is thought to be the area beneath the sarcolemma in the vicinity of the SR Ca^{2+} release channels, the junctional space. During Ca^{2+} release from the SR, $[Ca^{2+}]$ in this area is several-fold higher than in the bulk cytosol, and the inward NCX current is much larger than predicted from the global $[Ca^{2+}]_i$ (Lipp et al. 1990; Trafford et al. 1995; Weber et al. 2002). The second issue is that $[Ca^{2+}]_i$ has an allosteric regulatory effect, with a slow increase in current density with maintained elevation of $[Ca^{2+}]_i$ (Weber et al. 2001). These aspects are important for understanding the direction of the NCX current during the action potential.

2.2
$[Na^+]_i$ as a Regulator of Na/Ca Exchange Function

During a single cardiac cycle, transient increases in $[Na^+]_i$ due to Na^+ influx with the upstroke of the action potential could also influence the direction of the NCX current (Leblanc and Hume 1990). Computation of the expected changes in $[Na^+]_i$ due to the Na^+ current indicates that this influx could only have a substantial effect if it occurred in a restricted space (Lederer et al. 1990). Several experiments have attempted to measure the local $[Na^+]$ in this subsarcolemmal space, but the evidence for a high local $[Na^+]$ related to the Na^+ current remains equivocal (reviewed in Verdonck et al. 2004). On a longer term basis, on the other hand, there is ample evidence that an increase in cytosolic $[Na^+]$ shifts E_{NCX} to more negative values and induces net Ca^{2+}

gain (see e.g. Eisner et al. 1984; Harrison et al. 1992; Mubagwa et al. 1997). This results in a higher SR Ca^{2+} content and higher availability for release. Increased Ca^{2+} entry via the Na/Ca exchanger during depolarization could also provide additional Ca^{2+} for activation of the ryanodine receptor to trigger Ca^{2+} release from the SR. Indeed, with high $[Na^+]_i$, reverse mode NCX can by itself induce Ca^{2+} release from the SR, albeit with lower efficiency than L-type Ca^{2+} current, I_{CaL} (Levi et al. 1994; Sham et al. 1995; Sipido et al. 1997). Most likely release is triggered primarily by I_{CaL}, and modulated by Ca^{2+} entry or removal through NCX (Goldhaber et al. 1999; Litwin et al. 1998; Su et al. 2001).

2.3
Can Other Mechanisms Replace NCX in Ca^{2+} Extrusion?

It has long been known that another pathway, in addition to NCX, exists to remove Ca^{2+} from cardiac cells. This plasma membrane Ca-ATPase (referred to as PMCA) is found in essentially all cells in the body. It uses the energy provided by hydrolysis of ATP to expel Ca^{2+} ions from the cell. It is generally thought to transport protons into the cell, and the stoichiometry may be 2 H^+ per Ca^{2+}, thus making the Ca-ATPase electroneutral (Schwiening et al. 1993), although this point is controversial (Salvador et al. 1998). The PMCA exists in four isoforms which differ in their tissue distribution (Strehler and Zacharias 2001). One problem with studying the PMCA is that there are no specific inhibitors for it. Eosin and its derivatives such as carboxyeosin have been used (Gatto and Milanik 1993; Bassani et al. 1995; Choi and Eisner 1999a).

Most early work on the PMCA was performed on vesicles or purified preparations as opposed to measuring fluxes in intact tissues. It was found that, in contrast to NCX, the PMCA has a high affinity (low K_m) and low V_{max} (Caroni and Carafoli 1981). The idea therefore grew that the PMCA was responsible for regulating resting $[Ca^{2+}]_i$, whereas the NCX was responsible for reducing Ca^{2+} following an elevation (Carafoli 1987). While this may be the case in nerve fibres (DiPolo and Beaugé 1979), one should note that the cardiac cell never rests, and it is therefore unclear why it would have a mechanism for dealing with "resting" Ca^{2+} fluxes.

Evidence suggesting a functional role for the PMCA in cardiac muscle has been obtained by inhibiting other Ca^{2+} removal processes. For example, it is known that removal of external Na^+ produces an increase of $[Ca^{2+}]_i$ as Ca^{2+} enters the cell on reverse mode NCX. However, $[Ca^{2+}]_i$ then decays to a level only somewhat greater than control (Allen et al. 1983). This implies that something other than NCX must be capable of removing Ca^{2+} from the cytoplasm. This NCX-independent Ca^{2+} removal persists when SR function is disabled (Allen et al. 1983), and it therefore presumably results from either sequestration of Ca^{2+} by mitochondria or pumping of Ca^{2+} out of the cell by the PMCA.

The experiments described above do not provide quantitative data comparing NCX and NCX-independent Ca^{2+} removal mechanisms. Such data have been obtained in two different ways. (1) One approach, illustrated in Fig. 5, is to apply caffeine rapidly to release Ca^{2+} ions from the SR. This results in a transient increase of $[Ca^{2+}]_i$ until Ca^{2+} is removed from the cytoplasm by the combined effects of NCX, PMCA and mitochondria. If NCX is inhibited, the rate of decay of $[Ca^{2+}]_i$ is decreased and the ratio of the decreased rate to the control gives the ratio: (PMCA+mitochondria)/(NCX+PMCA+mitochondria). From this it has been estimated that the PMCA+mitochondria together account for up to 25% to 30% with estimates varying depending upon species and other conditions (Negretti et al. 1993; Bers et al. 1993). Separation between mitochondrial and PMCA fluxes has been performed either by inhibiting mitochondria (although here there is a worry of secondary changes of ATP concentration affecting the PMCA) or inhibiting the Ca-ATPase either by increasing external Ca^{2+} or using carboxyeosin. (2) The above approaches suffer from the problem that, in order to investigate pumping of calcium, one would like to measure the changes of *total* cell Ca^{2+}, whereas the indicators that are used measure free Ca^{2+}. A more quantitative approach, therefore, measures the buffering power of the cell for Ca^{2+} and calculates the actual transport rate. This also allows the flux to be calculated as a function of $[Ca^{2+}]_i$. It was found that the apparent affinity of the Ca^{2+} removal processes across the sarcolemma was the same whether or not Na^+ was present (Choi et al. 2000), in agreement with previous work on

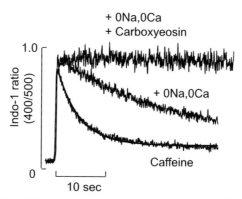

Fig. 5 Ca^{2+} removal by different systems in the absence of SR Ca^{2+} uptake. Application of caffeine induces Ca^{2+} release from the SR, and in the maintained presence of caffeine, Ca^{2+} removal from the cytosol is the result of the activity of the Na/Ca exchanger, the mitochondria and PMCA. Three applications of caffeine (10 mM; applied for the period shown by the *bar*) are shown superimposed. Caffeine was applied under the following conditions: control, all systems operational; 0 Na^+, 0 Ca^{2+}, mitochondria and PMCA, no Na/Ca exchange; 0 Na^+, 0 Ca^{2+} and carboxyeosin, no Na/Ca exchange and no PMCA. The continuous curves through the data are exponentials with rate constants of 0.36 and 0.079 s^{-1}, respectively. (Reproduced from Choi and Eisner 1999b)

intact muscles (Lamont and Eisner 1996). This is *not* what would be expected for a low-affinity NCX and a higher affinity PMCA.

Very recently, important data have been provided by the introduction of mice bred to contain no NCX in the heart (Henderson et al. 2004) (as opposed to global knockout of NCX, which is lethal). These animals appeared to have near normal cardiac function. More precisely, in these animals the majority of cells had no detectable NCX. However, 10%–20% of cells did have NCX. It is impossible to exclude the possibility that in the intact heart, cells with NCX create a diffusion gradient down which Ca^{2+} can diffuse from neighbours with no NCX. However, the authors studied single cells and found that, even in cells with no NCX, normal systolic $[Ca^{2+}]_i$ transients could be observed. They also found that the amplitude of the L-type Ca^{2+} current was reduced to 50% of control and suggested that the remainder of the Ca^{2+} current could be accommodated by non-NCX-mediated Ca^{2+} extrusion from the cell. One problem with this conclusion is that it requires a larger NCX-independent Ca^{2+} extrusion process than found in the work described above. In addition their study found that when caffeine was applied to cells in which NCX was knocked out, the increase of $[Ca^{2+}]_i$ was maintained (at least for the 1 s duration application), a result inconsistent with the idea that there is significant Ca^{2+} extrusion by a non-NCX mechanism.

Other recent work has suggested that the PMCA may be expressed in localized domains such as caveolae and, either by regulating local $[Ca^{2+}]_i$ or by direct interaction, may regulate the activity of NO synthase (Schuh et al. 2001).

In conclusion, although mechanisms other than NCX can produce measurable Ca^{2+} efflux from the cell, we feel that currently available data have not shown a clear physiological or pathological role for these fluxes. This conclusion serves to emphasize the importance of NCX.

3
Electrogenic Na/Ca Exchange Modulates the Action Potential and Generates Afterdepolarizations

3.1
Na/Ca Exchange Current During the Action Potential

As illustrated in Fig. 2 and discussed above, the Ca^{2+} transport by the exchanger is initially into the cell and later out of the cell during a single cardiac cycle. This implies that the NCX current has repolarizing as well as depolarizing effects. The net result on the action potential time course and duration is therefore complex and the effect of blocking the current much less predictable than for Na^+ or K^+ currents. Much depends on the balance between the initial outward and the subsequent inward current.

Many experimental studies have addressed this issue, using often indirect or complex approaches. Egan et al. examined NCX currents during the action potential by interposing voltage clamps (Egan et al. 1989). They incorporated their findings in a mathematical model which includes only a brief outward current during the initial action potential (Noble et al. 1991; see also Fig. 2). For increasing $[Na^+]_i$ up to 8 mM, the outward current increases, yet action potential duration increases as the later inward component also increases. A later study examined the effects of strongly buffering $[Ca^{2+}]_i$ with BAPTA (Janvier et al. 1997). In this study, the dominant effect of the NCX current was to prolong the action potential. In the Luo-Rudy model of the guinea-pig ventricular cell, the initial outward current is large and sustained for most of the plateau of the action potential, probably related to the high $[Na^+]_i$ in this model (14 mM at 1 Hz stimulation) (Luo and Rudy 1994). When simulating increased Na^+ loading, this current increases further and the action potential shortens primarily due to the NCX current (Faber and Rudy 2000).

More recently, Armoundas et al. used the small NCX inhibitory peptide XIP to examine the effect of blocking NCX on the action potential profile (Armoundas et al. 2003). From these data and further modelling, they conclude that the NCX current is a predominantly depolarizing current for $[Na^+]_i$ of 5 mM, and a predominantly repolarizing current when cytosolic $[Na^+]$ is high (10 mM or above). Weber et al. combined measurements of NCX current during the action potential using interpolated voltage clamps, estimates of subsarcolemmal $[Ca^{2+}]$ and modelling to derive the direction of the NCX current (Weber et al. 2003). These authors show that the NCX current is mostly inward during the plateau of the action potential. This result is also obtained in the most recent modelling data from this group, which incorporate the local ion concentrations and the allosteric regulation of the exchanger (Shannon et al. 2004). Taken together most of the data thus indicate that in normal tissue and without elevated $[Na^+]_i$, the Na/Ca exchanger is predominantly inward during the action potential plateau. This may change with disease (see Sect. 4.2).

3.2
Na/Ca Exchange and Delayed Afterdepolarizations During Ca^{2+} Overload

As an inward current, the NCX current can contribute to afterdepolarizations. Early afterdepolarizations occur on the late plateau or early repolarization phase of the action potential, which is usually prolonged; delayed afterdepolarizations occur after full repolarization and are related to Ca^{2+} overload (Fig. 6).

During the cardiac cycle, the amount of Ca^{2+} that enters the cell (largely via the L-type current) must be removed from the cell (largely via NCX). This requires that the $[Ca^{2+}]_i$ transient is of the correct amplitude such that the degree of activation of NCX produces exactly the right amount of Ca^{2+} efflux to balance the influx. If the amplitude of the Ca^{2+} transient is too small, then the efflux will be less than the influx. This will then result in net gain

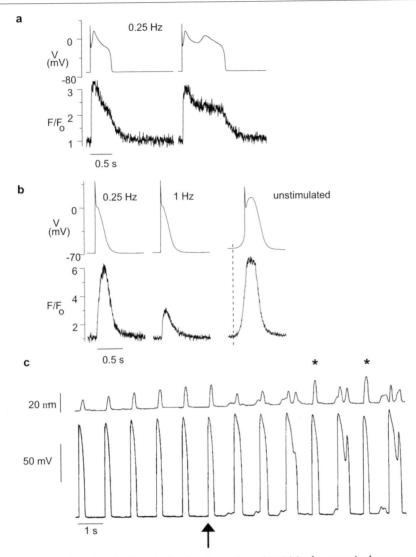

Fig. 6a–c Early (EAD) and delayed afterdepolarizations (DAD) in dog ventricular myocytes. **a**, EADs. In myocytes isolated from the left ventricle (LV) of the hypertrophied heart of dogs with complete atrioventricular (AV) node block for 4–6 weeks, action potentials are prolonged in particular at low frequencies of stimulation (Volders et al. 1998a). Spontaneous EADs are occasionally observed, as in this example (at 0.25 Hz) and become very prominent after application of almokalant (Volders et al. 1998). **b**, DADs When dialysed with a pipette solution with increased [Na] (20 mM), the action potentials are shorter, a prominent negative frequency response is present (Mubagwa et al. 1997) and at-rest spontaneous Ca^{2+} release occurs with a DAD triggering an action potential. **c**, EADs and DADs. Under adrenergic stimulation (20 nM isoproterenol), Ca^{2+} release increases (indicated by the increase of shortening) and spontaneous release occurs, with both early and delayed afterdepolarizations. (Modified after Volders et al. 1997)

of Ca^{2+} by the cell and therefore by the SR. As a consequence, the amplitude of the $[Ca^{2+}]_i$ transient will increase until the efflux balances the influx. (For further discussion of this issue see Eisner et al. 1998, 2000.) Under some conditions, however, the Ca^{2+} content of the SR increases to a level at which spontaneous release occurs. This condition is often referred to as "calcium overload". The release generally takes the form of Ca^{2+} waves which propagate along and between cells (Capogrossi et al. 1984; Mulder et al. 1989; Wier et al. 1987). The importance of these waves is that they can activate Ca^{2+}-dependent inward currents. Before discussing these currents, we will briefly review the mechanisms that produce the Ca^{2+}-overloaded situation.

3.2.1
Mechanisms of Calcium Overload

Perhaps the simplest way in which to study Ca^{2+} overload is by using electrically quiescent cells. Under normal conditions these have a constant resting $[Ca^{2+}]_i$ of the order of 100 nM. However, if Ca^{2+} influx into the cell is increased (by raising external Ca^{2+}) or efflux is decreased (by reducing the transmembrane Na^+ gradient) then spontaneous waves of Ca^{2+} release from the SR result. Simultaneous measurements of SR content show that as Ca^{2+} influx is increased there is, at first, an increase of SR content. As Ca^{2+} influx is further increased, spontaneous Ca^{2+} release develops. From this point there is no further increase of SR content, rather increasing Ca^{2+} influx results in an increased frequency of Ca^{2+} release (Díaz et al. 1997). These observations are consistent with the idea that Ca^{2+} release from the SR occurs when the SR content has reached a certain critical threshold level. Measurements of Ca^{2+} sparks have shown that Ca^{2+} waves are preceded by an increased frequency of these sparks (Cheng et al. 1996).

There are two possible causes of this apparent SR Ca content threshold for Ca^{2+} waves. (1) It could represent a threshold for wave *initiation*. As SR Ca^{2+} is elevated, the frequency and amplitude of Ca^{2+} sparks will increase and waves may be more likely to be initiated. (2) It could reflect a threshold for wave *propagation*. This requires that sufficient Ca^{2+} is released from one release site such that when it has diffused to the next release site then, even allowing for Ca^{2+} reuptake into the SR and pumping out of the cell, there is a sufficiently large trigger for Ca^{2+} release to ensure that the wave can continue to propagate. It should also be noted that some work has suggested that it is not SR content that determines whether or not a wave occurs but, rather cytoplasmic Ca^{2+} (Edgell et al. 2000).

Most of the work referred to above relates to experiments performed in quiescent cells, and the situation is more complicated when the heart is stimulated. Under these conditions, the normal systolic Ca^{2+} transient is followed by a delayed release which can activate an aftercontraction. As the degree of Ca^{2+} overload increases, the interval between the systolic and subsequent

Ca^{2+} releases decreases (Kass et al. 1978). This is important, inasmuch as if the subsequent release follows with an interval which is greater than the interval between heartbeats, it will not be seen.

3.2.2
Delayed Afterdepolarizations

The Ca^{2+} waves activate inward currents (Kass et al. 1978; Lederer and Tsien 1976) that can produce arrhythmogenic delayed afterdepolarizations (DADs). These DADs were first identified under conditions of calcium overload produced by digitalis intoxication when it was shown that the appearance of DADs correlated with arrhythmogenic changes in the ECG (Ferrier et al. 1973; Rosen et al. 1973). Subsequent work has also shown that DADs are associated with arrhythmias produced by ischaemia and reperfusion (reviewed in Carmeliet 1999).

Repetitive stimulation increases the amplitude of the DAD (Ferrier et al. 1973). When the threshold for activation of the Na^+ current is reached, an action potential is generated. This mechanism therefore has the right properties to account for the focal nature of some triggered tachyarrhythmias.

The nature of the Ca-activated current activated by Ca^{2+} waves has been investigated in several studies. The consensus is that the vast majority of this current results from activation of the NCX (Fedida et al. 1987; Mechmann and Pott 1986).

3.2.3
A Dual Role for NCX in Arrhythmogenesis

From the above it is clear that NCX plays two roles in the generation of DAD-dependent arrhythmias.

- Changes of NCX activity will affect the Ca^{2+} balance of the cell. Therefore, anything which decreases Ca^{2+} efflux on the exchanger or increases Ca^{2+} influx will make Ca^{2+} overload more likely.

- NCX actually carries the arrhythmogenic current, and therefore the greater the activity of NCX the larger the current that will be activated by a given Ca^{2+} wave.

In vitro Ca^{2+} overload and DADs can be induced in various ways: increasing external Ca^{2+}, reducing external Na^+, increasing intracellular Na^+ by inhibiting the Na/K ATPase or by enhancing Na^+ influx via the Na^+ channel, or increasing Ca^{2+} load by adrenergic stimulation. In vivo and from a clinical point of view, the conditions most commonly associated with DAD-dependent arrhythmias are increased intracellular Na^+, i.e. during digitalis intoxication (fortunately rare), during ischaemia/reperfusion, possibly in cardiac hypertrophy and failure (see Sect. 4.2), and conditions of excessive adrenergic stimulation. In most of these conditions, the Na/Ca exchanger indeed has the dual role described. With Ca^{2+} overload related to increased $[Na^+]_i$, action potentials are usually short as observed experimentally and during theoretical modelling (Armoundas et al. 2003; Faber and Rudy 2000; Mubagwa et al. 1997). This enhances the potential for DAD formation, as the diastolic interval is long. An illustration of a triggered action potential during spontaneous Ca^{2+} release with high $[Na^+]_i$ is shown in Fig. 6b.

3.3
Na/Ca Exchange and Early Afterdepolarizations

Less well studied is the role of inward NCX current in early afterdepolarizations (EADs), i.e. those occurring on the action potential plateau or during early repolarization (Volders et al. 2000). EADs are most often observed at low frequencies of stimulation and in the presence of action potential prolongation (Fig. 6a). Experimental studies have implicated reactivation of Ca^{2+} channels and window currents in the upstroke of the EAD (Hirano et al. 1992; January and Riddle 1989), as well as Na^+ window currents (Boutjdir et al. 1994). This depolarization occurs on top of a delayed repolarization that provides the conditioning phase. This conditioning phase can result from reduced repolarizing currents, such as decreased K^+ currents. This is typically illustrated by drug-induced or acquired long QT syndrome (LQTS), which is associated with polymorphic ventricular tachycardia (PVT). Experimentally EADs can be observed in isolated myocytes or multicellular preparations (e.g. Antzelevitch et al. 1996). Alternatively, the conditioning phase is predominantly due to increased depolarizing current such as Na^+ current (e.g. el Sherif et al. 1988), but also inward NCX current. Particularly in the presence of adrenergic stimulation, experimental evidence indicates that the exchanger is of major importance (Volders et al. 1997). Under these conditions, EADs can occur over a wide range of take-off potentials, and the mechanisms for DAD and EAD formation overlap (Fig. 6c).

4
Evidence for the Role of Na/Ca Exchange in Generating and/or Sustaining Arrhythmias

The level of evidence for the implication of NCX in arrhythmias varies from 'highly likely' to 'possible'. While in experimental conditions some evidence that is more direct can be provided, most of it remains indirect. The best case can be made for ventricular arrhythmias induced on reperfusion after an ischaemic period. During ischaemia the cells depolarize, acidify and accumulate Na^+ due to loss of the Na/K ATPase function (see Carmeliet 1999 for a review on ionic changes in ischaemia). On reperfusion, the mechanisms regulating pH, such as the Na/H exchanger, lead to a large influx of Na^+, which in turn leads to

Ca^{2+} overload via the Na/Ca exchanger. $[Ca^{2+}]_i$ oscillations have been documented in isolated tissues and NCX current oscillations in myocytes (Benndorf et al. 1991; Cordeiro et al. 1994). Arrhythmias are often focal (Pogwizd 2003). The interest in preventing Na^+ and Ca^{2+} overload in ischaemia and reperfusion injury is only partially driven by the anti-arrhythmic effects—as many of these arrhythmias are transient—and is largely related to potential salvage of myocardium from necrosis. The protection offered by Na/H inhibitors was reviewed in Avkiran and Marber (2002), but a recent clinical trial was not as successful as expected. The experience with NCX inhibition is reviewed Sect. 6.

In the following section we review in more detail the evidence for a role for NCX in generating or sustaining arrhythmias in a number of chronic diseases.

4.1
Na/Ca Exchange and Atrial Fibrillation

In chronic atrial fibrillation, structural remodelling and electrical remodelling of the atrial myocytes conspire to create an arrhythmogenic substrate (reviewed in Allessie et al. 2001, 2002; Nattel and Li 2000). NCX current is enhanced in atrial fibrillation secondary to heart failure (Li et al. 2000), and was shown to contribute to maintaining the action potential plateau and duration in human atrial myocytes (Benardeau et al. 1996). However, in chronic atrial fibrillation, typically a shorter action potential is found, primarily related to loss of Ca^{2+} current. Together with the slowing of conduction this favours re-entry. The actual mechanisms of atrial fibrillation generation, re-entry versus focal, remain a matter of debate. A mechanism proposed more recently gives a more prominent role to the Na/Ca exchanger. Ectopic activity originating in the muscular sleeves of the pulmonary veins could be an important triggering mechanism of the atrial tachyarrhythmia (Chen et al. 2001, 2002; Chen et al. 2003). Myocytes isolated from this area have indeed specific membrane properties (Ehrlich et al. 2003) and have been reported to be more prone to DADs than the typical atrial myocyte (Chen et al. 2002).

A last element in the pathogenesis of atrial fibrillation is the Ca^{2+} overload as a signal for remodelling. This has been proposed from in vitro studies of rapid stimulation of atrial myocytes (Sun et al. 1998). In the experimental situation, reducing Ca^{2+} influx with Ca^{2+} channel blockers favourably influenced remodelling (Tieleman et al. 1997). A clinical study using verapamil, however, has not confirmed this approach (Van Noord et al. 2001).

4.2
Na/Ca Exchange and Arrhythmias in Cardiac Hypertrophy and Heart Failure

During cardiac remodelling in response to increased load or consequent to myocardial infarction and ischaemic heart disease, many changes occur that will affect the contribution of the Na/Ca exchanger to electrical activity and potential arrhythmogenesis. The most immediately relevant changes are altered expression of the Na/Ca exchanger and modulation of exchanger activity due to altered $[Ca^{2+}]_i$ and $[Na^+]_i$. As recently reviewed, such changes are variable with the degree of hypertrophy, stage of decompensation and the stimulus leading to remodelling (Schillinger et al. 2003; Sipido et al. 2002; Verdonck et al. 2003b). This cautions against generalizing scenarios on the Na/Ca exchanger in hypertrophy and heart failure. Nevertheless, from data on some of the best-studied examples of remodelling, some important aspects can be highlighted.

4.2.1
Increased Ca^{2+} Removal via NCX in End-Stage Heart Failure

In myocytes from human hearts obtained at the time of transplantation, $[Ca^{2+}]_i$ transients are prolonged and Ca^{2+} uptake into the SR is reduced (Beuckelmann et al. 1992; Hasenfuss and Pieske 2002; Piacentino, III et al. 2003). This promotes removal of Ca^{2+} via the Na/Ca exchanger. Although the amount of charge extruded by the exchanger in steady state is determined by the amount of Ca^{2+} entry, not by the expression/activity levels of NCX (Bridge et al. 1990; Negretti et al. 1995), the latter will determine the kinetics of the current and time course of Ca^{2+} removal. Therefore, an increased inward current at the end of the action potential can be part of the prolongation of the action potential typically seen in human heart failure. Increased Ca^{2+} removal through NCX has also been observed in some well-characterized animal models of heart failure. In the dog with tachycardia-induced heart failure, the increased Ca^{2+} removal is the result of reduced SR Ca^{2+} uptake and increased activity and expression of the exchanger (Hobai and O'Rourke 2000; O'Rourke et al. 1999). Modelling of the electrical activity in this dog model, however, did not clearly indicate that this will contribute to action potential prolongation (Winslow et al. 1999). The rabbit with heart failure due to combined pressure and volume overload also has increased Ca^{2+} removal by the Na/Ca exchanger (Pogwizd et al. 1999). Enhanced expression and function result in larger inward and outward exchanger currents, and this is also seen in theoretical models of the NCX current during the action potential as illustrated in Fig. 7 (Pogwizd et al. 2003). The net effect on the action potential profile is therefore at present uncertain. What is clear, however, is that in this model the increased inward current is particularly important in the enhanced propensity for arrhythmias triggered by DADs (Pogwizd et al. 2001). For any given release of Ca^{2+} from the SR, the NCX current is larger in myocytes from the failing hearts. In combination with a reduced density of the repolarizing inward rectifier current, the probability that spontaneous Ca^{2+} release will effectively elicit an action potential is greatly enhanced.

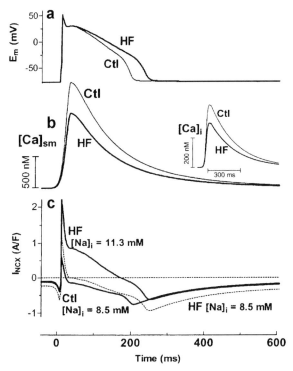

Fig. 7a–c Alterations in the amplitude and time course of the Na/Ca exchange current shown in panel **c**, in a rabbit model of heart failure (HF) with increased intracellular Na$^+$, as calculated from the measured changes in action potential (**a**), global Ca^{2+} transient (**b**) and calculated Ca^{2+} near the membrane (**b**, *inset*). Whereas the higher NCX expression simply increases the amplitude of the current (if Na$^+$ remains at 8.5 mM), the increase in Na$^+$ results in a large increase in the outward current component. (After Pogwizd et al. 2003)

4.2.2
Increased Ca^{2+} Influx via NCX in Heart Failure and Hypertrophy

The same rabbit model just mentioned was also studied extensively by Fiolet and co-workers. They found that $[Na^+]_i$ is increased though increased activity of the Na/H exchanger (Baartscheer et al. 2003b,c). The expected higher Ca^{2+} influx via the exchanger could explain the higher diastolic Ca^{2+} values that were observed and the larger fractional Ca^{2+} release, despite a lower SR Ca^{2+} content. Similar to the Pogwizd et al. studies, Baartscheer et al. reported a higher incidence of DADs, in particular under adrenergic stimulation. Inhibition of the Na/H exchanger reversed these changes, indicating that the primary event is the increase in Na$^+$ influx (Baartscheer et al. 2003c).

$[Na^+]_i$ is also increased in human heart failure (Pieske et al. 2002) and increased Ca^{2+} influx via the exchanger could contribute to the Ca^{2+} available for contraction (Dipla et al. 1999; Piacentino et al. 2003; Weber et al. 2002).

In heart failure in humans and the animal models mentioned above, the amplitude of the $[Ca^{2+}]_i$ transient is actually reduced at normal heart rates, as is the Ca^{2+} loading of the SR. This is not necessarily so in compensated hypertrophy (Shorofsky et al. 1999). This is exemplified by the dog with chronic atrioventricular block (AVB). Bradycardia and volume overload lead to biventricular hypertrophy with preserved and even enhanced contractile function, at least in the first 6 weeks (Vos et al. 1998). At the cellular level, NCX activity is increased and $[Na^+]_i$ is higher (Sipido et al. 2000; Verdonck et al. 2003a). Because there is no apparent reduction in SERCA function, these changes result in the SR Ca^{2+} content being larger, and spontaneous Ca^{2+} release is more likely to occur, accompanied by large inward exchanger currents (Sipido et al. 2000).

In the context of a balance of Ca^{2+} fluxes during the cardiac cycle, any additional Ca^{2+} influx during the action potential must result in increased exchanger efflux at a later time during the cycle. So even if functional emphasis is placed on increased Ca^{2+} influx due to higher $[Na^+]_i$, this will always be accompanied by enhanced efflux and inward exchanger currents that can contribute to arrhythmogenesis.

4.2.3
Incidence of Afterdepolarizations in Cardiac Hypertrophy and Failure

The occurrence of DADs has been documented in the above-mentioned animal models and in human heart failure. In compensated hypertrophy, this follows rather straightforwardly from an enhanced Ca^{2+} loading of the myocytes. In vivo, DADs can be elicited by pacing protocols that enhance contractility, further exacerbated by increased Na^+ loading under ouabain (de Groot et al. 2000; Fig. 8). In the case of heart failure with reduced SR Ca^{2+} loading, DADS and triggered action potentials still occur because of a lowered threshold with reduced inward rectifierents (Pogwizd et al. 2001; Pogwizd 2003), and/or because diastolic Ca^{2+} is elevated (Baartscheer et al. 2003a). Another important element is adrenergic stimulation, which may induce DADs with high incidence in myocytes from the failing heart, including human (Baartscheer et al. 2003a; Pogwizd et al. 2001; Verkerk et al. 2001). In addition, Purkinje fibres may be more sensitive to the development of DADs (Boyden et al. 2000), and a high incidence has been reported in the dog after myocardial infarction (Boutjdir et al. 1990).

As mentioned above, the role of the NCX current in EADs is in providing inward current during the priming phase, with a more prominent role for the exchanger as depolarizing current during adrenergic stimulation. There are currently few experimental data that have directly linked exchanger current to EADs in hypertrophy and failure. We have proposed such a role in the dog with chronic AVB (Sipido et al. 2000; Volders et al. 1998), and it could be hypothesized for the EADs observed in human failing myocytes under adrenergic stimulation (Veldkamp et al. 2001).

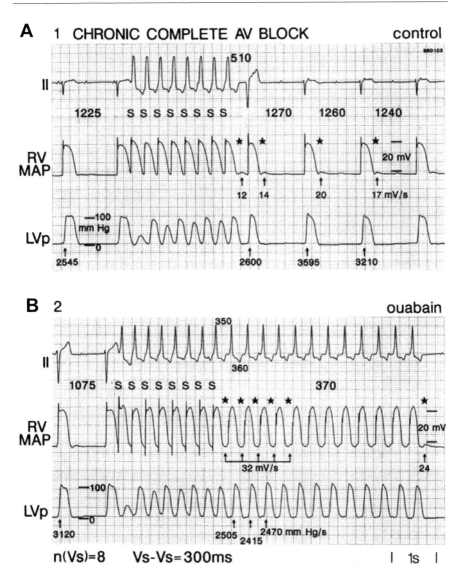

Fig. 8a,b Arrhythmias related to DADs in the dog with chronic AV block. **a** The dog with chronic AV block has increased loading of the SR (Sipido et al. 2000) and a train of 8 stimuli (S) results in an increase in DAD slope (*arrow*) and one DAD-related triggered beat (*) recorded here in vivo with a MAP catheter (de Groot et al. 2000). **b** In these dogs, $[Na^+]_i$ in the isolated myocytes is enhanced (Verdonck et al. 2003a) and in vivo the heart is more sensitive to ouabain. In the presence of ouabain, +LV dP/dt is higher, and pacing now results in ventricular tachycardia (12 beats). In MAP, DADs are visible during VT and directly after tachycardia terminates. (Reproduced from de Groot et al. 2000)

A last element is heterogeneity in exchanger current density between regions in the heart, which is already present at baseline (Zygmunt et al. 2000) and could be enhanced during remodelling (Sipido et al. 2000; Yoshiyama et al. 1997) thus contributing to dispersion of repolarization.

4.2.4
In Vivo Evidence for Na/Ca Exchange-Mediated Arrhythmias in Heart Failure

Sudden death and potentially lethal arrhythmias have been documented during the many recent ICD studies in heart failure. Ventricular tachycardia is often initiated by ectopic activity, but these types of ECG recordings do not allow us to distinguish between (micro) re-entry or abnormal focal activity underlying the extrasystoles. Yet there is evidence that triggered activity is an important mechanism underlying arrhythmias in patients with heart failure (Paulus et al. 1992; Pogwizd et al. 1992, 1995; and see review, Janse 2004). Though there is no direct evidence for the Na/Ca exchanger, in view of the discussions above, the role of the Na/Ca exchanger in afterdepolarizations is clear.

4.3
Na/Ca Exchange and Congenital Arrhythmias

Currently there are no known direct associations between mutations of the Na/Ca exchanger and arrhythmic disease. There are, however, indirect links to other congenital syndromes leading to PVT and associated with EADs and DADs. In congenital LQTS type 3, a late depolarizing Na^+ current contributes to action potential prolongation and provides the conditioning phase for EADs (Bennett et al. 1995; Clancy and Rudy 1999; Nuyens et al. 2001). In the D1790G mutation LQTS3, theoretical modelling indicates an important role for increased inward NCX current consequent on an increase in intracellular Ca^{2+} (Wehrens et al. 2000). In LQTS4, a mutation in ankyrin results in defective targeting of the Na/Ca exchanger to the T-tubular membrane and abnormal Ca^{2+} cycling (Mohler et al. 2003). In LQTS1, arrhythmias are occurring preferably under adrenergic stimulation and the Na/Ca exchanger could thus be indirectly involved. Adrenergic stimulation is also the trigger for often-lethal arrhythmias in the catecholaminergic polymorphic ventricular tachycardia (CPVT) syndrome that is associated with mutations in Ca^{2+}-handling proteins such as the ryanodine receptor (Laitinen et al. 2001; Priori et al. 2001) and calsequestrin (Lahat et al. 2001).

5
What Are the Expected Consequences of Na/Ca Exchange Block on Ca Handling?

5.1
General Considerations

A simple analysis suggests that inhibitors of NCX will have two classes of effects on the Ca-dependent arrhythmias discussed above. There will be effects due to (1) changes of the degree of Ca^{2+} (over)load of the cell and (2) changes of the membrane current and resulting arrhythmias produced by a given degree of Ca^{2+} overload. We will now consider these two effects.

Under normal conditions the NCX produces a net Ca^{2+} efflux. Therefore, partial inhibition of the exchanger produced either by stopping a fraction of the NCX completely or stopping all of the NCX partially would be expected to increase systolic $[Ca^{2+}]_i$ until the increased $[Ca^{2+}]_i$ compensates for the decrease of NCX sites and results in the same time-averaged Ca^{2+} efflux as in control. This effect by itself will be positively inotropic. However, the tendency to load the cell with calcium may also result in a state of Ca^{2+} overload and thence arrhythmias. On the other hand, the fact that NCX has been inhibited means that each Ca^{2+} wave will result in less Ca^{2+} efflux from the cell and thus in a smaller arrhythmogenic inward current. This would suggest that a given degree of inotropy produced by NCX inhibition will be accompanied by less arrhythmogenic problems than will be the result of producing the inotropy by other means of loading the cell with calcium. In heart failure with a low level of SR Ca^{2+} loading, inhibition of the Na/Ca exchanger may thus be an option. Indeed, in myocytes from the dog with tachycardia-induced cardiomyopathy, partial inhibition of the Na/Ca exchanger resulted in a positive inotropic effect, and afterdepolarizations were not observed (Hobai et al. 2004).

A different situation may exist when Ca^{2+} overload occurs by primary increase in Ca^{2+} influx via other pathways, as can occur during adrenergic stimulation. Inhibition of the Ca^{2+} removal pathway may then result in an unacceptable further increase of cellular Ca^{2+}.

5.2
Unidirectional Block of NCX

It has been reported that the drug KB-R7943 can inhibit reverse-mode NCX more effectively than forward mode (Iwamoto et al. 1996). This would be expected to decrease Ca^{2+} influx through the exchanger and this might be beneficial, e.g. during reperfusion. The relative lack of effect on forward-mode exchange would allow the exchanger to continue to pump Ca^{2+} out of the cell and thereby decrease the degree of Ca^{2+} overload. However, since the arrhythmias are produced by the forward mode of the exchange, any remaining overload would still result in arrhythmias. It is necessary, however, to be cautious about the results obtained with these drugs. This is because the reversal potential of NCX is determined by the transmembrane Na^+ and Ca^{2+} gradients. Therefore, at least near equilibrium, it is impossible to inhibit one direction of the pump more than the other, as this would change the direction of the net flux and thereby change the reversal potential.

If there is preferential drug binding in the presence of high internal $[Na^+]$, as suggested for SEA0400 in heterologous expression systems (Lee et al. 2004), such 'selectivity' might be an advantage.

6
Current Experience with NCX Blockers

NCX blockers as antiarrhythmic agents have been tested primarily in conditions where the arrhythmogenic mechanism was thought to be related to Ca^{2+} overload either related to the exchanger itself, as in the case of Na^+ overload, or through other channels, such as the repeated activation of Ca^{2+} channels in atrial fibrillation. One can examine these data in two ways: first, simply as an evaluation of the efficiency of a given compound; second, as a test for clarifying the contribution of NCX to the arrhythmias. Given the fact that, so far, the specificity of the compounds is rather poor, the first objective can be addressed, but the result of the second is rather uncertain.

In this part we will review the properties and selectivity of the available compounds and the results obtained.

6.1
'First Generation' of NCX Blockers

A large number of agents were initially used to inhibit NCX, but these had low potency and completely lacked selectivity. Amiloride (Siegl et al. 1984), a widely used diuretic, its derivatives (4,5)3′,4′-dichlorobenzamil (DCB) and 2′,4′-dimethylbenzamil (DMB), and an antiarrhythmic drug, bepridil, with strong Na^+ and Ca^{2+} channel-blocking properties, are typical examples (Kaczorowski et al. 1989). These agents were reported to be effective in different experimental arrhythmia models, but their low potency and lack of selectivity made them unsuitable to use as tools to test the role of NCX in arrhythmogenesis.

6.2
NCX Inhibitory Peptide

NCX inhibitory peptide (XIP) was developed (Chin et al. 1993; Li et al. 1991) based on the structure of NCX. It is a useful tool in patch-clamp experiments, where it can be applied through the patch pipette with IC_{50} of the submicromolar range. It is an excellent tool for evaluation of NCX function. However, since it does not cross the plasma membrane, it cannot be used in in vivo studies and has little potential therapeutic value in patients.

Hobai et al. recently reported that dialysis of myocytes from failing hearts increased SR Ca^{2+} loading (Hobai et al. 2004; Hobai and O'Rourke 2004), which is consistent with the inotropic effect of NCX inhibition postulated above. XIP was also used by these authors to probe the NCX current during the action potential (see Sect. 3.1).

6.3
KB-R7943

Recently, better drugs have been developed, even if still not optimally specific (Fig. 9). KB-R7943, an isothiourea derivate (Watano et al. 1996), has been reported as an effective blocker of NCX. In some studies it was described as a more potent reverse (Iwamoto et al. 1996) than forward mode inhibitor, but in other studies it was shown that both the forward and reverse mode of NCX was equally affected by the compound (Kimura et al. 1999; Tanaka et al. 2002). There is also great inconsistency of the reported IC_{50} values for NCX block by KB-R7943, ranging from 0.3 µM to 9.5 µM (Iwamoto et al. 1996; Takahashi et al. 2003; Tanaka et al. 2002; Watano et al. 1996). These differences probably reflect difficulties and variety in the methodology of measuring NCX in the cardiac muscle. Also, species differences may have importance, since rat has different Ca^{2+} handling and higher intracellular Na^+ concentration than other mammals.

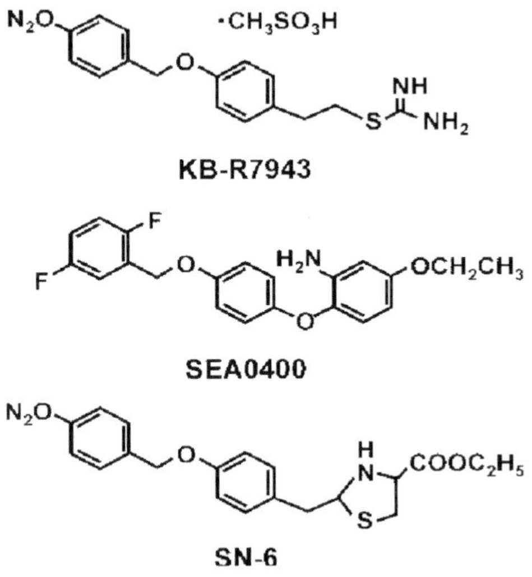

Fig. 9 Chemical structure of some newer NCX inhibitors

The major problem with KB-R7943 as a useful pharmacological tool seems its poor selectivity. KB-R7943 has been found to inhibit fast Na^+, L-type Ca^{2+}, inward rectifier and delayed K^+ currents in the low micromolar range (Fig. 10; Tanaka et al. 2002).

Despite this, KB-R7943 has been used in a number of cellular, multicellular and in vivo studies. In the in vitro studies, it was found to reduce Ca^{2+} overload induced by glycosides (Satoh et al. 2000) and during ischaemia/reperfusion (Baczko et al. 2003; Ladilov et al. 1999), as illustrated in Fig. 11. In the latter condition, cardioprotective and antiarrhythmic effects could be documented in multicellular or intact heart preparations (Mukai et al. 2000; Nakamura et al. 1998; Satoh et al. 2003). Further antiarrhythmic effects were shown in Na^+ overload induced by glycosides or Na-channel openers (Amran et al. 2004; Satoh et al. 2003). Administered in vivo, KB-R7943 also prevented the development of atrial fibrillation evoked by rapid pacing (Miyata et al. 2002), but failed to reduce arrhythmic death in an in vivo ischaemia study (Miyamoto et al. 2002). While the multiple effects on other transmembrane ion channels mentioned previously make it hard to use these results as indications for the role of NCX in the pathophysiology, useful information comes from it. First, the additional block of Na^+ and Ca^{2+} current may actually be an advantage for ischaemia-reperfusion related arrhythmias by decreasing the danger of

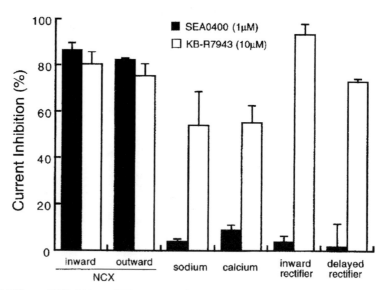

Fig. 10 Effects of SEA-0400 and KB-R7943 on ionic currents in isolated guinea-pig ventricular myocytes. The inward and outward NCX currents are measured at −80 mV and +30 mV, respectively, Na^+ current at −20 mV, L-type Ca^{2+} current at +10 mV, inward rectifying K^+ current at −60 mV and delayed rectifier K^+ current at +50 mV. Inhibition of current amplitudes in the presence of SEA-0400 (1 μM) and KB-R7943 (10 μM) is expressed as a percentage of the values in the absence of compounds. (Reproduced from Tanaka et al. 2002)

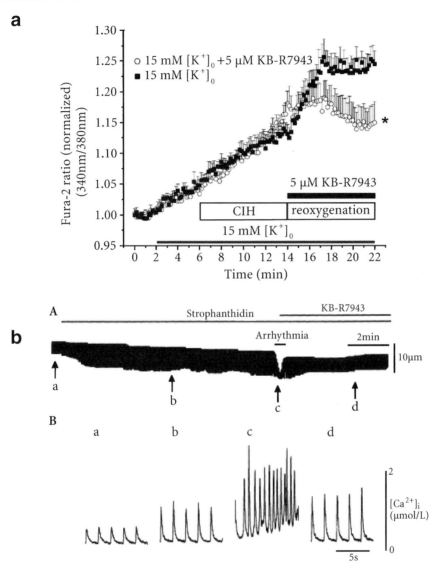

Fig. 11a,b Protection against Ca^{2+} overload by KB-R7943. **a** Chemically induced hypoxia and reoxygenation-induced Ca^{2+} overload in rat ventricular myocytes. $[Ca^{2+}]_i$ measurements in 15 mM $[K^+]_o$ (1) under baseline conditions, and (2) during 8 min chemically induced hypoxia, (3) followed by 8 min reoxygenation. KB-R7943 (5 µM), was applied during reoxygenation. $*p < 0.05$ vs control (15 mM $[K^+]_o$), $n = 6$ experiments, 20 and 29 cells. Reproduced from Baczko et al. (2003). **b** Block of strophanthidin-induced arrhythmia but not inotropy. *A*, Continuous recording of twitch cell shortening in a rat ventricular myocyte superfused with 50 µmol/L strophanthidin and 5 µmol/L KB-R7943 added as indicated by bars. *B*, Ca^{2+} transients recorded during control perfusion (*a*), 5 min after starting strophanthidin perfusion (*b*), when arrhythmia appeared (11 min, indicated by bar; *c*), and 3 min after addition of KB-R7943 (*d*). (Reproduced from Satoh et al. 2000)

possible Ca^{2+} overload due to diminished Ca^{2+} efflux caused by the forward NCX inhibition. This property of KB-R7943 may also have been important in preventing atrial fibrillation in the dog (Miyata et al. 2002). Second, on the down-side, the K^+ channel block by KB-R7943 (Tanaka et al. 2002) can further decrease the repolarization reserve in heart failure where several potassium channels like I_{to}, I_{K1} and I_{Ks} are downregulated. This latter may prolong action potential duration and enhance dispersion of repolarization, thereby increasing the risk of proarrhythmia in heart failure. In atrial fibrillation on the other hand, prolongation of the action potential would rather be an advantage.

6.4
SEA-0400

SEA-0400 is the most potent and selective inhibitor of NCX which is available and reported so far. Tanaka et al. (2002) showed in guinea-pig ventricular myocytes, by the patch-clamp technique, that SEA-0400 equally inhibited NCX in the forward and reverse mode with an IC_{50} value of 40 nM and 32 nM, respectively. When the same authors studied the effect of SEA-0400 on the fast Na^+, L-type Ca^{2+}, inward rectifier K^+ and delayed rectifier K^+ currents, it was found that even at 1 µM the compound affected these currents less than 10% (Fig. 10). Somewhat different results were obtained by Lee et al. (2004) who found a high potency (IC_{50} of 23–78 nM) of SEA-0400 in the reverse, but far less in the forward, mode. Also in this study, it was shown that SEA-0400 altered outward NCX peak current recovery, and the effect of the compound was strongly dependent on intracellular Na^+ and Ca^{2+} concentrations. The authors concluded that SEA-0400 acts by favouring Na^+-dependent inactivation of the NCX current. A similar conclusion was reported by Bouchard et al. (2004) and Iwamoto et al. (2004b) based on mutant NCX1 measurements. The discrepancy between the results of these latter studies (Bouchard et al. 2004; Iwamoto et al. 2004b; Lee et al. 2004) and that of Tanaka et al. (2002) are not clear, but may relate to the marked differences between the experimental conditions and preparations applied. Tanaka et al. (2002) used guinea-pig native ventricular myocytes at 35°C–36°C with more physiological pipette and extracellular solutions while Lee et al. (2004) carried out measurement with the giant excised patch technique in *Xenopus* laevis oocytes in which NCX1.1 was expressed. Also, in the latter study the intracellular Na^+ and Ca^{2+} was high, 100 mM and 3–10 µM, and the temperature was 30°C. The selectivity of SEA-0400 was questioned by Reuter et al. (2002) based on NCX knockout mice experiments where intracellular Ca^{2+} transient was markedly reduced by 1 µM SEA-0400. However, in native freshly isolated dog ventricular myocytes we did not observe a significant effect of 1 µM SEA-0400 on the intracellular Ca^{2+} transient measured by the fura-2 ratiometric technique (Nagy et al. 2004).

In spite of the mentioned inconsistencies concerning the published results to date, it seems that SEA-0400 is a better tool than KB-R7843 to investigate

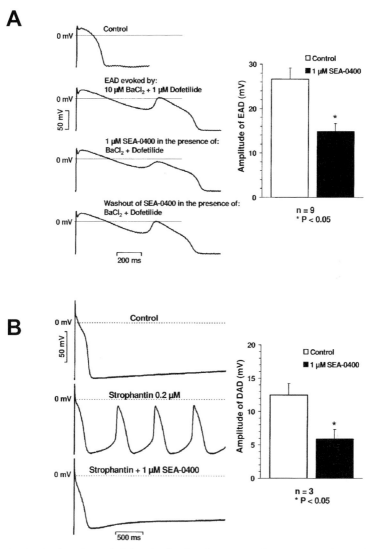

Fig. 12a,b Block of early and delayed afterdepolarizations by SEA-0400 in canine myocardium. **a** The effect of 1 μM SEA-0400 on EADs in right ventricular papillary muscles, stimulated at slow cycle lengths (1,500–3,000 ms) in the presence of 1 μM dofetilide plus 10 μM BaCl₂. On the *left*, the results of a representative experiment are shown, on the *right*, the average values of the amplitude of EADs are presented before (*open bars*) and after (*filled bars*) the administration of SEA-0400. **b** The effect of SEA-0400 on the delayed afterdepolarization (DAD) in canine cardiac Purkinje fibres, superfused with 0.2 μM strophantin. A train of 40 stimuli was applied at a cycle length of 400 ms, followed by a 20-s long stimulation-free period that generated DADs. On the *left*, results of a representative experiment are shown, on the *right*, average values of the amplitude of DADs are given before (*open bars*) and after (*filled bars*) the application of 1 μM SEA-0400. (Reproduced from Nagy et al. 2004)

the role of NCX inhibition in cardioprotection and arrhythmogenesis. SEA-0400 was reported to exert cardioprotective (Takahashi et al. 2003; Yoshiyama et al. 2004) and antiarrhythmic effects (Yoshiyama et al. 2004) after coronary ligation and reperfusion experiments, both in Langendorff perfused isolated rabbit heart and in in vivo rat experiments. These effects were explained by the inhibition of the reverse mode of the NCX by SEA-0400.

Recently we showed (Nagy et al. 2004) that the amplitude of DAD and EAD was significantly decreased by 1 µM SEA-0400 in dog Purkinje fibres and right ventricular papillary muscle, respectively (Fig. 12). Based on these results, we—like Pogwizd earlier (Pogwizd 2003)—also speculated that the inhibition of the forward mode of the NCX could also represent an antiarrhythmic mechanism, especially in certain situations, like in heart failure where K$^+$ channels are downregulated and NCX is upregulated. Also, this speculation can be extended to the atria and pulmonary vein, since in these preparations DADs and EADs were observed (Chen et al. 2000, 2001, 2003) and implicated in the mechanism of atrial fibrillation.

6.5
New and Other NCX Inhibitors

Two other compounds, CGP-37157, a mitochondrial NCX inhibitor (Cox et al. 1993), and a new compound, SN-6 (Iwamoto et al. 2004a), were shown to affect sarcolemmal NCX in the micromolar range (Omelchenko et al. 2003), but no data are available about their selectivity and antiarrhythmic activity.

Antisense approaches have been used to inhibit NCX in celluar experiments and may potentially be further developed (Lipp et al. 1995; Eigel & Hadley, 2001).

7
Conclusions and Perspectives

Based on our knowledge of mechanisms of arrhythmogenesis and the expected effects of NCX inhibition described above, one can theoretically and tentatively identify a number of situations that would constitute an indication for NCX inhibition (Table 1).

First, one can expect a potential benefit of selectively reducing the global NCX current (but never full inhibition) in a limited number of conditions, namely when the risk for Ca^{2+} overload is not very high. This could theoretically be postulated for heart failure, in particular under β-blockade as protection against Ca^{2+} overload.

Second, if a selective inhibition of reverse mode is possible, then conditions of Ca^{2+} overload that are related to Na$^+$ overload and Ca^{2+} influx via reverse mode NCX, would be a prime indication, such as ischaemia/reperfusion.

Table 1 Potential indications for selective or non-selective NCX inhibition and putative effects on Ca^{2+} homeostasis and action potential time course

	'Heart failure'	Na-dependent overload	Non-Na dependent Ca overload, e.g. adrenergic
Potential targets			
Action potential	Prolonged	Shortened	Slightly prolonged
Afterdepolarizations	EAD, DAD	DAD	DAD, 'early' DAD
SR Ca content	Decreased or normal	Increased	Increased
Type of invention			
Selective partial but bidirectional NCX inhibition	Could reduce DAD and have positive motropic effect	Would increase Ca overload	Would increase Ca overload
Selective unidirectional block of reverse mode	Could further reduce SR content	Would reduce Ca overload	Uncertain
NCX inhibition with combined L-type Ca channel inhibition	Uncertain	Could reduce Ca overload but further shorten action potential	Would reduce Ca overload and shorten action potential

Third, a less selective drug, that would also block Ca^{2+} channels and thereby further reduce Ca^{2+} influx, may actually be a better choice for conditions with a high risk of Ca^{2+} overload, perhaps as under adrenergic stimulation or in compensated hypertrophy.

To establish and verify these propositions, we need drugs with different profiles that are well-characterized at the cellular and molecular level. The data from the cell lab must be integrated with in vivo studies in relevant (large) animal models covering a spectrum of disease. Considering the potential of such drugs certainly encourages further research and investment in developing selective, as well as less selective, Na/Ca exchange inhibitors.

References

Allen DG, Eisner DA, Lab MJ, Orchard CH (1983) The effects of low sodium solutions on intracellular calcium concentration and tension in ferret ventricular muscle. J Physiol (Lond) 345:391–407

Allessie M, Ausma J, Schotten U (2002) Electrical, contractile and structural remodeling during atrial fibrillation. Cardiovasc Res 54:230–246

Allessie MA, Boyden PA, Camm AJ, Kleber AG, Lab MJ, Legato MJ, Rosen MR, Schwartz PJ, Spooner PM, Van Wagoner DR, Waldo AL (2001) Pathophysiology and prevention of atrial fibrillation. Circulation 103:769–777

Amran MS, Hashimoto K, Homma N (2004) Effects of sodium-calcium exchange inhibitors, KB-R7943 and SEA0400, on aconitine-induced arrhythmias in guinea pigs in vivo, in vitro, and in computer simulation studies. J Pharmacol Exp Ther 310:83–89

Antzelevitch C, Sun ZQ, Zhang ZQ, Yan GX (1996) Cellular and ionic mechanisms underlying erythromycin-induced long QT intervals and torsade de pointes. J Am Coll Cardiol 28:1836–1848

Armoundas AA, Hobai IA, Tomaselli GF, Winslow RL, O'Rourke B (2003) Role of sodium-calcium exchanger in modulating the action potential of ventricular myocytes from normal and failing hearts. Circ Res 93:46–53

Avkiran M, Marber MS (2002) $Na^{(+)}/H^{(+)}$ exchange inhibitors for cardioprotective therapy: progress, problems and prospects. J Am Coll Cardiol 39:747–753

Baartscheer A, Schumacher CA, Belterman CN, Coronel R, Fiolet JW (2003a) SR calcium handling and calcium after-transients in a rabbit model of heart failure. Cardiovasc Res 58:99–108

Baartscheer A, Schumacher CA, Belterman CN, Coronel R, Fiolet JW (2003b) $[Na^+]i$ and the driving force of the Na^+/Ca^{2+}-exchanger in heart failure. Cardiovasc Res 57:986–995

Baartscheer A, Schumacher CA, van Borren MM, Belterman CN, Coronel R, Fiolet JW (2003c) Increased Na^+/H^+-exchange activity is the cause of increased $[Na^+]i$ and underlies disturbed calcium handling in the rabbit pressure and volume overload heart failure model. Cardiovasc Res 57:1015–1024

Baczko I, Giles WR, Light PE (2003) Resting membrane potential regulates $Na^{(+)}-Ca^{2+}$ exchange-mediated Ca^{2+} overload during hypoxia-reoxygenation in rat ventricular myocytes. J Physiol 550:889–898

Bassani RA, Bassani JWM, Bers DM (1995) Relaxation in ferret ventricular myocytes: role of the sarcolemmal Ca ATPase. Pflugers Arch 430:573–578

Benardeau A, Hatem SN, Rucker Martin C, Le Grand B, Mace L, Dervanian P, Mercadier JJ, Coraboeuf E (1996) Contribution of Na^+/Ca^{2+} exchange to action potential of human atrial myocytes. Am J Physiol 271:H1151–H1161

Benndorf K, Friedrich M, Hirche H (1991) Reoxygenation-induced arrhythmogenic transient inward currents in isolated cells of the guinea-pig heart. Pflügers Arch 418:248–260

Bennett PB, Yazawa K, Makita N, George AL Jr (1995) Molecular mechanism for an inherited cardiac arrhythmia. Nature 376:683–685

Bers DM (2002) Cardiac excitation-contraction coupling. Nature 415:198–205

Bers DM, Bassani JWM, Bassani RA (1993) Competition and redistribution among calcium transport systems in rabbit cardiac myocytes. Cardiovasc Res 27:1772–1777

Beuckelmann DJ, Nabauer M, Erdmann E (1992) Intracellular calcium handling in isolated ventricular myocytes from patients with terminal heart failure. Circulation 85:1046–1055

Blaustein MP, Lederer WJ (1999) Sodium/calcium exchange: its physiological implications. Physiol Rev 79:763–854

Bokhari F, Newman D, Greene M, Korley V, Mangat I, Dorian P (2004) Long-term comparison of the implantable cardioverter defibrillator versus amiodarone: eleven-year follow-up of a subset of patients in the Canadian Implantable Defibrillator Study (CIDS). Circulation 110:112–116

Bouchard R, Omelchenko A, Le HD, Choptiany P, Matsuda T, Baba A, Takahashi K, Nicoll DA, Philipson KD, Hnatowich M, Hryshko LV (2004) Effects of SEA0400 on mutant NCX1.1 Na^+-Ca^{2+} exchangers with altered ionic regulation. Mol Pharmacol 65:802–810

Boutjdir M, El-Sherif N, Gough WB (1990) Effects of caffeine and ryanodine on delayed afterdepolarizations and sustained rhythmic activity in 1-day-old myocardial infarction in the dog. Circulation 81:1393–1400

Boutjdir M, Restivo M, Wei Y, Stergiopoulos K, el Sherif N (1994) Early afterdepolarization formation in cardiac myocytes: analysis of phase plane patterns, action potential, and membrane currents. J Cardiovasc Electrophysiol 5:609–620

Boyden PA, Pu J, Pinto J, Keurs HE (2000) $Ca^{(2+)}$ transients and $Ca^{(2+)}$ waves in Purkinje cells: role in action potential initiation. Circ Res 86:448–455

Bridge JHB, Smolley JR, Spitzer KW (1990) The relationship between charge movements associated with ICa and INa-Ca in cardiac myocytes. Science 248:376–378

Capogrossi MC, Kort AA, Spurgeon HA, Suárez-Isla BA, Lakatta EG (1984) Spontaneous contractile waves and stimulated contractions exhibit the same Ca^{2+} and species dependence in isolated cardiac myocytes and papillary muscles. Biophys J 45:94a

Carafoli E (1987) Intracellular calcium homeostasis. Annu Rev Biochem 56:395–433

The Cardiac Arrhythmia Suppression Trial (CAST) Investigators (1989) Preliminary report: effect of encainide and flecainide on mortality in a randomized trial of arrhythmia suppression after myocardial infarction. N Engl J Med 321:406–412

The Cardiac Arrhythmia Suppression Trial II Investigators (1992) Effect of the antiarrhythmic agent moricizine on survival after myocardial infarction. N Engl J Med 327:227–233

Carmeliet E (1999) Cardiac ionic currents and acute ischemia: from channels to arrhythmias. Physiol Rev 79:917–1017

Caroni P, Carafoli E (1981) The Ca^{2+}-pumping ATPase of heart sarcolemma. J Biol Chem 256:3263–3270

Chen SA, Chen YJ, Yeh HI, Tai CT, Chen YC, Lin CI (2003) Pathophysiology of the pulmonary vein as an atrial fibrillation initiator. Pacing Clin Electrophysiol 26:1576–1582

Chen YJ, Chen SA, Chang MS, Lin CI (2000) Arrhythmogenic activity of cardiac muscle in pulmonary veins of the dog: implication for the genesis of atrial fibrillation. Cardiovasc Res 48:265–273

Chen YJ, Chen SA, Chen YC, Yeh HI, Chan P, Chang MS, Lin CI (2001) Effects of rapid atrial pacing on the arrhythmogenic activity of single cardiomyocytes from pulmonary veins: implication in initiation of atrial fibrillation. Circulation 104:2849–2854

Chen YJ, Chen YC, Yeh HI, Lin CI, Chen SA (2002) Electrophysiology and arrhythmogenic activity of single cardiomyocytes from canine superior vena cava. Circulation 105:2679–2685

Cheng H, Lederer MR, Lederer WJ, Cannell MB (1996) Calcium sparks and $[Ca^{2+}]i$ waves in cardiac myocytes. Am J Physiol 270:C148–C159

Chin T, Spitzer KW, Philipson KD, Bridge JHB (1993) The effect of exchanger inhibitory peptide (XIP) on sodium-calcium exchange current in guinea pig ventricular cells. Circ Res 72:497–503

Choi HS, Eisner DA (1999a) The effects of inhibition of the sarcolemmal Ca-ATPase on systolic calcium fluxes and intracellular calcium concentration in rat ventricular myocytes. Pflugers Arch 437:966–971

Choi HS, Eisner DA (1999b) The role of sarcolemmal Ca^{2+}-ATPase in the regulation of resting calcium concentration in rat ventricular myocytes. J Physiol (Lond) 515:109–118

Choi HS, Trafford AW, Eisner DA (2000) Measurement of calcium entry and exit in quiescent rat ventricular myocytes. Pflugers Arch 440:600–608

Clancy CE, Rudy Y (1999) Linking a genetic defect to its cellular phenotype in a cardiac arrhythmia. Nature 400:566–569

Cordeiro JM, Howlett SE, Ferrier GR (1994) Simulated ischaemia and reperfusion in isolated guinea pig ventricular myocytes. Cardiovasc Res 28:1794–1802

Cox DA, Conforti L, Sperelakis N, Matlib MA (1993) Selectivity of inhibition of $Na^{(+)}$-Ca^{2+} exchange of heart mitochondria by benzothiazepine CGP-37157. J Cardiovasc Pharmacol 21:595–599

de Groot SH, Schoenmakers M, Molenschot MM, Leunissen JD, Wellens HJ, Vos MA (2000) Contractile adaptations preserving cardiac output predispose the hypertrophied canine heart to delayed afterdepolarization-dependent ventricular arrhythmias. Circulation 102:2145–2151

Díaz ME, Trafford AW, O'Neill SC, Eisner DA (1997) Measurement of sarcoplasmic reticulum Ca^{2+} content and sarcolemmal Ca^{2+} fluxes in isolated rat ventricular myocytes during spontaneous Ca^{2+} release. J Physiol (Lond) 501:3–16

Dipla K, Mattiello JA, Margulies KB, Jeevanandam V, Houser SR (1999) The sarcoplasmic reticulum and the Na^+/Ca^{2+} exchanger both contribute to the Ca^{2+} transient of failing human ventricular myocytes. Circ Res 84:435–444

DiPolo R, Beaugé L (1979) Physiological role of ATP-driven calcium pump in squid axon. Nature 278:271–273

Edgell RM, De Souza AI, MacLeod KT (2000) Relative importance of SR load and cytoplasmic calcium concentration in the genesis of aftercontractions in cardiac myocytes. Cardiovasc Res 47:769–777

Egan TM, Noble D, Noble SJ, Powell T, Spindler AJ, Twist VW (1989) Sodium-calcium exchange during the action potential in guinea-pig ventricular cells. J Physiol (Lond) 411:639–661

Ehara T, Matsuoka S, Noma A (1989) Measurement of reversal potential of Na^+-Ca^{2+} exchange current in single guinea-pig ventricular cells. J Physiol (Lond) 410:227–249

Ehrlich JR, Cha TJ, Zhang L, Chartier D, Melnyk P, Hohnloser SH, Nattel S (2003) Cellular electrophysiology of canine pulmonary vein cardiomyocytes: action potential and ionic current properties. J Physiol (Lond) 551:801–813

Eigel BN, Hadley RW (2001) Antisense inhibition of Na^+/Ca^{2+} exchange during anoxia/reoxygenation in ventricular myocytes. Am J Physiol Heart Circ Physiol 281:H2184–H2190

Eisner DA, Lederer WJ, Vaughan Jones RD (1984) The quantitative relationship between twitch tension and intracellular sodium activity in sheep cardiac Purkinje fibres. J Physiol (Lond) 355:251–266

Eisner DA, Trafford AW, Diaz ME, Overend CL, O'Neill SC (1998) The control of Ca release from the cardiac sarcoplasmic reticulum: regulation versus autoregulation. Cardiovasc Res 38:589–604

Eisner DA, Choi HS, Diaz ME, O'Neill SC, Trafford AW (2000) Integrative analysis of calcium cycling in cardiac muscle. Circ Res 87:1087–1094

el Sherif N, Zeiler RH, Craelius W, Gough WB, Henkin R (1988) QTU prolongation and polymorphic ventricular tachyarrhythmias due to bradycardia-dependent early afterdepolarizations. Afterdepolarizations and ventricular arrhythmias. Circ Res 63:286–305

Epstein AE (2004) An update on implantable cardioverter-defibrillator guidelines. Curr Opin Cardiol 19:23–25

Faber GM, Rudy Y (2000) Action potential and contractility changes in $[Na^{(+)}](i)$ overloaded cardiac myocytes: a simulation study. Biophys J 78:2392–2404

Fedida D, Noble D, Rankin AC, Spindler AJ (1987) The arrhythmogenic transient inward current ITI and related contraction in isolated guinea-pig ventricular myocytes. J Physiol (Lond) 392:523–542

Ferrier GR, Saunders JH, Mendez C (1973) A cellular mechanism for the generation of ventricular arrhythmias by acetylstrophanthidin. Circ Res 32:600–609

Fujioka Y, Komeda M, Matsuoka S (2000) Stoichiometry of Na^+-Ca^{2+} exchange in inside-out patches excised from guinea-pig ventricular myocytes. J Physiol (Lond) 523:339–351

Gatto C, Milanik MA (1993) Inhibition of red blood cell calcium pump by eosin and other fluorescein analogues. Am J Physiol 264:C1577–C1586

Goldhaber JI, Lamp ST, Walter DO, Garfinkel A, Fukumoto GH, Weiss JN (1999) Local regulation of the threshold for calcium sparks in rat ventricular myocytes: role of sodium-calcium exchange. J Physiol (Lond) 520:431–438

Harrison SM, McCall E, Boyett MR (1992) The relationship between contraction and intracellular sodium in rat and guinea-pig ventricular myocytes. J Physiol (Lond) 449:517–550

Hasenfuss G, Pieske B (2002) Calcium cycling in congestive heart failure. J Mol Cell Cardiol 34:951–969

Henderson SA, Goldhaber JI, So JM, Han T, Motter C, Ngo A, Chantawansri C, Ritter MR, Friedlander M, Nicoll DA, Frank JS, Jordan MC, Roos KP, Ross RS, Philipson KD (2004) Functional adult myocardium in the absence of Na^+-Ca^{2+} exchange: cardiac specific knockout of NCX1. Circ Res 95:604–611

Hirano Y, Moscucci A, January CT (1992) Direct measurement of L-type Ca^{2+} window current in heart cells. Circ Res 70:

Hobai IA, O'Rourke B (2000) Enhanced Ca^{2+}-activated Na^+-Ca^{2+} exchange activity in canine pacing-induced heart failure. Circ Res 87:690–698

Hobai IA, O'Rourke B (2004) The potential of Na^+/Ca^{2+} exchange blockers in the treatment of cardiac disease. Expert Opin Investig Drugs 13:653–664

Hobai IA, Maack C, O'Rourke B (2004) Partial inhibition of sodium/calcium exchange restores cellular calcium handling in canine heart failure. Circ Res 95:292–299

Iwamoto T, Watano T, Shigekawa M (1996) A novel isothiourea derivative selectively inhibits the reverse mode of Na^+/Ca^{2+} exchange in cells expressing NCX1. J Biol Chem 271:22391–22397

Iwamoto T, Inoue Y, Ito K, Sakaue T, Kita S, Katsuragi T (2004a) The exchanger inhibitory peptide region-dependent inhibition of Na^+/Ca^{2+} exchange by SN-6 [2-[4-(4-nitrobenzyloxy)benzyl]thiazolidine-4-carboxylic acid ethyl ester], a novel benzyloxyphenyl derivative. Mol Pharmacol 66:45–55

Iwamoto T, Kita S, Uehara A, Imanaga I, Matsuda T, Baba A, Katsuragi T (2004b) Molecular determinants of Na^+/Ca^{2+} exchange (NCX1) inhibition by SEA0400. J Biol Chem 279:7544–7553

Janse MJ (2004) Electrophysiological changes in heart failure and their relationship to arrhythmogenesis. Cardiovasc Res 61:208–217

January CT, Riddle JM (1989) Early afterdepolarizations: mechanism of induction and block. A role for L-type Ca^{2+} current. Circ Res 64:977–990

Janvier NC, Boyett MR (1996) The role of Na-Ca exchange current in the cardiac action potential. Cardiovasc Res 32:69–84

Janvier NC, Harrison SM, Boyett MR (1997) The role of inward Na^+-Ca^{2+} exchange current in the ferret ventricular action potential. J Physiol (Lond) 498:611–625

Kaczorowski GJ, Slaughter RS, King VF, Garcia ML (1989) Inhibitors of sodium-calcium exchange: identification and development of probes of transport activity. Biochim Biophys Acta 988:287–302

Kang TM, Hilgemann DW (2004) Multiple transport modes of the cardiac Na^+/Ca^{2+} exchanger. Nature 427:544–548

Kass RS, Lederer WJ, Tsien RW, Weingart R (1978) Role of calcium ions in transient inward currents and aftercontractions induced by strophanthidin in cardiac Purkinje fibres. J Physiol (Lond) 281:187–208

Kendall MJ (2000) Clinical trial data on the cardioprotective effects of beta-blockade. Basic Res Cardiol 95 Suppl 1:I25–I30

Kimura J, Watano T, Kawahara M, Sakai E, Yatabe J (1999) Direction-independent block of bi-directional Na^+/Ca^{2+} exchange current by KB-R7943 in guinea-pig cardiac myocytes. Br J Pharmacol 128:969–974

Ladilov Y, Haffner S, Balser-Schafer C, Maxeiner H, Piper HM (1999) Cardioprotective effects of KB-R7943: a novel inhibitor of the reverse mode of Na^+/Ca^{2+} exchanger. Am J Physiol 276:H1868–H1876

Lahat H, Pras E, Olender T, Avidan N, Ben Asher E, Man O, Levy-Nissenbaum E, Khoury A, Lorber A, Goldman B, Lancet D, Eldar M (2001) A missense mutation in a highly conserved region of CASQ2 is associated with autosomal recessive catecholamine-induced polymorphic ventricular tachycardia in Bedouin families from Israel. Am J Hum Genet 69:1378–1384

Laitinen PJ, Brown KM, Piippo K, Swan H, Devaney JM, Brahmbhatt B, Donarum EA, Marino M, Tiso N, Viitasalo M, Toivonen L, Stephan DA, Kontula K (2001) Mutations of the cardiac ryanodine receptor (RyR2) gene in familial polymorphic ventricular tachycardia. Circulation 103:485–490

Lamont C, Eisner DA (1996) The sarcolemmal mechanisms involved in the control of diastolic intracellular calcium in isolated rat cardiac trabeculae. Pflugers Arch 432:961–969

Leblanc N, Hume JR (1990) Sodium current-induced release of calcium from cardiac sarcoplasmic reticulum. Science 248:372–376

Lederer WJ, Tsien RW (1976) Transient inward current underlying arrhythmogenic effects of cardiotonic steroids in Purkinje fibers. J Physiol (Lond) 263:73–100

Lederer WJ, Niggli E, Hadley RW (1990) Sodium-calcium exchange in excitable cells: fuzzy space. Science 248:283

Lee C, Visen NS, Dhalla NS, Le HD, Isaac M, Choptiany P, Gross G, Omelchenko A, Matsuda T, Baba A, Takahashi K, Hnatowich M, Hryshko LV (2004) Inhibitory profile of SEA0400 [2-[4-[(2,5-difluorophenyl)methoxy]phenoxy]-5-ethoxyaniline] assessed on the cardiac Na^+-Ca^{2+} exchanger, NCX1.1. J Pharmacol Exp Ther 311:748–757

Levi AJ, Spitzer KW, Kohmoto O, Bridge JHB (1994) Depolarization-induced Ca entry via Na-Ca exchange triggers SR release in guinea pig cardiac myocytes. Am J Physiol 266:H1422–H1433

Li D, Melnyk P, Feng J, Wang Z, Petrecca K, Shrier A, Nattel S (2000) Effects of experimental heart failure on atrial cellular and ionic electrophysiology. Circulation 101:2631–2638

Li Z, Nicoll DA, Collins A, Hilgemann DW, Filoteo AG, Penniston JT, Weiss JN, Tomich JM, Philipson KD (1991) Identification of a peptide inhibitor of the cardiac sarcolemmal $Na^{(+)}$-Ca^{2+} exchanger. J Biol Chem 266:1014–1020

Lipp P, Pott L, Callewaert G, Carmeliet E (1990) Simultaneous recording of Indo-1 fluorescence and Na^+/Ca^{2+} exchange current reveals two components of Ca^{2+}-release from sarcoplasmic reticulum of cardiac atrial myocytes. FEBS Lett 275:181–184

Lipp P, Schwaller B, Niggli E (1995) Specific inhibition of NA–Ca exchange function by antisense oligodeoxynucleotides. FEBS Lett 364:198–202

Litwin SE, Li J, Bridge JH (1998) Na-Ca exchange and the trigger for sarcoplasmic reticulum Ca release: studies in adult rabbit ventricular myocytes. Biophys J 75:359–371

Luo CH, Rudy Y (1994) A dynamic model of the cardiac ventricular action potential. I. Simulations of ionic currents and concentration changes. Circ Res 74:1071–1096

Mechmann S, Pott L (1986) Identification of Na-Ca exchange current in single cardiac myocytes. Nature 319:597–599

Members of the Sicilan Gambit (2001) New approaches to antiarrhythmic therapy: emerging therapeutic applications of the cell biology of cardiac arrhythmias. Cardiovasc Res 52:345–360

Miyamoto S, Zhu BM, Kamiya K, Nagasawa Y, Hashimoto K (2002) KB-R7943, a Na^+/Ca^{2+} exchange inhibitor, does not suppress ischemia/reperfusion arrhythmias nor digitalis arrhythmias in dogs. Jpn J Pharmacol 90:229–235

Miyata A, Zipes DP, Hall S, Rubart M (2002) Kb-R7943 prevents acute, atrial fibrillation-induced shortening of atrial refractoriness in anesthetized dogs. Circulation 106:1410–1419

Mohler PJ, Schott JJ, Gramolini AO, Dilly KW, Guatimosim S, duBell WH, Song LS, Haurogne K, Kyndt F, Ali ME, Rogers TB, Lederer WJ, Escande D, Le Marec H, Bennett V (2003) Ankyrin-B mutation causes type 4 long-QT cardiac arrhythmia and sudden cardiac death. Nature 421:634–639

Mubagwa K, Wei Lin, Sipido KR, Bosteels S, Flameng W (1997) Monensin-induced reversal of positive force-frequency relationship in cardiac muscle: role of intracellular sodium in rest-dependent potentiation of contraction. J Mol Cell Cardiol 29:977–989

Mukai M, Terada H, Sugiyama S, Satoh H, Hayashi H (2000) Effects of a selective inhibitor of Na^+/Ca^{2+} exchange, KB-R7943, on reoxygenation-induced injuries in guinea pig papillary muscles. J Cardiovasc Pharmacol 35:121–128

Mulder BJM, de Tombe PP, ter Keurs HE (1989) Spontaneous and propagated contractions in rat cardiac trabeculae. J Gen Physiol 93:943–961

Myerburg RJ, Mitrani R, Interian A Jr, Castellanos A (1998) Interpretation of outcomes of antiarrhythmic clinical trials: design features and population impact. Circulation 97:1514–1521

Nagy ZA, Virag L, Toth A, Biliczki P, Acsai K, Banyasz T, Nanasi P, Papp JG, Varro A (2004) Selective inhibition of sodium-calcium exchanger by SEA-0400 decreases early and delayed afterdepolarization in canine heart. Br J Pharmacol 143:827–831

Nakamura A, Harada K, Sugimoto H, Nakajima F, Nishimura N (1998) [Effects of KB-R7943, a novel Na^+/Ca^{2+} exchange inhibitor, on myocardial ischemia/reperfusion injury]. Nippon Yakurigaku Zasshi 111:105–115

Nattel S, Li D (2000) Ionic remodeling in the heart: pathophysiological significance and new therapeutic opportunities for atrial fibrillation. Circ Res 87:440–447

Nattel S, Yue L, Wang Z (1999) Cardiac ultrarapid delayed rectifiers: a novel potassium current family o f functional similarity and molecular diversity. Cell Physiol Biochem 9:217–226

Negretti N, O'Neill SC, Eisner DA (1993) The relative contributions of different intracellular and sarcolemmal systems to relaxation in rat ventricular myocytes. Cardiovasc Res 27:1826–1830

Negretti N, Varro A, Eisner DA (1995) Estimate of net calcium fluxes and sarcoplasmic reticulum calcium content during systole in rat ventricular myocytes. J Physiol (Lond) 486:581–591

Noble D, Noble SJ, Bett GC, Earm YE, Ho WK, So IK (1991) The role of sodium-calcium exchange during the cardiac action potential. Ann N Y Acad Sci 639:334–353

Nuyens D, Stengl M, Dugarmaa S, Rossenbacker T, Compernolle V, Rudy Y, Smits JF, Flameng W, Clancy CE, Moons L, Vos MA, Dewerchin M, Benndorf K, Collen D, Carmeliet E, Carmeliet P (2001) Abrupt rate accelerations or premature beats cause life-threatening arrhythmias in mice with long-QT3 syndrome. Nat Med 7:1021–1027

O'Rourke B, Kass DA, Tomaselli GF, Kaab S, Tunin R, Marban E (1999) Mechanisms of altered excitation-contraction coupling in canine tachycardia-induced heart failure, I: experimental studies. Circ Res 84:562–570

Omelchenko A, Bouchard R, Le HD, Choptiany P, Visen N, Hnatowich M, Hryshko LV (2003) Inhibition of canine (NCX1.1) and Drosophila (CALX1.1) $Na^{(+)}$-Ca(2+) exchangers by 7-chloro-3,5-dihydro-5-phenyl-1H-4,1-benzothiazepine-2-one (CGP-37157). J Pharmacol Exp Ther 306:1050–1057

Paulus WJ, Goethals MA, Sys SU (1992) Failure of myocardial inactivation: a clinical assessment in the hypertrophied heart. Basic Res Cardiol 87 Suppl 2:145–161

Piacentino V III, Weber CR, Chen X, Weisser-Thomas J, Margulies KB, Bers DM, Houser SR (2003) Cellular basis of abnormal calcium transients of failing human ventricular myocytes. Circ Res 92:651–658

Pieske B, Maier LS, Piacentino V, Weisser J, Hasenfuss G, Houser S (2002) Rate dependence of $[Na^+]i$ and contractility in nonfailing and failing human myocardium. Circulation 106:447–453

Pogwizd SM (1995) Nonreentrant mechanisms underlying spontaneous ventricular arrhythmias in a model of nonischemic heart failure in rabbits. Circulation 92:1034–1048

Pogwizd SM (2003) Clinical potential of sodium-calcium exchanger inhibitors as antiarrhythmic agents. Drugs 63:439–452

Pogwizd SM, Hoyt RH, Saffitz JE, Corr PB, Cox JL, Cain ME (1992) Reentrant and focal mechanisms underlying ventricular tachycardia in the human heart. Circulation 86:1872–1887

Pogwizd SM, Qi M, Yuan W, Samarel AM, Bers DM (1999) Upregulation of Na^+/Ca^{2+} exchanger expression and function in an arrhythmogenic rabbit model of heart failure. Circ Res 85:1009–1019

Pogwizd SM, Schlotthauer K, Li L, Yuan W, Bers DM (2001) Arrhythmogenesis and contractile dysfunction in heart failure: roles of sodium-calcium exchange, inward rectifier potassium current, and residual β-adrenergic responsiveness. Circ Res 88:1159–1167

Pogwizd SM, Sipido KR, Verdonck F, Bers DM (2003) Intracellular Na in animal models of hypertrophy and heart failure: contractile function and arrhythmogenesis. Cardiovasc Res 57:887–896

Priori SG, Napolitano C, Tiso N, Memmi M, Vignati G, Bloise R, Sorrentino V, Danieli GA (2001) Mutations in the cardiac ryanodine receptor gene (hRyR2) underlie catecholaminergic polymorphic ventricular tachycardia. Circulation 103:196–200

Reuter H, Henderson SA, Han T, Matsuda T, Baba A, Ross RS, Goldhaber JI, Philipson KD (2002) Knockout mice for pharmacological screening: testing the specificity of Na^+-Ca^{2+} exchange inhibitors. Circ Res 91:90–92

Rosen MR, Gelband H, Hoffman BF (1973) Correlation between effects of ouabain on the canine electrocardiogram and transmembrane potentials of isolated Purkinje fibers. Circulation 47:65–72

Salvador JM, Inesi G, Rigaud J-L, Mata AM (1998) Ca^{2+} transport by reconstituted synaptosomal ATPase is associated with H^+ countertransport and net charge displacement. J Biol Chem 273:18230–18234

Satoh H, Ginsburg KS, Qing K, Terada H, Hayashi H, Bers DM (2000) KB-R7943 block of Ca^{2+} influx via Na^+/Ca^{2+} exchange does not alter twitches or glycoside inotropy but prevents Ca^{2+} overload in rat ventricular myocytes. Circulation 101:1441–1446

Satoh H, Mukai M, Urushida T, Katoh H, Terada H, Hayashi H (2003) Importance of Ca^{2+} influx by Na^+/Ca^{2+} exchange under normal and sodium-loaded conditions in mammalian ventricles. Mol Cell Biochem 242:11–17

Schillinger W, Fiolet JW, Schlotthauer K, Hasenfuss G (2003) Relevance of Na^+-Ca^{2+} exchange in heart failure. Cardiovasc Res 57:921–933

Schuh K, Uldrijan S, Telkamp M, Rothlein N, Neyses L (2001) The plasmamembrane calmodulin-dependent calcium pump: a major regulator of nitric oxide synthase I. J Cell Biol 155:201–205

Schwiening CJ, Kennedy HJ, Thomas RC (1993) Calcium-hydrogen exchange by the plasma membrane Ca-ATPase of voltage-clamped snail neurons. Proc R Soc Lond B Biol Sci 253:285–289

Sham JSK, Cleemann L, Morad M (1995) Functional coupling of Ca^{2+} channels and ryanodine receptors in cardiac myocytes. Proc Natl Acad Sci U S A 92:121–125

Shannon TR, Wang F, Puglisi J, Weber C, Bers DM (2004) A mathematical treatment of integrated Ca dynamics within the ventricular myocyte. Biophys J 87:3351–3371

Shorofsky SR, Aggarwal R, Corretti M, Baffa JM, Strum JM, Al-Seikhan BA, Kobayashi YM, Jones LR, Wier WG, Balke CW (1999) Cellular mechanisms of altered contractility in the hypertrophied heart. Circ Res 84:424–434

Siegl PK, Cragoe EJJ, Trumble MJ, Kaczorowski GJ (1984) Inhibition of Na^+/Ca^{2+} exchange in membrane vesicle and papillary muscle preparations from guinea pig heart by analogs of amiloride. Proc Natl Acad Sci U S A 81:3238–3242

Sipido KR, Callewaert G, Porciatti F, Vereecke J, Carmeliet E (1995) $[Ca^{2+}]i$-dependent membrane currents in guinea-pig ventricular cells in the absence of Na/Ca exchange. Pflügers Arch 430:871–878

Sipido KR, Maes MM, Van de Werf F (1997) Low efficiency of Ca^{2+} entry through the Na/Ca exchanger as trigger for Ca^{2+} release from the sarcoplasmic reticulum. Circ Res 81:1034–1044

Sipido KR, Volders PGA, de Groot SH, Verdonck F, Van de Werf F, Wellens HJ, Vos MA (2000) Enhanced Ca^{2+} release and Na/Ca exchange activity in hypertrophied canine ventricular myocytes: a potential link between contractile adaptation and arrhythmogenesis. Circulation 102:2137–2144

Sipido KR, Volders PG, Vos MA, Verdonck F (2002) Altered Na/Ca exchange activity in cardiac hypertrophy and heart failure: a new target for therapy? Cardiovasc Res 53:782–805

Strehler EE, Zacharias DA (2001) Role of alternative splicing in generating isoform diversity among plasma membrane calcium pumps. Physiol Rev 81:21–50

Su Z, Sugishita K, Ritter M, Li F, Spitzer KW, Barry WH (2001) The sodium pump modulates the influence of I(Na) on $[Ca^{2+}]i$ transients in mouse ventricular myocytes. Biophys J 80:1230–1237

Sun H, Gaspo R, Leblanc N, Nattel S (1998) Cellular mechanisms of atrial contractile dysfunction caused by sustained atrial tachycardia. Circulation 98:719–727

Takahashi K, Takahashi T, Suzuki T, Onishi M, Tanaka Y, Hamano-Takahashi A, Ota T, Kameo K, Matsuda T, Baba A (2003) Protective effects of SEA0400, a novel and selective inhibitor of the Na^+/Ca^{2+} exchanger, on myocardial ischemia-reperfusion injuries. Eur J Pharmacol 458:155–162

Tanaka H, Nishimaru K, Aikawa T, Hirayama W, Tanaka Y, Shigenobu K (2002) Effect of SEA0400, a novel inhibitor of sodium-calcium exchanger, on myocardial ionic currents. Br J Pharmacol 135:1096–1100

Tieleman RG, De Langen C, Van Gelder IC, De Kam PJ, Grandjean J, Bel KJ, Wijffels MC, Allessie MA, Crijns HJ (1997) Verapamil reduces tachycardia-induced electrical remodeling of the atria. Circulation 95:1945–1953

Trafford AW, Eisner DA (2002) Excitation contraction coupling in cardiac muscle. In: So-
laro RJ, Moss RL (eds) Molecular control in striated muscle contraction. Kluwer, Dordrecht, pp 49–89

Trafford AW, Diaz ME, O'Neill SC, Eisner DA (1995) Comparison of subsarcolemmal and bulk
calcium concentration during spontaneous calcium release in rat ventricular myocytes.
J Physiol (Lond) 488:577–586

Trafford AW, Diaz ME, O'Neill SC, Eisner DA (2002) Integrative analysis of calcium signalling
in cardiac muscle. Front Biosci 7:D843–D852

Van Noord T, Van Gelder IC, Tieleman RG, Bosker HA, Tuinenburg AE, Volkers C, Veeger NJ,
Crijns HJ (2001) VERDICT: the Verapamil versus Digoxin Cardioversion Trial: a randomized study on the role of calcium lowering for maintenance of sinus rhythm after
cardioversion of persistent atrial fibrillation. J Cardiovasc Electrophysiol 12:766–769

Varro A, Biliczki P, Iost N, Virag L, Hala O, Kovacs P, Matyus P, Papp JG (2004) Theoretical
possibilities for the development of novel antiarrhythmic drugs. Curr Med Chem 11:1–11

Veldkamp MW, Verkerk AO, van Ginneken AC, Baartscheer A, Schumacher C, de Jonge N,
de Bakker JM, Opthof T (2001) Norepinephrine induces action potential prolongation
and early afterdepolarizations in ventricular myocytes isolated from human end-stage
failing hearts. Eur Heart J 22:955–963

Verdonck F, Volders PGA, Vos MA, Sipido KR (2003a) Increased Na^+ concentration and
altered Na/K pump activity in hypertrophied canine ventricular cells. Cardiovasc Res
57:1035–1043

Verdonck F, Volders PGA, Vos MA, Sipido KR (2003b) Intracellular Na^+ and altered Na^+
transport mechanisms in cardiac hypertrophy and failure. J Mol Cell Cardiol 35:5–25

Verdonck F, Mubagwa K, Sipido KR (2004) [$Na^{(+)}$] in the subsarcolemmal 'fuzzy' space and
modulation of [$Ca(2+)$](i) and contraction in cardiac myocytes. Cell Calcium 35:603–612

Verkerk AO, Veldkamp MW, Baartscheer A, Schumacher CA, Klopping C, van Ginneken AC,
Ravesloot JH (2001) Ionic mechanism of delayed afterdepolarizations in ventricular cells
isolated from human end-stage failing hearts. Circulation 104:2728–2733

Volders PG, Vos MA, Szabo B, Sipido KR, de Groot SH, Gorgels AP, Wellens HJ, Lazzara R
(2000) Progress in the understanding of cardiac early afterdepolarizations and torsades
de pointes: time to revise current concepts. Cardiovasc Res 46:376–392

Volders PGA, Kulcsar A, Vos MA, Sipido KR, Wellens HJ, Lazzara R, Szabo B (1997) Similarities between early and delayed afterdepolarizations induced by isoproterenol in canine
ventricular myocytes. Cardiovasc Res 34:348–359

Volders PGA, Sipido KR, Vos MA, Kulcsar A, Verduyn SC, Wellens HJJ (1998b) Cellular
basis of biventricular hypertrophy and arrhythmogenesis in dogs with chronic complete
atrioventricular block and acquired torsade de pointes. Circulation 98:1136–1147

Vos MA, de Groot SH, Verduyn SC, van der Zande J, Leunissen HD, Cleutjens JP, van Bilsen M,
Daemen MJ, Schreuder JJ, Allessie MA, Wellens HJ (1998) Enhanced susceptibility for
acquired torsade de pointes arrhythmias in the dog with chronic, complete AV block is
related to cardiac hypertrophy and electrical remodeling. Circulation 98:1125–1135

Waldo AL, Camm AJ, deRuyter H, Friedman PL, MacNeil DJ, Pauls JF, Pitt B, Pratt CM,
Schwartz PJ, Veltri EP (1996) Effect of d-sotalol on mortality in patients with left ventricular dysfunction after recent and remote myocardial infarction. The SWORD Investigators. Survival With Oral d-Sotalol. Lancet 348:7–12

Watano T, Kimura J, Morita T, Nakanishi H (1996) A novel antagonist, No. 7943, of the
Na^+/Ca^{2+} exchange current in guinea-pig cardiac ventricular cells. Br J Pharmacol
119:555–563

Weber CR, Ginsburg KS, Philipson KD, Shannon TR, Bers DM (2001) Allosteric regulation of Na/Ca exchange current by cytosolic Ca in intact cardiac myocytes. J Gen Physiol 117:119–132

Weber CR, Piacentino V, Ginsburg KS, Houser SR, Bers DM (2002) $Na^{(+)}$-Ca(2+) exchange current and submembrane [Ca(2+)] during the cardiac action potential. Circ Res 90:182–189

Weber CR, Ginsburg KS, Bers DM (2003) Cardiac submembrane [Na^+] transients sensed by Na^+-Ca^{2+} exchange current. Circ Res 92:950–952

Wehrens XH, Abriel H, Cabo C, Benhorin J, Kass RS (2000) Arrhythmogenic mechanism of an LQT-3 mutation of the human heart $Na^{(+)}$ channel alpha-subunit: a computational analysis. Circulation 102:584–590

Wier WG, Cannell MB, Berlin JR, Marban E, Lederer WJ (1987) Cellular and subcellular heterogeneity of [Ca^{2+}]i in single heart cells revealed by fura-2. Science 235:325–328

Winslow RL, Rice JJ, Jafri S, Marban E, O'Rourke B (1999) Mechanisms of altered excitation-contraction coupling in canine tachycardia-induced heart failure, II: model studies. Circ Res 84:571–586

Yoshiyama M, Takeuchi K, Hanatani A, Kim S, Omura T, Toda I, Teragaki M, Akioka K, Iwao H, Yoshikawa J (1997) Differences in expression of sarcoplasmic reticulum Ca^{2+}-ATPase and Na^+-Ca^{2+} exchanger genes between adjacent and remote noninfarcted myocardium after myocardial infarction. J Mol Cell Cardiol 29:255–264

Yoshiyama M, Hayashi T, Nakamura Y, Omura T, Izumi Y, Matsumoto R, Takeuchi K, Kitaura Y, Yoshikawa J (2004) Effects of cellular cardiomyoplasty on ventricular remodeling assessed by Doppler echocardiography and topographic immunohistochemistry. Circ J 68:580–586

Zipes DP, Wellens HJJ (1998) Sudden cardiac death. Circulation 98:2334–2351

Zygmunt AC, Goodrow RJ, Antzelevitch C (2000) INaCa contributes to electrical heterogeneity within the canine ventricle. Am J Physiol 278:H1671–H1678

HEP (2006) 171:201–220
© Springer-Verlag Berlin Heidelberg 2006

A Role for Calcium/Calmodulin-Dependent Protein Kinase II in Cardiac Disease and Arrhythmia

T. J. Hund[1] (✉) · Y. Rudy[2]

[1] Department of Pathology and Immunology,
Washington University in Saint Louis School of Medicine, 660 S. Euclid Ave.,
Campus Box 8118, Saint Louis MO, 63118, USA
thund@wustl.edu

[2] Cardiac Bioelectricity and Arrhythmia Center and Department of Biomedical
Engineering, Washington University in Saint Louis, One Brookings Dr., Campus Box 1097,
Saint Louis MO, 63130-4899, USA

Abstract More than 20 years have passed since the discovery that a collection of specific calcium/calmodulin-dependent phosphorylation events is the result of a single multifunctional kinase. Since that time, we have learned a great deal about this multifunctional and ubiquitous kinase, known today as calcium/calmodulin-dependent protein kinase II (CaMKII). CaMKII is interesting not only for its widespread distribution and broad specificity but also for its biophysical properties, most notably its activation by the critical second messenger complex calcium/calmodulin and its autophosphorylating capability. A central role for CaMKII has been identified in regulating a diverse array of fundamental cellular activities. Furthermore, altered CaMKII activity profoundly impacts function in the brain and heart. Recent findings that CaMKII expression in the heart changes during hypertrophy, heart failure, myocardial ischemia, and infarction suggest that CaMKII may be a viable therapeutic target for patients suffering from common forms of heart disease.

Keywords Calcium/calmodulin-dependent protein kinase II · Cardiac · Electrophysiology ·
Calcium/calmodulin · Heart disease · Arrhythmia

1
Introduction

Calcium/calmodulin-dependent protein kinase II (CaMKII) is a ubiquitous and
multifunctional holoenzyme belonging to a superfamily of serine/threonine
calcium/calmodulin-dependent protein kinases that also includes myosin light
chain kinase (MLCK), phosphorylase kinase, eEf2 kinase (also known as
CaMKIII), CaMKI, and CaMKIV (for reviews, see Schulman 1988; Schulman
et al. 1992; Braun and Schulman 1995; Hook and Means 2001). CaMKII was initially discovered as a series of seemingly independent calcium/calmodulin-de-
pendent phosphorylations (Le Peuch et al. 1979; Kennedy and Greengard 1981;
Woodgett et al. 1982). Similarity between a calcium/calmodulin-dependent
kinase that phosphorylates synapsin I in the brain and glycogen synthase
kinase in skeletal muscle led one group to postulate the existence of a single
calcium/calmodulin-dependent kinase with broad specificity (termed
"calcium/calmodulin-dependent multi-protein kinase"; McGuinness et al.
1983). The two decades since these early pioneering studies have seen mounting experimental and theoretical evidence that CaMKII indeed mediates a vast
array of critical cellular behaviors in many different tissues. This chapter dis-
cusses the electrophysiological function of CaMKII under normal and pathological conditions. The focus is primarily on the role of CaMKII in the heart,
although the brain is also discussed in some detail due to the fact that CaMKII
has been extensively characterized in neuronal tissue.

2
Background

2.1
CaMKII Structure

CaMKII consists of multiple subunits assembled in an homomultimeric or
heteromultimeric structure, resembling a pinwheel with the N-terminal association domains forming a central hub (Kanaseki et al. 1991; Kolodziej et al.
2000; Hoelz et al. 2003; Gaertner et al. 2004). Electron micrographs initially
revealed the holoenzyme to be a hub-and-spoke assembly of 8 or 10 units
(Kanaseki et al. 1991) (Fig. 1). More recent three-dimensional reconstructions indicate a dodecameric assembly (Kolodziej et al. 2000; Gaertner et al.
2004), while the crystal structure of the truncated association domain reveals
a tetradecameric structure with a 50-Å pore (Hoelz et al. 2003). Each subunit

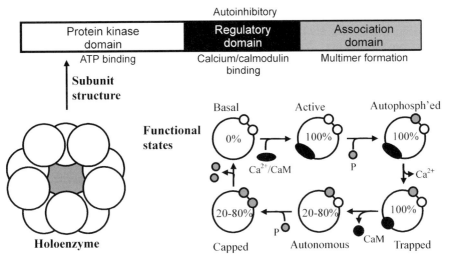

Fig. 1 Schematic of calcium/calmodulin-dependent protein kinase II, illustrating the structure, functional domains, and major kinetic states of the enzyme

is between 50 kDa and 60 kDa, and consists of three functional domains: (1) a N-terminal kinase domain with high homology to other protein kinases, including the ATP-binding consensus sequence; (2) a regulatory domain containing overlapping autoinhibitory and calmodulin-binding domains; (3) an association domain in the C-terminal region involved in assembly of subunits or association of holoenzyme with other proteins (Schulman 1988; Fig. 1). These core domains are approximately 85% homologous across isoforms (Tobimatsu and Fujisawa 1989). A variable domain located past the C-terminus of the regulatory domain accounts for most of the divergence among isoforms, along with a second variable insert at the C-terminus of δ-isoforms (Braun and Schulman 1995).

Four genes α, β, δ, and γ encode at least 30 CaMKII isoforms (Hudmon and Schulman 2002) with every cell type containing at least one isoform (Hook and Means 2001). The α- and β-isoforms are found only in nervous tissue while γ and δ are distributed in most tissues (Tobimatsu and Fujisawa 1989). To date, at least eleven δ-isoforms have been identified (Tobimatsu and Fujisawa 1989; Mayer et al. 1993; Schworer et al. 1993; Edman and Schulman 1994; Mayer et al. 1995; Hagemann et al. 1999; Hoch et al. 1999) with the δ_C and δ_B (or δ_2 and δ_3, respectively) isoforms being the most prevalent in the heart (Schworer et al. 1993; Edman and Schulman 1994). The δ_B splice variant has an 11-amino-acid localization sequence in its variable domain, which targets it to the nucleus (Srinivasan et al. 1994).

2.2
Functional States and Kinetics

Binding of Ca^{2+}/calmodulin to the regulatory domain displaces the autoinhibitory segment and activates a kinase subunit. The kinase undergoes autophosphorylation when an active subunit phosphorylates a neighboring subunit in the same assembly (Bennett et al. 1983; Kuret and Schulman 1985; Hanson et al. 1994). Autophosphorylation occurs at a specific threonine residue, Threonine[286] (in CaMKIIα) or Threonine[287] (in other isoforms), and increases the affinity of the kinase for calmodulin 1,000-fold, thereby trapping bound calmodulin (Lai et al. 1987; Schworer et al. 1988; Lou and Schulman 1989; Meyer et al. 1992). The kinase remains fully active in the trapped state even after Ca^{2+} returns to resting values. Eventually calmodulin unbinds from the kinase; however, even in this autonomous state the kinase retains 20%–80% of its maximal activity. Unbinding of calmodulin triggers Ca^{2+}-independent autophosphorylation at specific residues (Lou and Schulman 1989), which reduces the affinity for calmodulin and caps kinase activity (even in the presence of Ca^{2+}/calmodulin) to 20%–80% of maximal activity. The kinase returns to its basal state once complete dephosphorylation has occurred.

2.3
Cellular Substrates

Neuronal targets for CaMKII are many and include glutamate receptors (Derkach et al. 1999), gap junctions (Pereda et al. 1998), tyrosine hydroxylase (Griffith and Schulman 1988), synapsin I (Kennedy and Greengard 1981; Greengard et al. 1993), nitric oxide synthase (Nakane et al. 1991), and ion channels (Barrett et al. 2000; Wang et al. 2002). In smooth muscle cells, CaMKII regulates muscle contraction by targeting caldesmon (Ikebe et al. 1990) to relieve its inhibition of myosin ATPase and MLCK to decrease its Ca^{2+}/calmodulin sensitivity. In epithelial cells, CaMKII affects secretion and volume regulation via chloride channels and nonselective cation channels. CaMKII has been shown to regulate organelle transport during mitosis in *Xenopus* melanophores by phosphorylating myosin-V (Karcher et al. 2001). CaMKII isoforms also regulate the expression of several genes, including c-fos (target for α-isoform), interleukin-2 (γ-isoform target), and atrial natriuretic factor ($δ_B$ target) (Dash et al. 1991; Nghiem et al. 1994; Ramirez et al. 1997).

In cardiac cells, CaMKII regulates the cardiac sarcoplasmic reticulum (SR), the organelle responsible for intracellular storage and release of calcium during the cardiac cycle. This effect was first discovered as a calcium/calmodulin-dependent and cAMP-independent phosphorylation of phospholamban (PLB) in SR vesicles isolated from canine hearts (Le Peuch et al. 1979; Bilezikjian et al. 1981). PLB binds to and inhibits the SR Ca^{2+}-ATPase (SERCA2a) responsible for reuptake of calcium into the SR (MacLennan and Kranias 2003). CaMKII phosphorylates PLB at Threonine[17] (Wegener et al. 1989; Hagemann et al. 2000), thereby relieving inhibition of SERCA2a (Odermatt et al. 1996). There is experimental evidence that CaMKII also phosphorylates SERCA2a directly. Toyofuku et al. identified Serine[38] as the site on SERCA2a phosphorylated by CaMKII (Toyofuku et al. 1994). They and others report an increase in the maximum uptake rate of calcium into the SR in response to CaMKII phosphorylation of SERCA2a (Hawkins et al. 1994; Mattiazzi et al. 1994; Toyofuku et al. 1994), a finding which has been disputed (Odermatt et al. 1996; Reddy et al. 1996). Odermatt et al. showed that incubation of control cells in the presence of ethyleneglycoltetraacetic acid (EGTA) destabilizes the cells, producing apparent CaMKII-dependent changes in the SR calcium uptake rate (Odermatt et al. 1996). However, since then, Xu et al. have confirmed that CaMKII phosphorylation enhances the SR calcium uptake rate in vitro (Xu and Narayanan 1999) and have reported phosphorylation of SERCA2a at Serine[38] in vivo (Xu et al. 1999), while a different group has measured a decreased uptake rate in transgenic mice expressing a CaMKII inhibitory peptide (Ji et al. 2003). Therefore, while controversy remains, mounting experimental evidence (in vitro and in vivo) supports the earlier findings that CaMKII phosphorylates SERCA2a to increase calcium uptake (Hawkins et al. 1994; Toyofuku et al. 1994).

The cardiac ryanodine receptor (RyR2) is located in the SR membrane and releases Ca^{2+} from internal stores in response to calcium influx through sarcolemmal L-type Ca^{2+} channels. Studies on canine SR vesicles show that phosphorylation of RyR2 by CaMKII activates the channel (Witcher et al. 1991). Studies using phospho-specific antibodies initially identified Serine[2809] as the residue phosphorylated by CaMKII on RyR2 (Witcher et al. 1991; Rodriguez et al. 2003). However, site-directed mutagenesis has recently revealed Serine[2815] to be the CaMKII-specific phosphorylation site (Wehrens et al. 2004). While some studies have found a decrease in RyR2 activity in response to increased CaMKII activity (Lokuta et al. 1995; Wu et al. 2001), drug studies (Netticadan et al. 1996) and studies where the SR calcium content is tightly controlled (Li et al. 1997) provide further evidence for a positive regulation of RyR2 by CaMKII. Recently, it has been shown that CaMKIIδ coimmunoprecipitates with RyR2 and the specific CaMKII inhibitor AIP decreases RyR2 Ca^{2+} spark frequency, duration, and width in rabbit hearts (Currie et al. 2004). Furthermore, CaMKII phosphorylation of recombinant RyR2 in planar lipid bilayers has been shown to increase channel open probability, while having no effect on RyR2^{S2815A} mutant channels (Wehrens et al. 2004).

CaMKII phosphorylates ion channels in the sarcolemmal membrane as well. The best characterized of these targets is the L-type Ca^{2+} channel (Anderson et al. 1994; Xiao et al. 1994; Yuan and Bers 1994; Dzhura et al. 2000), which serves as the trigger for SR Ca^{2+} release and is responsible for maintaining the prominent action potential plateau. Repetitive stimulation using voltage pulses from a hyperpolarized potential increases, or facilitates, the peak current carried by L-type Ca^{2+} channels ($I_{Ca(L)}$) (Marban and Tsien 1982; Lee 1987; Fedida et al. 1988). This facilitation occurs via CaMKII phosphorylation of the channel, which promotes a gating mode characterized by long openings (Yuan and Bers 1994; Dzhura et al. 2000). Experiments indicate that CaMKII phosphorylation increases $I_{Ca(L)}$ by between 40% and 50% at rapid pacing (Li et al. 1997; Zuhlke et al. 1999).

3
Functional Roles of CaMKII

3.1
Experimental and Theoretical Tools for Studying CaMKII Function

Pharmacological intervention with CaMKII-specific inhibitors such as KN-62 (Tokumitsu et al. 1990) and KN-93 (Sumi et al. 1991), or the calmodulin antagonist W-7 has been used extensively to study the function of CaMKII in a variety of tissues. While these pharmacological agents have proved useful in the study of CaMKII function, nonspecific effects have been discovered in some cases. For example, KN-93 blocks K^+ channels while KN-62 has been found to slow Ca^{2+} channel recovery from inactivation independent of CaMKII in cardiac myocytes (Yuan and Bers 1994; Anderson et al. 1998). Synthetic peptides corresponding to the autoinhibitory segment of CaMKII have also been developed to study CaMKII structure and function (Kelly et al. 1988; Payne et al. 1988; Malinow et al. 1989), as well as the highly specific and potent synthetic peptide autocamtide-2-related inhibitory peptide (AIP) (Ishida et al. 1995).

Genetic engineering has produced several valuable tools for use in CaMKII research. Among the first of these was the CaMKIIα knockout mouse developed by Silva and colleagues to study the molecular basis of long-term potentiation (LTP) in the hippocampus (Silva et al. 1992ab). Constitutively active CaMKIIα has been expressed through viral infection of hippocampal slices (Pettit et al. 1994) and via direct injection (Lledo et al. 1995). A transgenic mouse with a point mutation in Threonine[286] to aspartate, which mimics autophosphorylation, also expresses a constitutively active CaMKII (Mayford et al. 1995; Mayford et al. 1996). In contrast, mutation of Threonine[286] to alanine prevents autophosphorylation in mice (Cho et al. 1998; Giese et al. 1998). Transgenic overexpression of the δ_B and δ_C CaMKII isoforms have been used to study cardiac hypertrophy and heart failure in mice (Zhang et al. 2002; Maier et al. 2003), while transgenic expression of the synthetic CaMKII inhibitor AIP with a SR localization sequence has been used to study the role of CaMKII in cardiac function (Ji et al. 2003).

Mathematical modeling has been another useful tool in understanding CaMKII function (Lisman and Goldring 1988; Hanson et al. 1994; Michelson and Schulman 1994; Matsushita et al. 1995; Dosemeci and Albers 1996;

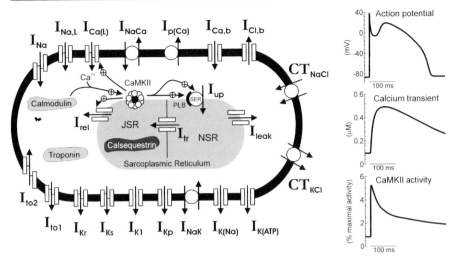

Fig. 2 Hund-Rudy dynamic (HRd) model of the canine epicardial myocyte. The model action potential, intracellular calcium transient, and CaMKII activity are shown after pacing to steady state at a pacing cycle length of 2,000 ms

Coomber 1998; Kubota 1999; Zhabotinsky 2000; Kubota and Bower 2001; Kikuchi et al. 2003; Hund and Rudy 2004). Early modeling work concluded that CaMKII in postsynaptic densities could theoretically store information in a stable manner required for long-term memory (Lisman 1985; Lisman and Goldring 1988). Hanson and colleagues later developed a set of differential equations describing CaMKII activity in response to a train of square-pulse calcium signals (Hanson et al. 1994). Since then, more advanced state-based models of CaMKII activity have been developed. Notably, Zhabotinsky has used one such model to examine the role of phosphatases in regulating CaMKII activity (Zhabotinsky 2000). Recently, we have incorporated CaMKII and its participation in rate-dependent cellular processes into a mathematical model of the canine epicardial action potential (Hund and Rudy 2004; Fig. 2). Whole-cell cardiac myocyte models have been used successfully to study myocardial ischemia (Shaw and Rudy 1997; Ch'en et al. 1998), heart failure (Winslow et al. 1999), and the molecular basis of congenital syndromes linked to sudden cardiac death (Clancy and Rudy 1999; Viswanathan and Rudy 2000).

3.2
CaMKII Function in Neurons

CaMKII is highly concentrated in the forebrain including the hippocampus, where it constitutes 2% of total protein (Erondu and Kennedy 1985). Excitatory pathways in the hippocampus and other regions show sustained enhancement of synaptic transmission in response to high-frequency stimulation, a property known as LTP and thought to underlie some forms of memory (see reviews: Bliss and Collingridge 1993; Lynch 2004). It is now generally accepted that CaMKII activation in postsynaptic densities is important for LTP induction (Malenka et al. 1989; Malinow et al. 1989; Silva et al. 1992a,b; Fukunaga et al. 1993; Pettit et al. 1994; Lledo et al. 1995). Early modeling studies concluded that the switch-like nature of CaMKII could theoretically encode long-term memory (Lisman and Goldring 1988). Experimental evidence for this first came from the fact that CaMKII inhibitors impair LTP in CA1 hippocampal cells (Malenka et al. 1989; Malinow et al. 1989). Subsequently, it was found that both LTP and spatial learning are severely impaired in mutant mice lacking CaMKIIα (Silva et al. 1992a,b). In fact, eliminating autophosphorylation by point mutation is enough to eliminate spatial learning (Cho et al. 1998; Giese et al. 1998).

The ability of CaMKII to encode LTP and memory depends on the enzyme's sensitivity to calcium/calmodulin and its unique regulatory properties, most notably autophosphorylation. Computer modeling and experimental studies alike have shown that the unique properties of CaMKII allow the enzyme to detect the frequency of calcium oscillations (Hanson et al. 1994; Michelson and Schulman 1994; De Koninck and Schulman 1998).

3.3
CaMKII Function in Cardiomyocytes

Calcium is an important second messenger in cardiac cells. Upon membrane depolarization during the cardiac action potential, calcium enters the cytosol primarily through L-type calcium channels, which are concentrated in T-tubules in close proximity to SR ryanodine receptor (RyR) calcium release channels. Local elevation of calcium concentration triggers a much greater calcium release from SR stores via RyR channels, giving rise to the calcium transient and ultimately myofibril contraction. In addition to being the primary signal for myocardial contraction, intracellular calcium regulates the transduction of electrical activation to mechanical function (excitation–contraction coupling) (see Bers 2001 for review). Over a century ago, it was discovered that myocardial contraction is stronger at faster pacing rates (staircase phenomenon or positive force-frequency relationship; Bowditch 1992). It has also been observed that the rate of muscle relaxation increases with pacing rate (frequency-dependent acceleration of relaxation, FDAR). In the heart, a role for CaMKII in regulating intracellular calcium cycling has been identified. It has been hypothesized that CaMKII underlies FDAR (Schouten 1990). Consistent with this hypothesis, CaMKII inhibitors have a dramatic effect on frequency-dependent acceleration of relaxation (Bassani et al. 1994) and excitation-contraction (EC) coupling (Li et al. 1997). Recently, we have used a computational approach to show that increased CaMKII activity at fast pacing rates enhances EC coupling gain and promotes a positive calcium transient-frequency relationship observed in normal myocytes (Hund and Rudy 2004; Fig. 2). Consistent with this simulation, CaMKII activity increases with pacing rate in isolated perfused rabbit hearts (Wehrens et al. 2004); the parallel increase in myocardial contractility with pacing rate is blocked by the CaMKII inhibitor KN-93.

The L-type Ca^{2+} current is important for phase 4 depolarization and automaticity of pacemaker cells from the sinoatrial (SA) node. Due to its facilitation of $I_{Ca(L)}$, CaMKII has been hypothesized to regulate SA node pacemaker activity and heart rate (Vinogradova et al. 2000). Support for this hypothesis comes from the fact that the CaMKII inhibitors AIP and KN-93 arrest spontaneous activity of cells isolated from rabbit SA node (Vinogradova et al. 2000).

4
Role of CaMKII in Cardiac Disease

Changes in CaMKII activity and/or expression have been documented in several animal models of cardiac disease. Hypertrophy, heart failure, myocardial ischemia, and infarction have all been associated with either an upregulation or downregulation of CaMKII activity (summarized in Table 1). In this section, we discuss the experimental findings regarding CaMKII alteration in cardiac disease and how changes in CaMKII may compromise cardiac function and promote the initiation of potentially fatal cardiac arrhythmias.

Table 1 CaMKII regulation in cardiac disease

Condition	Model	CaMKII change	Reference(s)
Hypertrophy	Transient aortic constriction in mice	Upregulation	Colomer et al. 2003
	Spontaneously hypertensive rat	Upregulation	Boknik et al. 2001; Hagemann et al. 2001; Hempel et al. 2002
Heart failure	Coronary artery ligation in rabbit	Upregulation	Currie and Smith 1999a,b
	Human	Upregulation	Hoch et al. 1999
	Microembolization in dog	Downregulation	Mishra et al. 2003
	Coronary artery occlusion in rat	Downregulation	Netticadan et al. 2000
Myocardial ischemia	Perfused rat heart	Downregulation	Osada et al. 1998; Netticadan et al. 1999
		Reduced autophosphorylation	Uemura et al. 2002
		No change	Vittone et al. 2002

4.1
Hypertrophy and Heart Failure

CaMKII upregulation in ventricular hypertrophy and heart failure has been observed by several groups. Myocytes isolated from hypertrophied myocardium 8 weeks after coronary artery ligation show increased levels of PLB phosphorylation and CaMKIIδ upregulation in a rabbit model of heart failure (Currie and Smith 1999a,b). Cardiac hypertrophy induced by transverse aortic constriction is associated with upregulated CaMKII activity, attributable to increased mRNA and protein levels (Colomer et al. 2003). Hypertensive rats show increased expression of the fetal δ_4 isoform compared to control (Hagemann et al. 2001; Hempel et al. 2002). Upregulation of δ_3 (or δ_B) has been identified in failing human myocardium (Hoch et al. 1999) and overexpression of CaMKIIδ$_B$ induces hypertrophy with decreased cardiac function in transgenic mice (Zhang et al. 2002). It has also been shown that mice overexpressing the cytosolic splice variant of cardiac CaMKII, δ_C, develop hypertrophy and heart failure (Zhang et al. 2003). Pressure overload may be a signal for altered CaMKII expression (Colomer et al. 2003), although the exact mechanism remains unknown.

A downregulation in CaMKII activity has been measured in other models of heart failure (Netticadan et al. 2000; Mishra et al. 2003). Intracoronary microembolization in the dog produces heart failure and decreased CaMKII activity (Mishra et al. 2003). Similarly, heart failure after myocardial infarction in the rat leads to CaMKII downregulation and reduced phosphorylation of SR substrates (Netticadan et al. 2000).

4.2
Ischemia-Reperfusion and Preconditioning

Transient global ischemia has been shown to decrease CaMKII activity in the brain (Aronowski et al. 1992; Churn et al. 1992; Hiestand et al. 1992; Westgate et al. 1994; Shackelford et al. 1995), which may be the result of a posttranslational modification in ATP binding to the kinase (Churn et al. 1992). Calcium channel blockers and the calmodulin antagonist, trifluoperazine, prevent CaMKII inactivation during ischemia in guinea pigs (Hiestand et al. 1992).

Myocardial ischemia results in abnormal contractile function and cardiac arrhythmia within minutes (Wit and Janse 1992; Mubagwa 1995). Paradoxically, reperfusion of the ischemic heart leads to further myocardial damage, which may be prevented by brief preconditioning cycles of ischemia-reperfusion before the onset of sustained ischemia. Ischemia-reperfusion has been shown to decrease cardiac function, SR Ca^{2+} uptake, and CaMKII-dependent phosphorylation of RyR, SERCA2a, and PLB in isolated perfused rat hearts (Osada et al. 1998; Netticadan et al. 1999). In the same preparation, preconditioning protects the myocardium from damage and eliminates differences in CaMKII activity before and after ischemia and CaMKII-mediated phosphorylation of the SR (Osada et al. 1998). The protective effects of preconditioning are greatly reduced by pretreatment with KN-93 (Osada et al. 2000). Reduced CaMKII phosphorylation of the SR is observed up to 4 weeks after myocardial infarction in a rat model of heart failure (Netticadan et al. 2000). Consistent with these findings, another group has measured translocation and reduced autophosphorylation of CaMKII in ischemic rat heart compared to control (Uemura et al. 2002). Others have observed no change in PLB phosphorylation at Threonine[17] after ischemia-reperfusion compared to pre-ischemic levels (Vittone et al. 2002). However, rats pretreated with KN-93 showed slowed recovery as did PLB[T17A] mutant mice (Said et al. 2003), suggesting an important role for CaMKII-mediated phosphorylation in recovery after ischemia.

Ischemia-reperfusion involves a number of physiological changes including hyperkalemia, acidosis, anoxia, and calcium overload. Reperfusion after prolonged ischemia results in severe calcium overload (Lee et al. 1987; Marban et al. 1987; Steenbergen et al. 1987), which has been found to inhibit CaMKII activity in cardiac cells (Netticadan et al. 2002). Therefore, calcium overload during ischemia may be one mechanism by which ischemia leads to altered CaMKII activity. In contrast, acidosis has been shown to activate CaMKII (Komukai et al. 2001; Nomura et al. 2002). It is not clear which effect, Ca^{2+} overload or acidosis, has a greater impact on CaMKII in vivo. Further investigation is necessary to fully understand the time course, mechanism, and impact of CaMKII changes during ischemia-reperfusion and ischemic preconditioning.

4.3
Cardiac Arrhythmia

Of the roughly 500,000 people that die from coronary heart disease each year, ventricular fibrillation is the immediate cause of death in the majority of cases (American Heart Association 2004). While a host of therapeutic strategies, including pharmaceuticals and medical devices, are available to prevent fibrillation and sudden cardiac death, clearly the need is great for a better understanding of what happens at the cellular level to predispose an ailing heart to life-threatening arrhythmias. Recently, several groups have examined the role of CaMKII in promoting cardiac arrhythmia. Early afterdepolarizations (EAD) are secondary depolarizations during the plateau or repolarization phase of the action potential that may serve as a triggering events for cardiac arrhythmia (see review: Volders et al. 2000). The CaMKII inhibitor KN-93 diminishes the inducibility of EADs by clofilium in isolated Langendorff-perfused rabbit hearts (Anderson et al. 1998). The calmodulin antagonist W-7 reduces the inducibility of torsades de pointes in an in vivo rabbit model (Mazur et al. 1999).

More recently, CaMKII-dependent arrhythmias have been identified in transgenic mouse models of cardiac hypertrophy (Wu et al. 2002; Kirchhof et al. 2004). Transgenic mice overexpressing constitutively active CaMKIV show reduced systolic function, enhanced CaMKII activity, and a greater number of arrhythmias compared to wild-type littermates at baseline and after isoproterenol, which were prevented by treatment with KN-93 (Wu et al. 2002). Cells isolated from transgenic mouse hearts showed a greater number of EADs, which were eliminated by the CaMKII inhibitory peptide, AC3-I (Wu et al. 2002). Knockout mice lacking the gene for the atrial natriuretic peptide receptor also show cardiac hypertrophy, increased CaMKII expression, and increased incidence of polymorphic ventricular tachycardia preceded by triggered activity (Kirchhof et al. 2004). Both W-7 and KN-93 greatly reduced the incidence of arrhythmia, as did the L-type Ca^{2+} channel blocker, verapamil. These experimental data support the hypothesis that CaMKII overexpression promotes arrhythmia by enhancing the inducibility of EADs.

EADs are generated by reactivation of the L-type Ca^{2+} current during the action potential (January and Riddle 1989; Zeng and Rudy 1995). Phosphorylation of L-type Ca^{2+} channels by CaMKII increases open channel probability and facilitates the current (discussed in Sect. 2.3), which may provide a mechanistic link between CaMKII overexpression and the induction of EADs and arrhythmias. Consistent with this hypothesis, Kirchhof and colleagues measured increased L-type open channel probability and CaMKII activity in their mouse model of cardiac hypertrophy (Kirchhof et al. 2004).

5
Summary and Conclusions

The importance of CaMKII as a ubiquitous "memory" macromolecule is becoming increasingly clear. Its sensitivity to the widespread second messenger calcium and its autophosphorylating capability make CaMKII ideally suited for mediating a number of important tasks in the body. Notably, its unique biophysical properties enable CaMKII to detect the frequency of calcium oscillations, making the enzyme a biochemical transducer of cell activity. While CaMKII has been thoroughly investigated in the nervous system, its importance in regulating cell function has only recently been appreciated in the heart. Several groups have established altered CaMKII activity and/or expression in heart failure, myocardial ischemia, and infarction. However, the data are incomplete and many questions remain to be answered. Clearly, more information is needed on the role of CaMKII in cardiac disease, the triggering events for alterations in CaMKII expression, and the relative importance of the different CaMKII targets in transducing the enzyme activity and its alteration by disease. This knowledge may lead to anti-arrhythmic therapeutic strategies to treat patients suffering from coronary heart disease.

References

American Heart Association (2004) Heart disease and stroke statistics—2004 update. American Heart Association

Anderson ME, Braun AP, Schulman H, Premack BA (1994) Multifunctional Ca^{2+}/calmodulin-dependent protein kinase mediates Ca^{2+}-induced enhancement of the L-type Ca^{2+} current in rabbit ventricular myocytes. Circ Res 75:854–861

Anderson ME, Braun AP, Wu Y, Lu T, Schulman H, Sung RJ (1998) KN-93, an inhibitor of multifunctional Ca^{2+}/calmodulin-dependent protein kinase, decreases early after depolarizations in rabbit heart. J Pharmacol Exp Ther 287:996–1006

Aronowski J, Grotta JC, Waxham MN (1992) Ischemia-induced translocation of Ca^{2+}/calmodulin-dependent protein kinase II: potential role in neuronal damage. J Neurochem 58:1743–1753

Barrett PQ, Lu HK, Colbran R, Czernik A, Pancrazio JJ (2000) Stimulation of unitary T-type Ca^{2+} channel currents by calmodulin-dependent protein kinase II. Am J Physiol Cell Physiol 279:C1694–1703

Bassani JW, Bassani RA, Bers DM (1994) Relaxation in rabbit and rat cardiac cells: species-dependent differences in cellular mechanisms. J Physiol (Lond) 476:279–293

Bennett MK, Erondu NE, Kennedy MB (1983) Purification and characterization of a calmodulin-dependent protein kinase that is highly concentrated in brain. J Biol Chem 258:12735–12744

Bers DM (2001) Excitation-contraction coupling and cardiac contractile force. Kluwer Academic Publishers, Dordrecht

Bilezikjian LM, Kranias EG, Potter JD, Schwartz A (1981) Studies on phosphorylation of canine cardiac sarcoplasmic reticulum by calmodulin-dependent protein kinase. Circ Res 49:1356–1362

Bliss TV, Collingridge GL (1993) A synaptic model of memory: long-term potentiation in the hippocampus. Nature 361:31–39

Boknik P, Heinroth-Hoffmann I, Kirchhefer U, Knapp J, Linck B, Luss H, Muller T, Schmitz W, Brodde O, Neumann J (2001) Enhanced protein phosphorylation in hypertensive hypertrophy. Cardiovasc Res 51:717–728

Bowditch HP (1992) On the peculiarities of excitability which the fibres of cardiac muscle show. In: Noble MIM, Seed WA (eds) The interval-force relationship of the heart: Bowditch revisited. Cambridge University Press, Cambridge, UK, pp 3–30

Braun AP, Schulman H (1995) The multifunctional calcium/calmodulin-dependent protein kinase: from form to function. Annu Rev Physiol 57:417–445

Ch'en FF, Vaughan-Jones RD, Clarke K, Noble D (1998) Modelling myocardial ischaemia and reperfusion. Prog Biophys Mol Biol 69:515–538

Cho YH, Giese KP, Tanila H, Silva AJ, Eichenbaum H (1998) Abnormal hippocampal spatial representations in αCaMKIIT286A and $CREB^{\alpha\Delta-}$ mice. Science 279:867–869

Churn SB, Taft WC, Billingsley MS, Sankaran B, DeLorenzo RJ (1992) Global forebrain ischemia induces a posttranslational modification of multifunctional calcium- and calmodulin-dependent kinase II. J Neurochem 59:1221–1232

Clancy CE, Rudy Y (1999) Linking a genetic defect to its cellular phenotype in a cardiac arrhythmia. Nature 400:566–569

Colomer JM, Mao L, Rockman HA, Means AR (2003) Pressure overload selectively up-regulates Ca^{2+}/calmodulin-dependent protein kinase II in vivo. Mol Endocrinol 17:183–192

Coomber CJ (1998) Site-selective autophosphorylation of Ca^{2+}/calmodulin-dependent protein kinase II as a synaptic encoding mechanism. Neural Comput 10:1653–1678

Currie S, Smith GL (1999a) Calcium/calmodulin-dependent protein kinase II activity is increased in sarcoplasmic reticulum from coronary artery ligated rabbit hearts. FEBS Lett 459:244–248

Currie S, Smith GL (1999b) Enhanced phosphorylation of phospholamban and down-regulation of sarco/endoplasmic reticulum Ca^{2+} ATPase type 2 (SERCA 2) in cardiac sarcoplasmic reticulum from rabbits with heart failure. Cardiovasc Res 41:135–146

Currie S, Loughrey CM, Craig MA, Smith GL (2004) Calcium/calmodulin-dependent protein kinase IIδ associates with the ryanodine receptor complex and regulates channel function in rabbit heart. Biochem J 377:357–366

Dash PK, Karl KA, Colicos MA, Prywes R, Kandel ER (1991) cAMP response element-binding protein is activated by Ca^{2+}/calmodulin- as well as cAMP-dependent protein kinase. Proc Natl Acad Sci U S A 88:5061–5065

De Koninck P, Schulman H (1998) Sensitivity of CaM kinase II to the frequency of Ca^{2+} oscillations. Science 279:227–230

Derkach V, Barria A, Soderling TR (1999) Ca^{2+}/calmodulin-kinase II enhances channel conductance of α-amino-3-hydroxy-5-methyl-4-isoxazolepropionate type glutamate receptors. Proc Natl Acad Sci U S A 96:3269–3274

Dosemeci A, Albers RW (1996) A mechanism for synaptic frequency detection through autophosphorylation of CaM kinase II. Biophys J 70:2493–2501

Dzhura I, Wu Y, Colbran RJ, Balser JR, Anderson ME (2000) Calmodulin kinase determines calcium-dependent facilitation of L-type calcium channels. Nat Cell Biol 2:173–177

Edman CF, Schulman H (1994) Identification and characterization of δB-CaM kinase and δC-CaM kinase from rat heart, two new multifunctional Ca^{2+}/calmodulin-dependent protein kinase isoforms. Biochim Biophys Acta 1221:89–101

Erondu NE, Kennedy MB (1985) Regional distribution of type II Ca^{2+}/calmodulin-dependent protein kinase in rat brain. J Neurosci 5:3270–3277

Fedida D, Noble D, Spindler AJ (1988) Use-dependent reduction and facilitation of Ca^{2+} current in guinea-pig myocytes. J Physiol (Lond) 405:439–460

Fukunaga K, Stoppini L, Miyamoto E, Muller D (1993) Long-term potentiation is associated with an increased activity of Ca^{2+}/calmodulin-dependent protein kinase II. J Biol Chem 268:7863–7867

Gaertner TR, Kolodziej SJ, Wang D, Kobayashi R, Koomen JM, Stoops JK, Waxham MN (2004) Comparative analyses of the three-dimensional structures and enzymatic properties of α, β, γ and δ isoforms of Ca^{2+}-calmodulin-dependent protein kinase II. J Biol Chem 279:12484–12494

Giese KP, Fedorov NB, Filipkowski RK, Silva AJ (1998) Autophosphorylation at Thr286 of the α calcium-calmodulin kinase II in LTP and learning. Science 279:870–873

Greengard P, Valtorta F, Czernik AJ, Benfenati F (1993) Synaptic vesicle phosphoproteins and regulation of synaptic function. Science 259:780–785

Griffith LC, Schulman H (1988) The multifunctional Ca^{2+}/calmodulin-dependent protein kinase mediates Ca^{2+}-dependent phosphorylation of tyrosine hydroxylase. J Biol Chem 263:9542–9549

Hagemann D, Hoch B, Krause EG, Karczewski P (1999) Developmental changes in isoform expression of Ca^{2+}/calmodulin-dependent protein kinase II δ-subunit in rat heart. J Cell Biochem 74:202–210

Hagemann D, Kuschel M, Kuramochi T, Zhu W, Cheng H, Xiao RP (2000) Frequency-encoding Thr17 phospholamban phosphorylation is independent of Ser16 phosphorylation in cardiac myocytes. J Biol Chem 275:22532–22536

Hagemann D, Bohlender J, Hoch B, Krause EG, Karczewski P (2001) Expression of Ca^{2+}/calmodulin-dependent protein kinase II δ-subunit isoforms in rats with hypertensive cardiac hypertrophy. Mol Cell Biochem 220:69–76

Hanson PI, Meyer T, Stryer L, Schulman H (1994) Dual role of calmodulin in autophosphorylation of multifunctional CaM kinase may underlie decoding of calcium signals. Neuron 12:943–956

Hawkins C, Xu A, Narayanan N (1994) Sarcoplasmic reticulum calcium pump in cardiac and slow twitch skeletal muscle but not fast twitch skeletal muscle undergoes phosphorylation by endogenous and exogenous Ca^{2+}/calmodulin-dependent protein kinase. Characterization of optimal conditions for calcium pump phosphorylation. J Biol Chem 269:31198–31206

Hempel P, Hoch B, Bartel S, Karczewski P (2002) Hypertrophic phenotype of cardiac calcium/calmodulin-dependent protein kinase II is reversed by angiotensin converting enzyme inhibition. Basic Res Cardiol 97 Suppl 1:I96–101

Hiestand DM, Haley BE, Kindy MS (1992) Role of calcium in inactivation of calcium/calmodulin dependent protein kinase II after cerebral ischemia. J Neurol Sci 113:31–37

Hoch B, Meyer R, Hetzer R, Krause EG, Karczewski P (1999) Identification and expression of δ-isoforms of the multifunctional Ca^{2+}/calmodulin-dependent protein kinase in failing and nonfailing human myocardium. Circ Res 84:713–721

Hoelz A, Nairn AC, Kuriyan J (2003) Crystal structure of a tetradecameric assembly of the association domain of Ca^{2+}/calmodulin-dependent kinase II. Mol Cell 11:1241–1251

Hook SS, Means AR (2001) Ca^{2+}/CaM-dependent kinases: from activation to function. Annu Rev Pharmacol Toxicol 41:471–505

Hudmon A, Schulman H (2002) Neuronal Ca^{2+}/calmodulin-dependent protein kinase II: the role of structure and autoregulation in cellular function. Annu Rev Biochem 71:473–510

Hund TJ, Rudy Y (2004) Rate dependence and regulation of action potential and calcium transient in a canine cardiac ventricular cell model. Circulation 110:3168–3174

Ikebe M, Reardon S, Scott-Woo GC, Zhou Z, Koda Y (1990) Purification and characterization of calmodulin-dependent multifunctional protein kinase from smooth muscle: isolation of caldesmon kinase. Biochemistry 29:11242–11248

Ishida A, Kameshita I, Okuno S, Kitani T, Fujisawa H (1995) A novel highly specific and potent inhibitor of calmodulin-dependent protein kinase II. Biochem Biophys Res Commun 212:806–812

January CT, Riddle JM (1989) Early afterdepolarizations: mechanism of induction and block. A role for L-type Ca^{2+} current. Circ Res 64:977–990

Ji Y, Li B, Reed TD, Lorenz JN, Kaetzel MA, Dedman JR (2003) Targeted inhibition of Ca^{2+}/calmodulin-dependent protein kinase II in cardiac longitudinal sarcoplasmic reticulum results in decreased phospholamban phosphorylation at threonine 17. J Biol Chem 278:25063–25071

Kanaseki T, Ikeuchi Y, Sugiura H, Yamauchi T (1991) Structural features of Ca^{2+}/calmodulin-dependent protein kinase II revealed by electron microscopy. J Cell Biol 115:1049–1060

Karcher RL, Roland JT, Zappacosta F, Huddleston MJ, Annan RS, Carr SA, Gelfand VI (2001) Cell cycle regulation of myosin-V by calcium/calmodulin-dependent protein kinase II. Science 293:1317–1320

Kelly PT, Weinberger RP, Waxham MN (1988) Active site-directed inhibition of Ca^{2+}/calmodulin-dependent protein kinase type II by a bifunctional calmodulin-binding peptide. Proc Natl Acad Sci U S A 85:4991–4995

Kennedy MB, Greengard P (1981) Two calcium/calmodulin-dependent protein kinases, which are highly concentrated in brain, phosphorylate protein I at distinct sites. Proc Natl Acad Sci U S A 78:1293–1297

Kikuchi S, Fujimoto K, Kitagawa N, Fuchikawa T, Abe M, Oka K, Takei K, Tomita M (2003) Kinetic simulation of signal transduction system in hippocampal long-term potentiation with dynamic modeling of protein phosphatase 2A. Neural Netw 16:1389–1398

Kirchhof P, Fabritz L, Kilic A, Begrow F, Breithardt G, Kuhn M (2004) Ventricular arrhythmias, increased cardiac calmodulin kinase II expression, and altered repolarization kinetics in ANP receptor deficient mice. J Mol Cell Cardiol 36:691–700

Kolodziej SJ, Hudmon A, Waxham MN, Stoops JK (2000) Three-dimensional reconstructions of calcium/calmodulin-dependent (CaM) kinase IIα and truncated CaM kinase IIα reveal a unique organization for its structural core and functional domains. J Biol Chem 275:14354–14359

Komukai K, Pascarel C, Orchard CH (2001) Compensatory role of CaMKII on ICa and SR function during acidosis in rat ventricular myocytes. Pflugers Arch 442:353–361

Kubota Y (1999) Decoding time-varying calcium signals by the postsynaptic biochemical network: computer simulations of molecular kinetics. Neurocomputing 26:29–38

Kubota Y, Bower JM (2001) Transient versus asymptotic dynamics of CaM kinase II: possible roles of phosphatase. J Comput Neurosci 11:263–279

Kuret J, Schulman H (1985) Mechanism of autophosphorylation of the multifunctional Ca^{2+}/calmodulin-dependent protein kinase. J Biol Chem 260:6427–6433

Lai Y, Nairn AC, Gorelick F, Greengard P (1987) Ca^{2+}/calmodulin-dependent protein kinase II: identification of autophosphorylation sites responsible for generation of Ca^{2+}/calmodulin-independence. Proc Natl Acad Sci U S A 84:5710–5714

Le Peuch CJ, Haiech J, Demaille JG (1979) Concerted regulation of cardiac sarcoplasmic reticulum calcium transport by cyclic adenosine monophosphate dependent and calcium/calmodulin-dependent phosphorylations. Biochemistry 18:5150–5157

Lee HC, Smith N, Mohabir R, Clusin WT (1987) Cytosolic calcium transients from the beating mammalian heart. Proc Natl Acad Sci U S A 84:7793–7797

Lee KS (1987) Potentiation of the calcium-channel currents of internally perfused mammalian heart cells by repetitive depolarization. Proc Natl Acad Sci U S A 84:3941–3945

Li L, Satoh H, Ginsburg KS, Bers DM (1997) The effect of Ca^{2+}-calmodulin-dependent protein kinase II on cardiac excitation-contraction coupling in ferret ventricular myocytes. J Physiol (Lond) 501:17–31

Lisman JE (1985) A mechanism for memory storage insensitive to molecular turnover: a bistable autophosphorylating kinase. Proc Natl Acad Sci U S A 82:3055–3057

Lisman JE, Goldring MA (1988) Feasibility of long-term storage of graded information by the Ca^{2+}/calmodulin-dependent protein kinase molecules of the postsynaptic density. Proc Natl Acad Sci U S A 85:5320–5324

Lledo PM, Hjelmstad GO, Mukherji S, Soderling TR, Malenka RC, Nicoll RA (1995) Calcium/calmodulin-dependent kinase II and long-term potentiation enhance synaptic transmission by the same mechanism. Proc Natl Acad Sci U S A 92:11175–11179

Lokuta AJ, Rogers TB, Lederer WJ, Valdivia HH (1995) Modulation of cardiac ryanodine receptors of swine and rabbit by a phosphorylation-dephosphorylation mechanism. J Physiol (Lond) 487:609–622

Lou LL, Schulman H (1989) Distinct autophosphorylation sites sequentially produce autonomy and inhibition of the multifunctional Ca^{2+}/calmodulin-dependent protein kinase. J Neurosci 9:2020–2032

Lynch MA (2004) Long-term potentiation and memory. Physiol Rev 84:87–136

MacLennan DH, Kranias EG (2003) Phospholamban: a crucial regulator of cardiac contractility. Nat Rev Mol Cell Biol 4:566–577

Maier LS, Zhang T, Chen L, DeSantiago J, Brown JH, Bers DM (2003) Transgenic CaMKIIδC overexpression uniquely alters cardiac myocyte Ca^{2+} handling: reduced SR Ca^{2+} load and activated SR Ca^{2+} release. Circ Res 92:904–911

Malenka RC, Kauer JA, Perkel DJ, Mauk MD, Kelly PT, Nicoll RA, Waxham MN (1989) An essential role for postsynaptic calmodulin and protein kinase activity in long-term potentiation. Nature 340:554–557

Malinow R, Schulman H, Tsien RW (1989) Inhibition of postsynaptic PKC or CaMKII blocks induction but not expression of LTP. Science 245:862–866

Marban E, Tsien RW (1982) Enhancement of calcium current during digitalis inotropy in mammalian heart: positive feed-back regulation by intracellular calcium? J Physiol (Lond) 329:589–614

Marban E, Kitakaze M, Kusuoka H, Porterfield JK, Yue DT, Chacko VP (1987) Intracellular free calcium concentration measured with 19F NMR spectroscopy in intact ferret hearts. Proc Natl Acad Sci U S A 84:6005–6009

Matsushita T, Moriyama S, Fukai T (1995) Switching dynamics and the transient memory storage in a model enzyme network involving Ca^{2+}/calmodulin-dependent protein kinase II in synapses. Biol Cybern 72:497–509

Mattiazzi A, Hove-Madsen L, Bers DM (1994) Protein kinase inhibitors reduce SR Ca transport in permeabilized cardiac myocytes. Am J Physiol Heart Circ Physiol 267:H812–820

Mayer P, Mohlig M, Schatz H, Pfeiffer A (1993) New isoforms of multifunctional calcium/calmodulin-dependent protein kinase II. FEBS Lett 333:315–318

Mayer P, Mohlig M, Idlibe D, Pfeiffer A (1995) Novel and uncommon isoforms of the calcium sensing enzyme calcium/calmodulin dependent protein kinase II in heart tissue. Basic Res Cardiol 90:372–379

Mayford M, Wang J, Kandel ER, O'Dell TJ (1995) CaMKII regulates the frequency-response function of hippocampal synapses for the production of both LTD and LTP. Cell 81:891–904

Mayford M, Bach ME, Huang YY, Wang L, Hawkins RD, Kandel ER (1996) Control of memory formation through regulated expression of a CaMKII transgene. Science 274:1678–1683

Mazur A, Roden DM, Anderson ME (1999) Systemic administration of calmodulin antagonist W-7 or protein kinase A inhibitor H-8 prevents torsade de pointes in rabbits. Circulation 100:2437–2442

McGuinness TL, Lai Y, Greengard P, Woodgett JR, Cohen P (1983) A multifunctional calmodulin-dependent protein kinase. Similarities between skeletal muscle glycogen synthase kinase and a brain synapsin I kinase. FEBS Lett 163:329–334

Meyer T, Hanson PI, Stryer L, Schulman H (1992) Calmodulin trapping by calcium-calmodulin-dependent protein kinase. Science 256:1199–1202

Michelson S, Schulman H (1994) CaM Kinase: a model for its activation and dynamics. J Theor Biol 171:281–290

Mishra S, Sabbah HN, Jain JC, Gupta RC (2003) Reduced Ca^{2+}-calmodulin-dependent protein kinase activity and expression in LV myocardium of dogs with heart failure. Am J Physiol Heart Circ Physiol 284:H876–883

Mubagwa K (1995) Sarcoplasmic reticulum function during myocardial ischaemia and reperfusion. Cardiovasc Res 30:166–175

Nakane M, Mitchell J, Forstermann U, Murad F (1991) Phosphorylation by calcium calmodulin-dependent protein kinase II and protein kinase C modulates the activity of nitric oxide synthase. Biochem Biophys Res Commun 180:1396–1402

Netticadan T, Xu A, Narayanan N (1996) Divergent effects of ruthenium red and ryanodine on Ca^{2+}/calmodulin-dependent phosphorylation of the Ca^{2+} release channel (ryanodine receptor) in cardiac sarcoplasmic reticulum. Arch Biochem Biophys 333:368–376

Netticadan T, Temsah R, Osada M, Dhalla NS (1999) Status of Ca^{2+}/calmodulin protein kinase phosphorylation of cardiac SR proteins in ischemia-reperfusion. Am J Physiol Cell Physiol 277:C384–391

Netticadan T, Temsah RM, Kawabata K, Dhalla NS (2000) Sarcoplasmic reticulum Ca^{2+}/Calmodulin-dependent protein kinase is altered in heart failure. Circ Res 86:596–605

Netticadan T, Temsah RM, Kawabata K, Dhalla NS (2002) Ca^{2+}-overload inhibits the cardiac SR Ca^{2+}-calmodulin protein kinase activity. Biochem Biophys Res Commun 293:727–732

Nghiem P, Ollick T, Gardner P, Schulman H (1994) Interleukin-2 transcriptional block by multifunctional Ca^{2+}/calmodulin kinase. Nature 371:347–350

Nomura N, Satoh H, Terada H, Matsunaga M, Watanabe H, Hayashi H (2002) CaMKII-dependent reactivation of SR Ca^{2+} uptake and contractile recovery during intracellular acidosis. Am J Physiol Heart Circ Physiol 283:H193–203

Odermatt A, Kurzydlowski K, MacLennan DH (1996) The vmax of the Ca^{2+}-ATPase of cardiac sarcoplasmic reticulum (SERCA2a) is not altered by Ca^{2+}/calmodulin-dependent phosphorylation or by interaction with phospholamban. J Biol Chem 271:14206–14213

Osada M, Netticadan T, Tamura K, Dhalla NS (1998) Modification of ischemia-reperfusion-induced changes in cardiac sarcoplasmic reticulum by preconditioning. Am J Physiol Heart Circ Physiol 274:H2025–2034

Osada M, Netticadan T, Kawabata K, Tamura K, Dhalla NS (2000) Ischemic preconditioning prevents I/R-induced alterations in SR calcium-calmodulin protein kinase II. Am J Physiol Heart Circ Physiol 278:H1791–1798

Payne ME, Fong YL, Ono T, Colbran RJ, Kemp BE, Soderling TR, Means AR (1988) Calcium/calmodulin-dependent protein kinase II. Characterization of distinct calmodulin binding and inhibitory domains. J Biol Chem 263:7190–7195

Pereda AE, Bell TD, Chang BH, Czernik AJ, Nairn AC, Soderling TR, Faber DS (1998) Ca^{2+}/calmodulin-dependent kinase II mediates simultaneous enhancement of gap-junctional conductance and glutamatergic transmission. Proc Natl Acad Sci U S A 95:13272–13277

Pettit DL, Perlman S, Malinow R (1994) Potentiated transmission and prevention of further LTP by increased CaMKII activity in postsynaptic hippocampal slice neurons. Science 266:1881–1885

Ramirez MT, Zhao XL, Schulman H, Brown JH (1997) The nuclear δB isoform of Ca^{2+}/calmodulin-dependent protein kinase II regulates atrial natriuretic factor gene expression in ventricular myocytes. J Biol Chem 272:31203–31208

Reddy LG, Jones LR, Pace RC, Stokes DL (1996) Purified, reconstituted cardiac Ca^{2+}-ATPase is regulated by phospholamban but not by direct phosphorylation with Ca^{2+}/calmodulin-dependent protein kinase. J Biol Chem 271:14964–14970

Rodriguez P, Bhogal MS, Colyer J (2003) Stoichiometric phosphorylation of cardiac ryanodine receptor on serine 2809 by calmodulin-dependent kinase II and protein kinase A. J Biol Chem 278:38593–38600

Said M, Vittone L, Mundina-Weilenmann C, Ferrero P, Kranias EG, Mattiazzi A (2003) Role of dual-site phospholamban phosphorylation in the stunned heart: insights from phospholamban site-specific mutants. Am J Physiol Heart Circ Physiol 285:H1198–1205

Schouten VJ (1990) Interval dependence of force and twitch duration in rat heart explained by Ca^{2+} pump inactivation in sarcoplasmic reticulum. J Physiol (Lond) 431:427–444

Schulman H (1988) The multifunctional Ca^{2+}/calmodulin-dependent protein kinase. Adv Second Messenger Phosphoprotein Res 22:39–112

Schulman H, Hanson PI, Meyer T (1992) Decoding calcium signals by multifunctional CaM kinase. Cell Calcium 13:401–411

Schworer CM, Colbran RJ, Keefer JR, Soderling TR (1988) Ca^{2+}/calmodulin-dependent protein kinase II. Identification of a regulatory autophosphorylation site adjacent to the inhibitory and calmodulin-binding domains. J Biol Chem 263:13486–13489

Schworer CM, Rothblum LI, Thekkumkara TJ, Singer HA (1993) Identification of novel isoforms of the δ subunit of Ca^{2+}/calmodulin-dependent protein kinase II. Differential expression in rat brain and aorta. J Biol Chem 268:14443–14449

Shackelford DA, Yeh RY, Hsu M, Buzsaki G, Zivin JA (1995) Effect of cerebral ischemia on calcium/calmodulin-dependent protein kinase II activity and phosphorylation. J Cereb Blood Flow Metab 15:450–461

Shaw RM, Rudy Y (1997) Electrophysiologic effects of acute myocardial ischemia: a theoretical study of altered cell excitability and action potential duration. Cardiovasc Res 35:256–272

Silva AJ, Paylor R, Wehner JM, Tonegawa S (1992a) Impaired spatial learning in α-calcium-calmodulin kinase II mutant mice. Science 257:206–211

Silva AJ, Stevens CF, Tonegawa S, Wang Y (1992b) Deficient hippocampal long-term potentiation in α-calcium-calmodulin kinase II mutant mice. Science 257:201–206

Srinivasan M, Edman CF, Schulman H (1994) Alternative splicing introduces a nuclear localization signal that targets multifunctional CaM kinase to the nucleus. J Cell Biol 126:839–852

Steenbergen C, Murphy E, Levy L, London RE (1987) Elevation in cytosolic free calcium concentration early in myocardial ischemia in perfused rat heart. Circ Res 60:700–707

Sumi M, Kiuchi K, Ishikawa T, Ishii A, Hagiwara M, Nagatsu T, Hidaka H (1991) The newly synthesized selective Ca^{2+}/calmodulin dependent protein kinase II inhibitor KN-93 reduces dopamine contents in PC12h cells. Biochem Biophys Res Commun 181:968–975

Tobimatsu T, Fujisawa H (1989) Tissue-specific expression of four types of rat calmodulin-dependent protein kinase II mRNAs. J Biol Chem 264:17907–17912

Tokumitsu H, Chijiwa T, Hagiwara M, Mizutani A, Terasawa M, Hidaka H (1990) KN-62, 1-[N,O-bis(5-isoquinolinesulfonyl)-N-methyl-L-tyrosyl]-4-phenylpiperazine, a specific inhibitor of Ca^{2+}/calmodulin-dependent protein kinase II. J Biol Chem 265:4315–4320

Toyofuku T, Curotto Kurzydlowski K, Narayanan N, MacLennan DH (1994) Identification of Ser38 as the site in cardiac sarcoplasmic reticulum Ca^{2+}-ATPase that is phosphorylated by Ca^{2+}/calmodulin-dependent protein kinase. J Biol Chem 269:26492–26496

Uemura A, Naito Y, Matsubara T (2002) Dynamics of Ca^{2+}/calmodulin-dependent protein kinase II following acute myocardial ischemia-translocation and autophosphorylation. Biochem Biophys Res Commun 297:997–1002

Vinogradova TM, Zhou YY, Bogdanov KY, Yang D, Kuschel M, Cheng H, Xiao RP (2000) Sinoatrial node pacemaker activity requires Ca^{2+}/calmodulin-dependent protein kinase II activation. Circ Res 87:760–767

Viswanathan PC, Rudy Y (2000) Cellular arrhythmogenic effects of congenital and acquired long-QT syndrome in the heterogeneous myocardium. Circulation 101:1192–1198

Vittone L, Mundina-Weilenmann C, Said M, Ferrero P, Mattiazzi A (2002) Time course and mechanisms of phosphorylation of phospholamban residues in ischemia-reperfused rat hearts. Dissociation of phospholamban phosphorylation pathways. J Mol Cell Cardiol 34:39–50

Volders PG, Vos MA, Szabo B, Sipido KR, de Groot SH, Gorgels AP, Wellens HJ, Lazzara R (2000) Progress in the understanding of cardiac early afterdepolarizations and torsades de pointes: time to revise current concepts. Cardiovasc Res 46:376–392

Wang Z, Wilson GF, Griffith LC (2002) Calcium/calmodulin-dependent protein kinase II phosphorylates and regulates the Drosophila Eag potassium channel. J Biol Chem 277:24022–24029

Wegener AD, Simmerman HK, Lindemann JP, Jones LR (1989) Phospholamban phosphorylation in intact ventricles. Phosphorylation of serine 16 and threonine 17 in response to beta-adrenergic stimulation. J Biol Chem 264:11468–11474

Wehrens XH, Lehnart SE, Reiken SR, Marks AR (2004) Ca^{2+}/calmodulin-dependent protein kinase II phosphorylation regulates the cardiac ryanodine receptor. Circ Res 94:e61–70

Westgate SA, Brown J, Aronowski J, Waxham MN (1994) Activity of Ca^{2+}/calmodulin-dependent protein kinase II following ischemia: a comparison between CA1 and dentate gyrus in a hippocampal slice model. J Neurochem 63:2217–2224

Winslow RL, Rice J, Jafri S, Marban E, O'Rourke B (1999) Mechanisms of altered excitation-contraction coupling in canine tachycardia-induced heart failure, II: model studies. Circ Res 84:571–586

Wit AL, Janse MJ (1992) The ventricular arrhythmias of ischemia and infarction: electrophysiological mechanisms. Futura Publishing, Mount Kisco

Witcher DR, Kovacs RJ, Schulman H, Cefali DC, Jones LR (1991) Unique phosphorylation site on the cardiac ryanodine receptor regulates calcium channel activity. J Biol Chem 266:11144–11152

Woodgett JR, Tonks NK, Cohen P (1982) Identification of a calmodulin-dependent glycogen synthase kinase in rabbit skeletal muscle, distinct from phosphorylase kinase. FEBS Lett 148:5–11

Wu Y, Colbran RJ, Anderson ME (2001) Calmodulin kinase is a molecular switch for cardiac excitation-contraction coupling. Proc Natl Acad Sci USA 98:2877–2881

Wu Y, Temple J, Zhang R, Dzhura I, Zhang W, Trimble R, Roden DM, Passier R, Olson EN, Colbran RJ, Anderson ME (2002) Calmodulin kinase II and arrhythmias in a mouse model of cardiac hypertrophy. Circulation 106:1288–1293

Xiao RP, Cheng H, Lederer WJ, Suzuki T, Lakatta EG (1994) Dual regulation of Ca^{2+}/calmodulin-dependent kinase II activity by membrane voltage and by calcium influx. Proc Natl Acad Sci U S A 91:9659–9663

Xu A, Narayanan N (1999) Ca^{2+}/calmodulin-dependent phosphorylation of the Ca^{2+}-ATPase, uncoupled from phospholamban, stimulates Ca^{2+}-pumping in native cardiac sarcoplasmic reticulum. Biochem Biophys Res Commun 258:66–72

Xu A, Netticadan T, Jones DL, Narayanan N (1999) Serine phosphorylation of the sarcoplasmic reticulum Ca^{2+}-ATPase in the intact beating rabbit heart. Biochem Biophys Res Commun 264:241–246

Yuan W, Bers DM (1994) Ca-dependent facilitation of cardiac Ca current is due to Ca-calmodulin-dependent protein kinase. Am J Physiol Heart Circ Physiol 267:H982–993

Zeng J, Rudy Y (1995) Early afterdepolarizations in cardiac myocytes: mechanism and rate dependence. Biophys J 68:949–964

Zhabotinsky AM (2000) Bistability in the Ca^{2+}/calmodulin-dependent protein kinase-phosphatase system. Biophys J 79:2211–2221

Zhang T, Johnson EN, Gu Y, Morissette MR, Sah VP, Gigena MS, Belke DD, Dillmann WH, Rogers TB, Schulman H, Ross J Jr, Brown JH (2002) The cardiac-specific nuclear δB isoform of Ca^{2+}/calmodulin-dependent protein kinase II induces hypertrophy and dilated cardiomyopathy associated with increased protein phosphatase 2A activity. J Biol Chem 277:1261–1267

Zhang T, Maier LS, Dalton ND, Miyamoto S, Ross J Jr, Bers DM, Brown JH (2003) The δC isoform of CaMKII is activated in cardiac hypertrophy and induces dilated cardiomyopathy and heart failure. Circ Res 92:912–919

Zuhlke R, Pitt G, Deisseroth K, Tsien R, Reuter H (1999) Calmodulin supports both inactivation and facilitation of L-type calcium channels. Nature 399:159–162

HEP (2006) 171:221–233
© Springer-Verlag Berlin Heidelberg 2006

AKAPs as Antiarrhythmic Targets?

S. O. Marx[1] (✉) · J. Kurokawa[2]

[1] Division of Cardiology, Department of Medicine and Pharmacology,
Columbia University College of Physicians and Surgeons, 630 W 168th St.,
New York NY, 10032, USA
sm460@columbia.edu

[2] Department of Bio-informational Pharmacology, Medical Research Institute,
Tokyo Medical and Dental University, 2-3-10 Kandasurugadai, Chiyoda-ku,
101-0062 Tokyo, Japan

Abstract Phosphorylation of ion channels plays a critical role in the modulation and amplification of biophysical signals. Kinases and phosphatases have broad substrate recognition
sequences. Therefore, the targeting of kinases and phosphatases to specific sites enhances
the regulation of diverse signaling events. Ion channel macromolecular complexes can be
formed by the association of A-kinase anchoring proteins (AKAPs) or other adaptor proteins
directly with the channel. The discovery that leucine/isoleucine zippers play an important
role in the recruitment of phosphorylation-modulatory proteins to certain ion channels has
permitted the elucidation of specific ion channel macromolecular complexes. Disruption of
signaling complexes by genetic defects can lead to abnormal physiological function. This
chapter will focus on evidence supporting the concept that ion channel macromolecular
complex formation plays an important role in regulating channel function in normal and
diseased states. Moreover, we demonstrate that abnormal complex formation may directly
lead to abnormal channel regulation by cellular signaling pathways, potentially leading to
arrhythmogenesis and cardiac dysfunction.

Keywords AKAPs · Leucine/isoleucine zippers · Ion channels ·
Macromolecular complexes · Phosphorylation

1
Introduction

Phosphorylation of ion channels plays a critical role in the modulation and amplification of biological signals. The specific activation and deactivation of signaling pathways by hormonal stimuli is enabled, in part, by phosphorylation. Kinases and phosphatases have broad substrate recognition sequences. Therefore, compartmentalization or targeting of signaling molecules represents an important modality to bring about specificity (Pawson and Scott 1997). Approximately 25 years ago, compartmentalization of cardiac cyclic AMP (cAMP)/protein kinase A (PKA) signaling was first proposed (Corbin et al. 1977). Recently, there have been significant advances in the elucidation of the mechanisms imparting specificity, providing supporting evidence for this hypothesis (Marx et al. 2000, 2001b, 2002; Zaccolo et al. 2002). This chapter will focus on evidence supporting the concept that ion channel macromolecular complex formation plays an important role in regulating channel function in normal and diseased states.

2
Protein Kinase A and A-Kinase Anchoring Proteins

PKA, a serine/threonine kinase, is a tetramer holoenzyme, comprising two catalytic (C) subunits and two regulatory (R) subunits (Corbin et al. 1977; Michel and Scott 2002). The association of the PKA catalytic subunit (PKA_c) with the R subunits maintains the holoenzyme in an inactive state. cAMP binding to the regulatory subunit permits dissociation of the catalytic subunits, relieving the inhibitory contact and thus brings about the phosphorylation of the appropriate target (Scott 1991; Theurkauf and Vallee 1982). A significant advance in the understanding of the modulation of PKA was the discovery of the role of A-kinase anchoring proteins (AKAPs), which anchor PKA_c to specific sites through binding to the R subunits (Feliciello et al. 2001; Pawson and Scott 1997). The catalytic subunits are encoded by three different genes (Cα, Cβ, and Cγ), whereas the regulatory subunits are encoded by four genes (RIα, RIβ, RIIα, RIIβ; Michel and Scott 2002; Scott 1991). PKA holoenzymes exist in two forms: type I (RIα and RIβ) are primarily cytoplasmic and are more sensitive to cAMP than type II (RIIα, RIIβ), which predominantly are associated with specific cellular proteins or structures (Michel and Scott 2002).

3
Scaffold Proteins

Microtubule-associated protein (MAP2) and AKAP75 were initially found to bind PKA (Sarkar et al. 1984; Theurkauf and Vallee 1982) through gel overlay, interaction cloning, yeast two hybrid screens and proteomic approaches. By classical definition, AKAPs associate with PKA because they contain an amphipathic helix that binds to the amino-terminus of the RII subunit (Carr et al. 1991). Recent evidence indicates that a few AKAPs also associate with the RI subunit (Michel and Scott 2002). AKAPs are localized to cellular compartments by specific sequences. Because of PKA_c association with the regulatory subunit, the PKA that is targeted by the AKAP is inactive, but can be activated in response to cAMP. In addition to PKA, AKAPs can also help to recruit a larger macromolecular complex. For instance, certain AKAPs associate with phosphodiesterases, protein phosphatases (PP), or both (Scott 1997). Muscle AKAP (mAKAP), which associates with ryanodine receptors (RyR)1 and RyR2, can recruit the PDE4D3 phosphodiesterase at the perinuclear region in rat cardiomyocytes (Dodge et al. 2001). Yotiao (AKAP9) can recruit PP1 in addition to PKA to the KCNQ1/KCNE1 ion channel (see Sect. 5). AKAPs serve as a multivalent scaffold to target kinases and phosphatases to specific compartments and play a major role in cardiovascular ion channel function in normal and diseased hearts. A recent study provided evidence that Yotiao also may serve as an effector in regulating the I_{Ks} channel (Kurokawa et al. 2004).

4
Ryanodine Receptor

The RyRs are the largest ion channels described to date (2.4 million daltons). They are tetrameric structures comprising four subunits, each approximately 600,000 Da. RyRs are ligand gated and are activated by micromolar Ca^{2+} and inhibited at millimolar Ca^{2+}. In cardiac muscle, RyR2 is activated by $Ca_V1.2$ (L-type Ca^{2+} channel)-mediated Ca^{2+} influx (Ca^{2+}-induced Ca^{2+} release; Fabiato and Fabiato 1979; Nabauer et al. 1989). The FK506-binding protein (FKBP12.6), a *cis–trans* peptidyl-prolyl isomerase, is associated with cardiac RyR2 and modulates its function by enhancing the cooperativity of the four subunits (Brillantes et al. 1994; Kaftan et al. 1996; Timerman et al. 1996). FKBP12.6 also influences neighboring channels through a process known as coupled gating (Marx et al. 1998, 2001a). Coupled gating provides a mechanism in which two or more physically connected channels gate simultaneously (Marx et al. 1998, 2001a).

FKBP12.6 binding to RyR2 can be physiologically regulated by PKA phosphorylation (Marx et al. 2000). In response to sympathetic stimulation, activation of the PKA pathway leads to dissociation of FKBP12.6 from the RyR2 complex. In failing hearts, PKA hyperphosphorylation of RyR leads to FKBP12.6 dissociation and abnormal channel function marked by increased Ca^{2+} sensitivity for activation, and elevated channel activity (probability of opening) associated with the appearance of subconductance states (Marx et al. 2000). Administration of metoprolol reverses the hyperphosphorylation, restoring the normal stoichiometry of the RyR macromolecular complex and normal channel function (Reiken et al. 2001; Reiken et al. 2003). The loss of FKBP12.6 (by genetic manipulation; FKBP12.6-null mice) leads to the development of exercise-induced arrhythmias and sudden cardiac death, due to aberrant Ca^{2+} release from the RyR (Wehrens et al. 2003). Heterozygous FKBP12.6-deficient mice also develop exercise-induced arrhythmias, due to a relative reduction in FKBP12.6, which is corrected by administration of the drug JTV-519 (Wehrens et al. 2004). These findings establish the critical role of the regulation of cardiac RyR phosphorylation.

The RyR contains a large cytosolic domain that regulates channel gating and serves as a scaffold for regulatory protein binding. RyRs contain leucine/isoleucine zippers (LIZ) that serve to recruit specific regulatory proteins. LIZs are α-helical structures that form coiled coils. They were originally found to mediate the binding of transcription factors to DNA (Landschulz et al. 1988). The sequence of coiled coils has been shown to contain heptad repeats $(abcdefg)_n$ in which hydrophobic residues occur at positions "a" and "d" and form the helix interface, while "b,c,e,f" and "g" are hydrophilic and form the solvent-exposed part of the coiled coil (Lupas 1996).

Prior to the discovery that LIZs play an important role in the recruitment of phosphorylation-modulatory proteins, they were found to be present in several ion channels including the human potassium channel hSK4 (hypothesized to play a role in the transduction of charge movement in Shaker potassium channel; McCormack et al. 1991) and in tetramer formation of the inositol triphosphate receptor (IP_3R) (Galvan et al. 1999). Moreover, the LIZ motif was shown to play an important role in the oligomerization of phospholamban, the phosphoprotein that regulates the SR Ca^{2+} ATPase (Arkin et al. 1994; Simmerman et al. 1996). We found LIZs in several ion channels including the RyR, IP_3R, $Ca_V1.2$ (L-type Ca^{2+} channel), and KCNQ1 (Hulme et al. 2002, 2003; Marx et al. 2000, 2001b, 2002; Tu et al. 2004).

The cardiac RyR2 contains three LIZs that serve to co-localize PP1, PP2A, and PKA to the channel (Marx et al. 2000, 2001b). The LIZs of RyR2 bind to LIZ in the targeting proteins spinophilin, PR130, and mAKAP (Fig. 1). By identifying the role of LIZs in mediating the formation of the RyR channel macromolecular complex, the isolation of the targeting proteins for the kinases and phosphatases was possible. mAKAP had been previously shown to co-localize with RyR based upon elegant immunostaining experiments (Yang et al. 1998) and was shown to bind to RyR2 based upon immunoprecipitation assays (Marx et al. 2000). A putative LIZ motif on RyR2 binds to a LIZ motif in mAKAP to mediate the association (Marx et al. 2001b). Disruption of the association of mAKAP/RII/PKA with the channel prevents cAMP-mediated phosphorylation of the channel and dissociation of FKBP12.6 (Marx et al. 2001b). Interestingly, mAKAP also binds to PDE4D3, potentially regulating the local concentration of cAMP around the cardiac RyR in vivo (Dodge et al. 2001). Control of local cAMP levels by an anchored PDE in the vicinity of

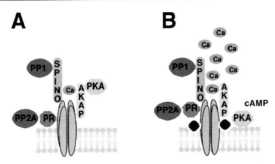

Fig. 1a,b Schematic representation of the RyR2 macromolecular complex. **a** PKA, PP1, and PP2A are targeted to the RyR by three anchoring proteins, mAKAP, spinophilin (*spino*), and PR130 (*PR*), respectively, via LIZ-mediated interactions betweens the anchoring proteins and the channel (Marx et al. 2001b). For illustrative purposes, only one scaffold protein is shown for each subunit, although each channel has four binding sites for mAKAP, spinophilin, and PR130. **b** Phosphorylation of S2809 on RyR2 (indicated by S) by PKA leads to increased channel open probability and increased Ca^{2+} release from SR

the RyR could potentially explain the differential regulation of PKA substrates seen in heart failure (phospholamban is hypophosphorylated whereas RyR is hyperphosphorylated; Huang et al. 1999; Mishra et al. 2002; Reiken et al. 2001; Schwinger et al. 1999).

In heart failure, the number of PP1 and PP2A catalytic subunits associated with the channel is reduced (Marx et al. 2000), potentially leading to pathological channel hyperphosphorylation. Like PKA, the localization of PP1 to subcellular targets has been shown to be mediated by anchoring proteins (Chisholm and Cohen 1988; Herzig and Neumann 2000; Stralfors et al. 1985). For instance, spinophilin/neurabin enables the binding of PP1 to post-synaptic density in neurons and RyR2 (Allen et al. 1997; Marx et al. 2000; McAvoy et al. 1999), and yotiao (AKAP9) enables the binding of PP1 to *N*-methyl-D-aspartate (NMDA) receptor and KCNQ1 (Lin et al. 1998; Marx et al. 2002). AKAP9 also binds to PKA, bringing both PKA and PP1 in close proximity to the ion channel.

5
I_{Ks} Channel

I_{Ks}, the slowly activating component of the human cardiac delayed rectifier K^+ current is a major contributor to repolarization of the cardiac action potential (AP) (Clancy et al. 2003; Kurokawa et al. 2001). Moreover, I_{Ks} is a dominant determinant of the physiological heart rate-dependent shortening of duration of AP (APD) (Zeng et al. 1995). The contribution of I_{Ks} to regulation of APD is augmented by the sympathetic nervous system (SNS). Stimulation of β-adrenergic receptor (β-AR) acts to increase the heart rate, and also results in a rate-dependent shortening of the APD (Kass and Wiegers 1982). The

I_{Ks} channel is one of several targets of PKA that occurs subsequent to β-AR stimulation. PKA-dependent phosphorylation of I_{Ks} channels results in increased I_{Ks} current amplitude and faster cardiac repolarization (Kurokawa et al. 2003). This increase in repolarization currents is essential to counter the stimulatory effects of PKA on L-type Ca^{2+} channels (Kass and Wiegers 1982). The result is that a balance of inward and outward membrane currents regulates the duration of ventricular APD, and consequently the Q-T interval, in response to SNS stimulation.

I_{Ks} results from the co-assembly of two subunits KCNQ1 (KvLQT1) and KCNE1 (minK) (Barhanin et al. 1996; Sanguinetti et al. 1996). The genes that encode the subunit components of the I_{Ks} channel, *KCNQ1* and *KCNE1*, have been shown to harbor mutations linked to the congenital long QT syndrome (LQTS). Mutations in *KCNQ1* cause LQT-1, and mutations in *KCNE1* channel cause LQT-5 (Splawski et al. 2000). In affected patients, triggers of arrhythmias are gene-specific, and those with mutations in either KCNQ1 or KCNE1 are at greatest risk of experiencing a fatal cardiac arrhythmia in the face of elevated SNS activity (Keating and Sanguinetti 2001; Priori et al. 1999; Schwartz et al. 2001). Unraveling the molecular links between the SNS and regulation of the KCNQ1/KCNE1 channel has direct implications for understanding the mechanistic basis of triggers of arrhythmias in LQTS.

The KCNQ1/KCNE1 channel forms a macromolecular signaling complex that is coordinated by binding of a targeting protein, yotiao (AKAP9) (Lin et al. 1998) via a LIZ motif in the C-terminus of KCNQ1, which in turn binds to and recruits PKA and PP1 to the channel (Marx et al. 2002). The complex then regulates the phosphorylation of Ser^{27} in the N-terminus of KCNQ1 (Marx et al. 2002; Fig. 2a). Reconstitution of PKA and PP1-mediated regulation of the KCNQ1/KCNE1 current in Chinese hamster ovary (CHO) cells requires co-expression of KCNQ1/KCNE1 and yotiao, and is ablated by mutation of the KCNQ1 LIZ which prevents yotiao binding to the channel, resulting in ablation of PKA phosphorylation of Ser^{27} (Marx et al. 2002).

Just as artificial mutations disrupt the LIZ motif, the naturally occurring *G589D* mutation at an "*e*" position in the LIZ motif of *hKCNQ1* disrupts targeting of yotiao to hKCNQ1 (Fig. 2b; Marx et al. 2002). The inherited *G589D* mutation has been linked to LQT-1 in Finnish families (Piippo et al. 2001). Moreover, the *KCNQ1-G589D* mutation, by virtue of the fact that it disrupts the LIZ motif in the C-terminus of KCNQ1 nullifies β-adrenergic-mediated regulation of the channel. The G589D mutation causes a defect in regulation of the channel by preventing assembly of the macromolecular complex that targets PKA and PP1 to the C-terminus of the channel. Affected LQTS patients suffer from dysfunctional regulation of QT duration during mental and physical stress (Paavonen et al. 2001) and are at risk of arrhythmia and sudden cardiac death during exercise (Piippo et al. 2001).

Kurokawa et al. (2003) demonstrated that cAMP-mediated functional regulation of KCNQ1/KCNE1 channels via PKA phosphorylation of the KCNQ1 N-

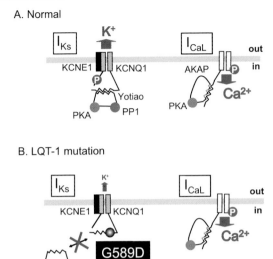

A. Normal

B. LQT-1 mutation

C. LQT-5 mutation

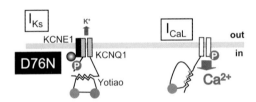

Fig. 2a–c Schematic diagrams of cardiac myocytes indicating signaling microdomains for I_{Ks} and I_{CaL} under SNS stimulation. **a** Normal (wildtype). In *left*, PKA and PP1 are targeted to the I_{Ks} (KCNQ1/KCNE1) channel by yotiao. In *right*, PKA is targeted to the L-type Ca^{2+} channel by AKAP15. In wildtype cells, elevated intracellular cAMP via β-AR stimulation leads to PKA-dependent phosphorylation of both K^+ and Ca^{2+} channels, resulting in enhancement of both channel currents. **b** LQT-1 mutation. Uncoupling yotiao via the LQT-1 *G589D* mutation (disruption of the LIZ motif) precludes I_{Ks}, but not I_{CaL}, channels from β-AR-mediated phosphorylation. **c** LQT-5 mutation. The LQT-5 *D76N* mutation does not uncouple I_{Ks} channels from PKA-mediated phosphorylation but ablates the functional response to β-AR stimulation. L-type Ca^{2+} channels are not affected by this mutation

terminus requires the expression of KCNQ1 with its auxiliary subunit KCNE1, although KCNE1 is not required for phosphorylation of KCNQ1 (Kurokawa et al. 2003). In other words, KCNQ1 phosphorylation is independent of co-assembly with KCNE1, but transduction of the phosphorylated channel into the physiologically essential increase in reserve channel activity requires the presence of KCNE1. In the absence of KCNE1, there is no significant effect of KCNQ1 phosphorylation on expressed channel activity.

The importance of KCNE1 association was revealed in a recent study that showed a point mutation in *KCNE1* linked to LQT-5, *D76N*, can severely disrupt the functional consequences of KCNQ1 phosphorylation (Kurokawa et al. 2003). This mutation reduces basal current density and would be expected to reduce repolarizing current, prolong cellular action potentials, and contribute to prolonged Q-T intervals in those expressing the mutation even in the absence of SNS stimulation (Bianchi et al. 1999). The *D76N* mutation also ablates functional regulation of the channels by cAMP (Fig. 2c; Kurokawa et al. 2003). The presumed consequence of the mutation is that in the face of SNS stimulation, there will be an insufficient reserve of K^+ channels to allow for appropriate shortening of the APD that is required at faster heart rates to allow for sufficiently long diastolic intervals required for ventricular filling.

Because PKA-dependent regulation of at least three key ion channels in the heart (RyR2, L-type Ca^{2+} channels, and KCNQ1/KCNE1 channels) requires assembly with AKAP-mediated macromolecular signaling complexes, it is clear that disruption of a subset of these complexes can lead to an imbalanced response to SNS stimulation. The LQT-1 mutation, *G589D*, is the first example of disease-associated disruption of a microdomain-signaling complex (Fig. 2b; Marx et al. 2002). The *D76N* mutation of KCNE1 represents a second, and additionally novel, mechanism of disrupting regulation of a local targeted ion channel (Fig. 2c; Kurokawa et al. 2003). In the latter case, however, it is functional uncoupling, and not biochemical uncoupling, that occurs.

6
Other Channels and Receptors in Heart

Activation of β-ARs and consequent phosphorylation by PKA increases the cardiac L-type Ca^{2+} current through $Ca_V 1.2$ channels. Recently, AKAP15 has been reported to target PKA to the $Ca_V 1.2$ channel in cardiac muscle via a LIZ motif (Hulme et al. 2003) as well as to the $Ca_V 1.1$ channel in skeletal muscle (Hulme et al. 2002).

Interaction between $β_2$-AR and gravin (AKAP250) (Fan et al. 2001; Tao et al. 2003) and between $Na^+–Ca^{2+}$ exchanger and mAKAP (Schulze et al. 2003) have also been reported.

7
Summary

It is now well established that macromolecular signaling complexes, coordinated by the binding of adaptor proteins to target proteins, are essential in creating micro signaling environments of many proteins, including ion channels (Marx et al. 2001, 2002; Tu et al. 2004; Hulme et al. 2002, 2003; Colledge et al. 2000; Hoshi et al. 2003; Westphal et al. 1999) and exchangers (Schulze et al. 2003). The identification of LIZ motifs in ion channels has provided a road map to elucidate new signaling pathways that modulate cardiac ion channels. Understanding how ion channels are modulated should represent a major focus, since phosphorylation can significantly modulate channel function and the cardiac action potential. Altered channel phosphorylation and function in disease states can lead to heart failure and arrhythmogenesis/sudden cardiac death. Disruption of signaling complexes by genetic defects can lead to sudden cardiac death (Paavonen et al. 2001; Piippo et al. 2001), kidney disease (Orellana et al. 2003), and cystic fibrosis (Sun et al. 2000), and genetic variation in AKAPs may raise the risk of susceptibility in complex diseases (Kammerer et al. 2003). Local signaling domains, coordinated by AKAPs, thus become important to our understanding of both the genesis of cardiac arrhythmias and of novel targets to treat and prevent them at the molecular level.

Acknowledgements S.O.M. is supported by National Heart, Lung and Blood Institute (HL-68093), American Heart Association Heritage Affiliate Grant-in-Aid and the Goldstein Family Fund.

References

Allen P, Ouimet C, Greengard P (1997) Spinophilin, a novel protein phosphatase 1 binding protein localized to dendritic spines. Proc Natl Acad Sci USA 94:9956–9961

Arkin IT, Adams PD, MacKenzie KR, Lemmon MA, Brunger AT, Engelman DM (1994) Structural organization of the pentameric transmembrane alpha-helices of phospholamban, a cardiac ion channel. EMBO J 13:4757–4764

Barhanin J, Lesage F, Guillemare E, Fink M, Lazdunski M, Romey G (1996) K(V)LQT1 and lsK (minK) proteins associate to form the I(Ks) cardiac potassium current. Nature 384:78–80

Bianchi L, Shen Z, Dennis AT, Priori SG, Napolitano C, Ronchetti E, Bryskin R, Schwartz PJ, Brown AM (1999) Cellular dysfunction of LQT5-minK mutants: abnormalities of IKs, IKr and trafficking in long QT syndrome. Hum Mol Genet 8:1499–1507

Brillantes A-MB, Ondrias K, Jayaraman T, Scott A, Kobrinsky E, Ehrlich BE, Marks A (1994) FKBP12 optimizes function of the cloned expressed calcium release channel (ryanodine receptor). Biophys J 66:A19

Carr DW, Stofko-Hahn RE, Fraser ID, Bishop SM, Acott TS, Brennan RG, Scott JD (1991) Interaction of the regulatory subunit (RII) of cAMP-dependent protein kinase with RII-anchoring proteins occurs through an amphipathic helix binding motif. J Biol Chem 266:14188–14192

Chisholm AA, Cohen P (1988) The myosin-bound form of protein phosphatase 1 (PP-1M) is the enzyme that dephosphorylates native myosin in skeletal and cardiac muscles. Biochim Biophys Acta 971:163–169

Clancy CE, Kurokawa J, Tateyama M, Wehrens XH, Kass RS (2003) K+ channel structure-activity relationships and mechanisms of drug-induced QT prolongation. Annu Rev Pharmacol Toxicol 43:441–461

Colledge M, Dean RA, Scott GK, Langeberg LK, Huganir RL, Scott JD (2000) Targeting of PKA to glutamate receptors through a MAGUK-AKAP complex. Neuron 27:107–119

Corbin JD, Sugden PH, Lincoln TM, Keely SL (1977) Compartmentalization of adenosine 3':5'-monophosphate and adenosine 3':5'-monophosphate-dependent protein kinase in heart tissue. J Biol Chem 252:3854–3861

Dodge KL, Khouangsathiene S, Kapiloff MS, Mouton R, Hill EV, Houslay MD, Langeberg LK, Scott JD (2001) mAKAP assembles a protein kinase A/PDE4 phosphodiesterase cAMP signaling module. EMBO J 20:1921–1930

Fabiato A, Fabiato F (1979) Calcium and cardiac excitation-contraction coupling. Annu Rev Physiol 41:473–484

Fan G, Shumay E, Wang HH, Malbon CC (2001) The scaffold protein gravin (AKAP250) binds the β2-adrenergic receptor via the receptor cytoplasmic R329 to L413 domain and provides a mobile scaffold during desensitization. J Biol Chem 276:24005–24014

Feliciello A, Gottesman ME, Avvedimento EV (2001) The biological functions of A-kinase anchor proteins. J Mol Biol 308:99–114

Galvan DL, Borrego-Diaz E, Perez PJ, Mignery GA (1999) Subunit oligomerization, and topology of the inositol 1,4,5-trisphosphate receptor. J Biol Chem 274:29483–29492

Herzig S, Neumann J (2000) Effects of serine/threonine protein phosphatases on ion channels in excitable membranes. Physiol Rev 80:173–210

Hoshi N, Zhang JS, Omaki M, Takeuchi T, Yokoyama S, Wanaverbecq N, Langeberg LK, Yoneda Y, Scott JD, Brown DA, Higashida H (2003) AKAP150 signaling complex promotes suppression of the M-current by muscarinic agonists. Nat Neurosci 6:564–571

Huang B, Wang S, Qin D, Boutjdir M, El-Sherif N (1999) Diminished basal phosphorylation level of phospholamban in the postinfarction remodeled rat ventricle: role of beta-adrenergic pathway, G(i) protein, phosphodiesterase, and phosphatases. Circ Res 85:848–855

Hulme JT, Ahn M, Hauschka SD, Scheuer T, Catterall WA (2002) A novel leucine zipper targets AKAP15 and cyclic AMP-dependent protein kinase to the C terminus of the skeletal muscle Ca2+ channel and modulates its function. J Biol Chem 277:4079–4087

Hulme JT, Lin TW, Westenbroek RE, Scheuer T, Catterall WA (2003) Beta-adrenergic regulation requires direct anchoring of PKA to cardiac CaV1.2 channels via a leucine zipper interaction with A kinase-anchoring protein 15. Proc Natl Acad Sci U S A 100:13093–13098

Kaftan E, Marks AR, Ehrlich BE (1996) Effects of rapamycin on ryanodine receptor/Ca2+-release channels from cardiac muscle. Circ Res 78:990–997

Kammerer S, Burns-Hamuro LL, Ma Y, Hamon SC, Canaves JM, Shi MM, Nelson MR, Sing CF, Cantor CR, Taylor SS, Braun A (2003) Amino acid variant in the kinase binding domain of dual-specific A kinase-anchoring protein 2: a disease susceptibility polymorphism. Proc Natl Acad Sci U S A 100:4066–4071

Kass RS, Wiegers SE (1982) The ionic basis of concentration-related effects of noradrenaline on the action potential of calf cardiac purkinje fibres. J Physiol 322:541–558

Keating MT, Sanguinetti MC (2001) Molecular and cellular mechanisms of cardiac arrhythmias. Cell 104:569–580

Kurokawa J, Abriel H, Kass RS (2001) Molecular basis of the delayed rectifier current I(ks)in heart. J Mol Cell Cardiol 33:873–882

Kurokawa J, Chen L, Kass RS (2003) Requirement of subunit expression for cAMP-mediated regulation of a heart potassium channel. Proc Natl Acad Sci U S A 100:2122–2127

Kurokawa J, Motoike HK, Rao J, Kass RS (2004) Regulatory action of the A-kinase anchoring protein Yotiao on a heart potassium channel downstream of PKA phosphorylation. Proc Natl Acad Sci U S A 101:16374–16378

Landschulz WH, Johnson PF, McKnight SL (1988) The leucine zipper: a hypothetical structure common to a new class of DNA binding proteins. Science 240:1759–1764

Lin JW, Wyszynski M, Madhavan R, Sealock R, Kim JU, Sheng M (1998) Yotiao, a novel protein of neuromuscular junction and brain that interacts with specific splice variants of NMDA receptor subunit NR1. J Neurosci 18:2017–2027

Lupas A (1996) Coiled coils: new structures and new functions. TIBS 21:375–382

Marx SO, Ondrias K, Marks AR (1998) Coupled gating between individual skeletal muscle Ca2+ release channels (ryanodine receptors). Science 281:818–821

Marx SO, Reiken S, Hisamatsu Y, Jayaraman T, Burkhoff D, Rosemblit N, Marks AR (2000) PKA phosphorylation dissociates FKBP12.6 from the calcium release channel (ryanodine receptor): defective regulation in failing hearts. Cell 101:365–376

Marx SO, Gaburjakova J, Gaburjakova M, Henrikson C, Ondrias K, Marks AR (2001a) Coupled gating between cardiac calcium release channels (ryanodine receptors). Circ Res 88:1151–1158

Marx SO, Reiken S, Hisamatsu Y, Gaburjakova M, Gaburjakova J, Yang YM, Rosemblit N, Marks AR (2001b) Phosphorylation-dependent regulation of ryanodine receptors: a novel role for leucine/isoleucine zippers. J Cell Biol 153:699–708

Marx SO, Kurokawa J, Reiken S, Motoike H, D'Armiento J, Marks AR, Kass RS (2002) Requirement of a macromolecular signaling complex for beta adrenergic receptor modulation of the KCNQ1-KCNE1 potassium channel. Science 295:496–499

McAvoy T, Allen PB, Obaishi H, Nakanishi H, Takai Y, Greengard P, Nairn AC, Hemmings HC Jr (1999) Regulation of neurabin I interaction with protein phosphatase 1 by phosphorylation. Biochemistry 38:12943–12949

McCormack K, Tanouye MA, Iverson LE, Lin JW, Ramaswami M, McCormack T, Campanelli JT, Mathew MK, Rudy B (1991) A role for hydrophobic residues in the voltage-dependent gating of Shaker K+ channels. Proc Natl Acad Sci U S A 88:2931–2935

Michel JJ, Scott JD (2002) AKAP mediated signal transduction. Annu Rev Pharmacol Toxicol 42:235–257

Mishra S, Gupta RC, Tiwari N, Sharov VG, Sabbah HN (2002) Molecular mechanisms of reduced sarcoplasmic reticulum Ca(2+) uptake in human failing left ventricular myocardium. J Heart Lung Transplant 21:366–373

Nabauer M, Callewaert G, Cleemann L, Morad M (1989) Regulation of calcium release is gated by calcium current, not gating charge, in cardiac myocytes. Science 244:800–803

Orellana SA, Quinones AM, Mandapat ML (2003) Ezrin distribution is abnormal in principal cells from a murine model of autosomal recessive polycystic kidney disease. Pediatr Res 54:406–412

Paavonen KJ, Swan H, Piippo K, Hokkanen L, Laitinen P, Viitasalo M, Toivonen L, Kontula K (2001) Response of the QT interval to mental and physical stress in types LQT1 and LQT2 of the long QT syndrome. Heart 86:39–44

Pawson T, Scott JD (1997) Signaling through scaffold, anchoring, and adaptor proteins. Science 278:2075–2080

Piippo K, Swan H, Pasternack M, Chapman H, Paavonen K, Viitasalo M, Toivonen L, Kontula K (2001) A founder mutation of the potassium channel KCNQ1 in long QT syndrome: implications for estimation of disease prevalence and molecular diagnostics. J Am Coll Cardiol 37:562–568

Priori SG, Barhanin J, Hauer RN, Haverkamp W, Jongsma HJ, Kleber AG, McKenna WJ, Roden DM, Rudy Y, Schwartz K, Schwartz PJ, Towbin JA, Wilde AM (1999) Genetic and molecular basis of cardiac arrhythmias: impact on clinical management parts I and II. Circulation 99:518–528

Reiken S, Gaburjakova M, Gaburjakova J, He Kl KL, Prieto A, Becker E, Yi Gh GH, Wang J, Burkhoff D, Marks AR (2001) β-Adrenergic receptor blockers restore cardiac calcium release channel (ryanodine receptor) structure and function in heart failure. Circulation 104:2843–2848

Reiken S, Gaburjakova M, Guatimosim S, Gomez AM, D'Armiento J, Burkhoff D, Wang J, Vassort G, Lederer WJ, Marks AR (2003) Protein kinase A phosphorylation of the cardiac calcium release channel (ryanodine receptor) in normal and failing hearts. Role of phosphatases and response to isoproterenol. J Biol Chem 278:444–453

Sanguinetti MC, Curran ME, Zou A, Shen J, Spector PS, Atkinson DL, Keating MT (1996) Coassembly of K(V)LQT1 and minK (IsK) proteins to form cardiac I(Ks) potassium channel. Nature 384:80–83

Sarkar D, Erlichman J, Rubin CS (1984) Identification of a calmodulin-binding protein that co-purifies with the regulatory subunit of brain protein kinase II. J Biol Chem 259:9840–9846

Schulze DH, Muqhal M, Lederer WJ, Ruknudin AM (2003) Sodium/calcium exchanger (NCX1) macromolecular complex. J Biol Chem 278:28849–28855

Schwartz PJ, Priori SG, Spazzolini C, Moss AJ, Vincent GM, Napolitano C, Denjoy I, Guicheney P, Breithardt G, Keating MT, Towbin JA, Beggs AH, Brink P, Wilde AA, Toivonen L, Zareba W, Robinson JL, Timothy KW, Corfield V, Wattanasirichaigoon D, Corbett C, Haverkamp W, Schulze-Bahr E, Lehmann MH, Schwartz K, Coumel P, Bloise R (2001) Genotype-phenotype correlation in the long-QT syndrome: gene-specific triggers for life-threatening arrhythmias. Circulation 103:89–95

Schwinger RH, Munch G, Bolck B, Karczewski P, Krause EG, Erdmann E (1999) Reduced Ca(2+)-sensitivity of SERCA 2a in failing human myocardium due to reduced serin-16 phospholamban phosphorylation. J Mol Cell Cardiol 31:479–491

Scott JD (1991) Cyclic nucleotide-dependent protein kinases. Pharmacol Ther 50:123–145

Scott JD (1997) Dissection of protein kinase and phosphatase targeting interactions. Soc Gen Physiol Ser 52:227–239

Simmerman HK, Kobayashi YM, Autry JM, Jones LR (1996) A leucine zipper stabilizes the pentameric membrane domain of phospholamban and forms a coiled-coil pore structure. J Biol Chem 271:5941–5946

Splawski I, Shen J, Timothy KW, Lehmann MH, Priori S, Robinson JL, Moss AJ, Schwartz PJ, Towbin JA, Vincent GM, Keating MT (2000) Spectrum of mutations in long-QT syndrome genes. KVLQT1, HERG, SCN5A, KCNE1, and KCNE2. Circulation 102:1178–1185

Stralfors P, Hiraga A, Cohen P (1985) The protein phosphatases involved in cellular regulation. Purification and characterisation of the glycogen-bound form of protein phosphatase-1 from rabbit skeletal muscle. Eur J Biochem 149:295–303

Sun F, Hug MJ, Bradbury NA, Frizzell RA (2000) Protein kinase A associates with cystic fibrosis transmembrane conductance regulator via an interaction with ezrin. J Biol Chem 275:14360–14366

Tao J, Wang H, Malbon CC (2003) Protein kinase A regulates AKAP250 (gravin) scaffold binding to the β2-adrenergic receptor. EMBO J 22:6419–6429

Theurkauf WE, Vallee RB (1982) Molecular characterization of the cAMP-dependent protein kinase bound to microtubule-associated protein 2. J Biol Chem 257:3284–3290

Timerman AP, Onoue H, Xin HB, Barg S, Copello J, Wiederrecht G, Fleischer S (1996) Selective binding of FKBP12.6 by the cardiac ryanodine receptor. J Biol Chem 271:20385–20391

Tu H, Tang TS, Wang Z, Bezprozvanny I (2004) Association of type 1 inositol 1,4,5-trisphosphate receptor with AKAP9 (Yotiao) and protein kinase A. J Biol Chem 279:19375–19382

Wehrens XH, Lehnart SE, Huang F, Vest JA, Reiken SR, Mohler PJ, Sun J, Guatimosim S, Song LS, Rosemblit N, D'Armiento JM, Napolitano C, Memmi M, Priori SG, Lederer WJ, Marks AR (2003) FKBP12.6 deficiency and defective calcium release channel (ryanodine receptor) function linked to exercise-induced sudden cardiac death. Cell 113:829–840

Wehrens XH, Lehnart SE, Reiken SR, Deng SX, Vest JA, Cervantes D, Coromilas J, Landry DW, Marks AR (2004) Protection from cardiac arrhythmia through ryanodine receptor-stabilizing protein calstabin2. Science 304:292–296

Westphal RS, Tavalin SJ, Lin JW, Alto NM, Fraser ID, Langeberg LK, Sheng M, Scott JD (1999) Regulation of NMDA receptors by an associated phosphatase-kinase signaling complex. Science 285:93–96

Yang J, Drazba JA, Ferguson DG, Bond M (1998) A-kinase anchoring protein 100 (AKAP100) is localized in multiple subcellular compartments in the adult rat heart. J Cell Biol 142:511–522

Zaccolo M, Magalhaes P, Pozzan T (2002) Compartmentalisation of cAMP and Ca(2+) signals. Curr Opin Cell Biol 14:160–166

Zeng J, Laurita KR, Rosenbaum DS, Rudy Y (1995) Two components of the delayed rectifier K+ current in ventricular myocytes of the guinea pig type. Theoretical formulation and their role in repolarization. Circ Res 77:140–152

HEP (2006) 171:235–266
© Springer-Verlag Berlin Heidelberg 2006

β-Blockers as Antiarrhythmic Agents

S. Zicha · Y. Tsuji · A. Shiroshita-Takeshita · S. Nattel (✉)

Montreal Heart Institute, 5000 Belanger East, Montreal Quebec, H1T 1C8, Canada
stanley.nattel@icm-mhi.org

Abstract Drugs that suppress β-adrenergic signaling by competitively inhibiting agonist binding to β-adrenergic receptors ("β-blockers") have important antiarrhythmic properties. They differ from most other antiarrhythmic agents by not directly modifying ion channel function; rather, they prevent the arrhythmia-promoting actions of β-adrenergic stimulation. β-Blockers are particularly useful in preventing sudden death due to ventricular tachyarrhythmias associated with acute myocardial ischemia, congenital long QT syndrome, and congestive heart failure. They are also quite valuable in controlling the ventricular rate in patients with atrial fibrillation. This chapter reviews the properties of β-adrenoceptor signaling, the basic mechanisms of cardiac arrhythmias on which β-blockers act, the ion channel mediators of β-adrenergic responses, the evidence for clinical antiarrhythmic indications for β-blocker therapy and the specific pharmacodynamic and pharmacokinetic properties of β-blockers that differentiate the various agents of this class.

Keywords β-Adrenoceptor antagonists · β-Adrenergic receptors · Sudden cardiac death · G protein-coupled receptors · Long QT syndrome

1
Introduction

Cardiovascular disease (CVD) remains the predominant cause of mortality in the world, with over 60 million people affected by some type of CVD in the USA alone (Heart Disease and Stroke—Statistics Update 2003, American Heart Association). Arrhythmias are an important contributor to CVD morbidity and mortality. The autonomic nervous system is an important regulator of cardiac electrical activity, and its function may be altered in CVD. The β-adrenergic signaling system is a crucial component of the autonomic control of cardiac electrical function. For these reasons, agents that alter β-adrenergic control of the heart by inhibiting β-adrenoceptor (AR) binding (known colloquially as "β-blockers") are an important component of the pharmacological treatment of arrhythmia.

Drugs that affect ionic currents directly by blocking voltage-gated ion channels include Vaughan Williams classes I (Na^+ channel blockers), III (K^+ channel blockers), and IV (Ca^{2+} channel blockers). The Cardiac Arrhythmia Suppression Trial (CAST) demonstrated in 1989 (CAST Investigators 1989) that blocking I_{Na} may increase, rather than decrease, arrhythmic mortality in patients at risk of sudden cardiac death due to ventricular tachyarrhythmias. The value of pure class III K$^+$ channel blockers has also been questioned with the realization that excessive prolongation of the cardiac action potential (AP) can lead to torsades de pointes (TdP) ventricular tachyarrhythmias, and with the publication of the Survival with Oral D-Sotalol (SWORD) trial (Waldo et al. 1996), which showed that class III agents may also increase mortality in at-risk patients.

β-Blocking agents were first discovered in 1958, with the identification of the β-blocking properties of the partial agonist dichloroisoproterenol, and the demonstration that dichloroisoproterenol could block adrenergic effects on the heart (Dresel 1960). The antiarrhythmic actions of β-blockers were characterized as "class II" antiarrhythmic properties by Singh and Vaughan Williams in 1970 (Singh and Vaughan Williams 1970). Class II antiarrhythmic agents are currently widely used for treating cardiac arrhythmias. Recent large trials on β-blocking agents such as the USCP (United States Carvedilol Program; Packer et al. 1996), CIBIS II (Cardiac Insufficiency Bisoprolol Study; Anonymous 1999b), MERIT-HF (Metoprolol CR/XL Randomised Intervention Trial in Heart Failure; Anonymous 1999a), and COPERNICUS (Carvedilol Prospective Randomized Cumulative Survival Trial; Eichhorn and Bristow 2001) all demonstrated that β-blockade prevents sudden death related to malignant arrhythmias in patients with congestive heart failure (CHF). In addition, β-blockers prevent lethal arrhythmias in patients with congenital long QT syndrome (LQTS), cardiac arrest survivors, some cases of ventricular tachycardia, and survivors of myocardial infarction. Inhibition of the effects of β-adrenergic stimulation contributes to the efficacy of drugs such as amiodarone or sotalol, which have β-blocking properties in addition to K$^+$ channel blocking capabilities. These observations highlight the importance of the β-adrenergic system in cardiac arrhythmias and the potentially important antiarrhythmic benefit from β-blockade.

2
β-Adrenergic Receptors in the Heart

The heart is under the influence of both the sympathetic and parasympathetic branches of the autonomic nervous system. The sympathetic nervous system (SNS) functions to increase heart rate and the force of contraction via adrenergic stimulation. SNS effects are usually balanced by those of the parasympathetic system, which decreases heart rate via muscarinic cholinergic receptors. Adrenergic effects were discovered in 1896 when Oliver and Shafer found that injected crude adrenal gland extracts could increase arterial pressure (Barcroft and Talbot 1968). Adrenaline was isolated as the active compound in 1900 by Farbwerke Hoechst and was marketed as a vasoconstrictor to stop bleeding and raise blood pressure in patients experiencing surgical shock and, more importantly, for the treatment of acute asthma (Sneader 2001). Over time, adrenaline was found to cause varied effects, including both vasoconstriction and vasodilation, but it wasn't until Ahlquist's discovery of α- and β-ARs in 1948 that these varying actions were understood (Ahlquist 1948). A principal target of cardiac sympathetic regulation is the sinoatrial node (SAN), which governs normal cardiac rate and is richly innervated by the SNS. However, all other regions of the heart receive sympathetic innervation, which can profoundly influence their electrical function.

2.1
Adrenergic Receptor Subtypes

Adrenergic receptors can be divided into α and β types based on their responses to antagonists and agonists such as noradrenaline, adrenaline, and isoproterenol (a synthetic derivative of adrenaline). β-Type ARs are more responsive to isoproterenol, while adrenaline acts more potently on α-type ARs. β-ARs are by far more prominent in the heart as compared to α-ARs, as demonstrated by a much greater inotropic effect of isoproterenol on cardiac tissue (Bohm et al. 1988; Brodde et al. 1998; Steinfath et al. 1992). This chapter will focus on β-adrenergic receptors, since α-receptor modulation is not used clinically as an antiarrhythmic intervention. α-ARs and β-ARs can be further divided into different subtypes based on their response to subtype-selective agonists and antagonists. To date, three subtypes of β-receptor have been identified in the heart: β_1, β_2, and β_3.

2.1.1
β_1 Adrenoceptors in the Heart

The β_1 subtype is the most prominent AR in the heart as determined by mRNA and protein quantification (Bristow et al. 1993; Bylund et al. 1994; Engelhardt et al. 1996; Ihl-Vahl et al. 1996; Ungerer et al. 1993). The ratio of β_1/β_2 receptors in the human heart is approximately 60/40 in the atrium and as high as 80/20 in the ventricle (Brodde 1991). β_1-ARs are stimulated by agonists such as dobutamine (Williams and Bishop 1981) and xamoterol (Nuttall and Snow 1982) and inhibited by selective antagonists such as practolol and metoprolol. The endogenous adrenergic neurotransmitter norepinephrine exerts its effects primarily through β_1 ARs (Kaumann et al. 1989; Motomura et al. 1990). The important inotropic (increased force of contraction) and chronotropic (increased heart rate) effects of norepinephrine are due to the stimulation of β_1-ARs, as demonstrated experimentally by the exogenous injection of norepinephrine or the release of endogenous norepinephrine by tyramine in healthy human subjects (Schafers et al. 1997) and in recent heart transplant patients who lack parasympathetic signaling (Leenen et al. 1998). The ability of β_1-AR signaling to initiate arrhythmia is illustrated by exercise-induced tachyarrhythmias related to norepinephrine release from sympathetic nerve endings (Brodde 1991; McDevitt 1989).

2.1.2
β₂-Adrenoreceptors in the Heart

The β_2-AR subtype is also found in the human heart (Brodde 1991), albeit to a lesser extent than β_1 ARs. Agonists for β_2-ARs include salbutamol (Kelman et al. 1969), terbutaline (Burnell and Maxwell 1971), and salmeterol (Ullman and Svedmyr 1988). β_2-ARs are more responsive to isoproterenol as compared to norepinephrine. When heart transplant patients are given isoproterenol, an increase in heart rate is observed, even in the presence of the highly selective β_1 antagonist bisoprolol (Hakim et al. 1997). Since the transplanted heart is not innervated, these observations cannot be due to reflex mechanisms. Interestingly, the greatest densities of β_2 ARs are found in the SAN (Rodefeld et al. 1996), which suggests a particularly important role in mediating adrenergic influences on heart rate. This has been demonstrated by in vivo studies: Despite the overall lesser amount of β_2-ARs in the heart, β_1 and β_2 AR-stimulation increases heart rate to an equal degree (Brodde 1991; McDevitt 1989). One of the concerns in designing β-blockers for the treatment of CVD is their deleterious effects on bronchial smooth muscle due to blockade of β_2-AR-mediated bronchodilation. Many early non-selective β-blockers caused bronchoconstriction, especially in asthmatic patients. Newer β-blockers have been designed to be more β_1 specific in order to avoid such complications.

The role of cardiac β_3 or β_4 ARs remains uncertain, although some recent evidence with the β_1/β_2 antagonist and β_3 agonist CGP 12177 does point to functional β_3-ARs in the heart (Arch and Kaumann 1993; Kaumann 1996).

2.2
β-Adrenoceptor Molecular Biology and Signaling in the Heart

All the β-adrenergic receptors are G protein-coupled receptors (GPCRs) with seven α-helix transmembrane spanning domains in the predominant α-subunit. GPCRs constitute the largest group of targets for pharmacological interventions on the market today. The β-AR was the first AR GPCR to be cloned (Dixon et al. 1986). GPCRs have a highly conserved structure, with the main differences occurring at the intracellular C-terminal, the extracellular N-terminal, and the G protein-binding long third cytoplasmic loop. G proteins are composed of three subunits, α, β, and γ. The α-subunit is capable of binding to guanine nucleotides (GTP) and catalyzing enzymatic conversion to guanosine diphosphate (GDP). In their inactive state, G proteins are found as an $\alpha\beta\gamma$ trimer bound to GDP. When an agonist binds to the receptor, the trimer is recruited to the intracellular loop region, resulting in the dissociation of GDP

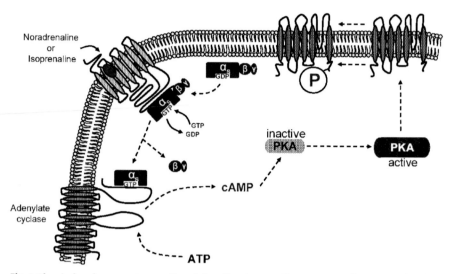

Fig. 1 Classic β_1 adrenoceptor-mediated signaling in a cardiomyocyte. After agonist binding (isoproterenol or noradrenaline), $G_{\alpha s}$ is activated and allows adenylate cyclase to convert ATP to the secondary messenger cAMP. In turn, cAMP activates protein kinase A (PKA), which can phosphorylate other proteins, including voltage-gated ion channels. The phosphorylation of ion channels alters their kinetic properties and can ultimately result in pathological changes to the action potential

and the subsequent binding of GTP. The G protein trimer then breaks up into its active α-GTP and βγ-subunit forms which diffuse into the cytosol to activate (or inactivate) enzymes or channel proteins. The β-subtypes differ in terms of the second messengers that transmit the adrenergic signal (see Dzimiri for an excellent review on β-AR signaling: Dzimiri 1999). The classic β-AR signaling pathway involves coupling to $G_{\alpha s}$, which in turn activates adenylyl cyclase that converts intracellular ATP to cyclic AMP (cAMP). This secondary effector activates the protein kinase A (PKA) pathway, which phosphorylates many proteins involved in ion channel and cellular contractile function (Fig. 1).

3
Normal Conduction to Arrhythmia: What Changes Are Happening in the Heart?

Normal electrical functioning of the heart requires an appropriate balance of inward and outward currents during the cardiac AP. An imbalance between inward and outward currents can lead to susceptibility to arrhythmia.

3.1
The Cardiac Action Potential

The phenomena underlying cardiac electrical activity are best appreciated at the cellular level by understanding the cardiac AP. Figure 2 shows a typical AP with the five different phases indicated. Inward (depolarizing) currents are shown in red and outward currents (bringing the cell back towards the resting potential) are shown in blue.

Phase 0 consists of a rapid depolarization from the resting membrane potential of approximately −80 to −90 mV to the "overshoot" of +40 mV and is caused by the opening of cardiac Na^+ channels, which carry a large Na^+ current (I_{Na}). Phase 0 depolarization is followed by the activation of various outward K^+ currents during phase 1, such as the transient outward current (I_{to}) and the ultra-rapid delayed rectifier (I_{Kur}). This early rapid repolarization phase is followed by the activation of phase 2 currents, which constitute fairly balanced inward and outward currents resulting in a phase of relatively constant transmembrane potential, the so-called "plateau" of the cardiac AP. During this phase, inward currents such as the L-type calcium current (I_{CaL}) and the late component of the sodium current ($I_{Na,L}$) balance outward currents such as I_{Kur} and the rapidly activating and slowly activating components of the delayed rectifier current, I_{Kr} and I_{Ks}. Ca^{2+} influx during the plateau phase is essential for electromechanical coupling, as Ca^{2+} ions trigger movement of the contractile filaments causing cardiac contraction. Phase 3 is the final rapid repolariza-

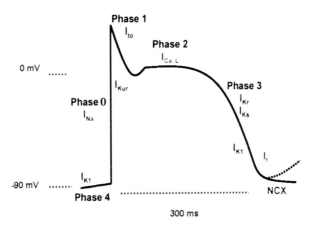

Fig. 2 Schematic of a typical cardiac action potential (AP), which is a recording of intracellular voltage as a function of time. The four phases are indicated, along with the ionic currents responsible for shaping the AP. Current direction is defined by the movement of positive ions. Ions that are at higher concentration in the extracellular space (like Na^+ and Ca^{2+}) move into the cell when the membrane allows them through, carrying depolarizing (inward) current. K^+, which is more concentrated inside the cell, tends to move out, carrying repolarizing (outward) current. *NCX*, Na^+,Ca^{2+}-exchanger current

tion phase of the AP and is dominated by the outward K^+ currents I_{Kr} and I_{Ks}. After repolarization is complete, maintenance of the resting membrane potential during phase 4 is controlled by the inward rectifier current (I_{K1}). In regions with pacemaker activity, such as the SAN, the hyperpolarization-activated current I_f is able to depolarize cells during phase 4, reaching the threshold for firing and producing pacemaker activity. The AP morphology varies in different regions of the heart because of heterogeneity in ion-current expression (Feng et al. 1998; Li et al. 2001; Wang et al. 1998). In addition to the currents mentioned above, other membrane currents such as I_{KATP}, I_{ClCa}, and the Na^+,Ca^{2+}-exchanger current (NCX) play a role. Disease states that predispose the myocardium to arrhythmia are often linked to changes in the expression of these currents (Nattel and Li 2000).

3.2
Basic Mechanisms of Arrhythmias

As mentioned previously, a disruption in the inward/outward current balance can lead to an arrhythmogenic state. CVDs that render the myocardium susceptible to arrhythmia include coronary artery disease, pericarditis, congenital heart disease, mitral valve disease, hypertension, ischemic heart disease, and congestive heart failure. Cardiac arrhythmias are believed to arise by four primary mechanisms: early afterdepolarizations (EADs), delayed afterdepolarizations (DADs), enhanced automaticity, and reentry.

3.2.1
Early Afterdepolarizations

Triggered activity that occurs before full repolarization of the AP is termed an EAD. This activity is the result of a spontaneous depolarization during a prolonged AP with increased duration (APD) and occurs more readily in Purkinje fibers (Nattel and Quantz 1988) and in ventricular mid-myocardial (M) cells (Sicouri and Antzelevitch 1995) than in other atrial or ventricular tissues. EADs can take place during the plateau (phase 2) or during phase 3 of the AP. The imbalance is most often the result of a decrease in outward currents; I_{Kr} and I_{Ks} are most directly implicated, but reductions in I_{K1} and I_{to} may also contribute (Beuckelmann et al. 1993; Han et al. 2001a; Kleiman and Houser 1989). APD prolongation accompanied by decreased outward currents allows inward currents such as I_{Ca} (for phase 2 EADs; De Ferrari et al. 1995; January and Riddle 1989; Luo and Rudy 1994; Nattel and Quantz 1988), reactivated fast I_{Na}, and the NCX for phase 3 EADs (Luo and Rudy 1994) to trigger depolarization. These afterdepolarizations may depolarize adjacent, repolarized cells to the threshold potential, triggering another depolarization (Cranefield 1977) and potentially initiating transmural reentry (Antzelevitch 2003).

EADs are central to arrhythmogenesis in patients exhibiting LQTS. Congenital LQTS patients may have a mutation in K^+ subunit genes including *KvLQT1*

(LQT1), *HERG* (LQT2), *mink* (LQT5) and *MiRP1* (LQT6). In addition to loss-of-function K^+ channel mutations, an inactivating defect in the Na^+ channel subunit $Na_v1.5$ (LQT3) and the membrane adaptor protein ankyrin-B (LQT4) have also been implicated in LQTS (for a review see Antzelevitch 2003). The result of these mutations is a prolongation of the APD and a disruption in the net current balance, predisposing the tissue to EADs. Pharmacological agents can also cause EADs by inducing "acquired" LQTS (Nattel 2000). Class III anti-arrhythmic agents which prolong APD by blocking delayed rectifier currents predispose patients to TdP arrhythmias (Hohnloser 1997; Roden 2004) by mimicking the functional defects of congenital LQTS. By increasing plateau Ca^{2+} current, β-adrenergic stimulation tends to prolong APD and promote EADs. This action is offset by activation of K^+ currents, especially I_{Ks} (see Sect. 4.1.1). When I_{Ks} is reduced by mutations in the DNA-encoding channel subunits or by cardiac disease, the tendency of β-adrenergic stimulation to promote EAD-related arrhythmias is enhanced and β-blockers may be beneficial in preventing arrhythmogenesis.

3.2.2
DADs

Delayed afterdepolarizations, another cause of abnormal impulse formation, occur after AP repolarization. DADs can occur in the ventricles, Purkinje fibers, and the atria and are typically caused by Ca^{2+} overload. When there is an increase in intracellular Ca^{2+}, a secondary diastolic release of Ca^{2+} from sarcoplasmic reticulum stores can occur after AP repolarization (Fabiato and Fabiato 1975; Lakatta 1992; Pogwizd and Bers 2004). This spontaneous release of Ca^{2+} results in extrusion of Ca^{2+} by NCX. The NCX exchanges one Ca^{2+} ion for three Na^+ ions, resulting in a net inward current in the direction of Na^+ transport, which depolarizes the cell. If these depolarizations reach threshold, an extrasystolic AP can be triggered. I_{Ca} is enhanced by β-adrenergic signaling, is the main source of cellular Ca^{2+} loading and plays a key role in the generation of DADs. Sympathetic nervous system activation in diseased myocardium leads to increased I_{Ca}, which can precipitate Ca^{2+} overloading (Belardinelli and Isenberg 1983; Malfatto et al. 1988; Wit and Cranefield 1976; Wit and Cranefield 1977). Arrhythmias due to digitalis toxicity are also frequently related to DADs (Vos et al. 1990; Zipes et al. 1974), since cardiac glycosides increase intracellular Ca^{2+} by inhibiting the Na^+/K^+ pump (Bigger 1985; Lee et al. 1980), causing DADs when Ca^{2+} loading becomes excessive.

3.2.3
Abnormal Automaticity

Under normal conditions, the SAN controls cardiac rate because of its faster intrinsic firing rate compared to other regions; however, regions like the AVN and the His-Purkinje system are capable of depolarizing spontaneously and displaying automaticity. Sympathetic over-stimulation of Purkinje fibers (Hauswirth et al. 1968) or cardiac pathology in other regions (including working atrium and ventricle; Lazzara and Scherlag 1988) can increase automatic discharge rates and generate cardiac arrhythmias. Resting membrane potential can become depolarized in diseased tissue (Cameron et al. 1983; Gelband and Bassett 1973; Nordin et al. 1989; Rossner and Sachs 1978; Wiederhold and Nilius 1986), possibly because of decreased inward rectifier current I_{K1} (Beuckelmann et al. 1993). In regions that do not normally exhibit automaticity, more positive resting membrane potential facilitates the initiation of an AP because less depolarization is needed to reach threshold voltage for firing (Janse and Wit 1989; Katzung and Morgenstern 1977), and contributes to the development of abnormal automaticity. In addition, the rate of phase 4 depolarization can be enhanced by reduced I_{K1}, increased I_f, or both.

3.2.4
Reentry

Reentry is a disorder of impulse conduction that is believed to cause many important clinical tachyarrhythmias (Cranefield et al. 1973). Once normal tissue is excited, the Na^+ channels become inactivated and another AP cannot be initiated until they recover from inactivation—a period of time termed the refractory period (RP). There are three main requirements for initiation of reentry: (1) two distinct pathways for AP propagation joined proximally and distally; (2) different RPs in the two pathways; and (3) development of unidirectional block, generally by premature activations exposing the RP differences. If a premature impulse encounters one pathway when it is refractory but the other can conduct, it can travel "antegradely" (in the normal direction) down the shorter-RP path and reach the distal end of the previously blocked pathway to a point at which excitability has been regained. It can then travel in the retrograde direction up this longer RP pathway and, if conditions are correct, re-excite the shorter RP pathway in the antegrade direction. This can lead to repetitive excitation that travels antegradely down the shorter RP path and retrogradely up the longer RP pathway.

The ability to maintain reentry depends on the relationship between circuit time (equivalent to tachycardia cycle length) and RP. For reentry to be sustained, circuit time has to be greater than RP—otherwise, the impulse will hit refractory tissue and extinguish. Thus, short RPs promote sustained reentry and long RPs prevent it. Circuit time is given by the length of the reentry circuit divided by conduction velocity, so slow conduction (which increases circuit time) favors reentry. Disrupting the balance of currents in the AP, especially during phase 2, has a profound effect on APD and therefore RP. For example, the downregulation of I_{Ca}, which can help prevent calcium overloading in the cell, also shortens APD and RP and promotes the maintenance of reentry (Nattel 2002). During fast atrial rates, as seen in atrial fibrillation (AF), I_{Ca} may decrease, reducing the RP and favoring arrhythmia perpetuation (Nattel 2002). In addition, AF is sometimes associated with decreased I_{Na}, which can lead to a decrease in atrial conduction velocity and also promote reentry (Gaspo et al. 1997). Increasing RP by increasing inward Ca^{2+} and Na^+ currents or reducing outward K^+ currents will have the opposite effect and suppress reentry.

4
Mechanisms of β-Blocker Action on Arrhythmias

Although the beneficial effects of β-blocker therapy have been known for quite a while (Singh and Vaughan Williams 1970), their exact mechanisms of action are still incompletely understood. The following sections will examine effects on ionic currents, as well as other fundamental properties of β-blockers that can contribute to antiarrhythmic actions.

4.1
Ionic Currents Affected by β-Adrenergic Signaling and β-Adrenoceptor Blockade

Many ionic currents are affected by β-adrenergic stimulation, resulting in complex effects on the AP. Generally, β-adrenergic stimulation results in rate-dependent APD shortening (Walsh and Kass 1991). Below is a summary of principal actions (Table 1).

Table 1 Currents affected by β-adrenergic signaling

Current	Effect of β-AR stimulation	Effect on AP	Mechanism of arrhythmia
I_{Ks}	↑	Shorten APD and refractory period	Reentry
I_{Kr}	↓	Prolong APD	EAD
I_f	↑	Increase the chance of premature depolarization	Automaticity
I_{K1}	↑	Resting membrane potential is more negative	Reentry
I_{Kur}	↑	Shorten the APD	
I_{NCX}	↑	Cause phase 4 depolarizations	DAD
$I_{CFTR, cardiac}$	↑	Depolarize resting membrane potential	Automaticity
$I_{Ca,L}$	↑	Prolong APD, contributes to Ca^{2+} overloading of the cell	DAD

4.1.1
The Slowly Activating Delayed Rectifier, I_{Ks}

The delayed rectifier currents regulate final repolarization of the AP, and both I_{Ks} and I_{Kr} components are influenced by adrenergic stimulation. I_{Ks} is a typical tetrameric voltage-gated K^+ channel composed of the KvLQT1 α-subunit and minK β-subunits encoded by *KCNQ1* and *KCNE1*, respectively (Barhanin et al. 1996; Sanguinetti et al. 1996). I_{Ks} expression is heterogeneous in the heart, especially across the left ventricular wall. Midmyocardial cells have a significantly longer APD, which is attributable in part to decreased I_{Ks} expression (Liu and Antzelevitch 1995), and I_{Ks} expression is greater in the right ventricle as compared to the left (Volders et al. 1999). β-Adrenergic stimulation increases I_{Ks} density three- to fivefold. This action prevents excessive APD prolongation in the face of sympathetic I_{Ca} augmentation (Han et al. 2001b). Incomplete deactivation of I_{Ks} may contribute to rate-dependent APD shortening (Stengl et al. 2003; Volders et al. 2003). Aside from gating effects of β-adrenergic stimulation that move I_{Ks} activation voltage negatively (towards plateau potentials), β-receptor stimulation has direct effects on I_{Ks} via the cAMP/PKA cascade. KvLQT1 is phosphorylated by PKA, which forms part of a macromolecular signaling complex with the protein yotiao (Marx et al. 2002). I_{Ks} phosphorylation increases current amplitude by increasing the rate of activation and decreasing the rate of deactivation. Because I_{Ks} stimulation by β-adrenergic activation is important to offset I_{Ca} augmentation and prevent excessive APD prolongation, in situations in which I_{Ks} is reduced, such as congenital LQTS types 1 and 5 (Priori and Napolitano 2004) and possibly acquired channelopathy due to cardiac remodeling (Li et al. 2002), β-adrenergic stimulation can lead to EADs and potentially malignant ventricular tachyarrhythmias. In such settings, β-AR antagonists may be particularly beneficial.

4.1.2
The Rapidly Activating Delayed Rectifier, I_{Kr}

I_{Kkr} repolarizes the cardiac AP during phase 3 and is particularly important in determining APD. I_{Kr} channels are made up of the α-subunit HERG and possibly MiRP1 β-subunits (Abbott et al. 1999). In LQTS type 2, HERG mutations decrease I_{Kr}, produce excess QT-prolongation and potentially lethal EAD-related arrhythmias. HERG has four PKA phosphorylation consensus sites, S238 (N-terminus), S890, S895, and S1137 (all C-terminus). The cAMP-mediated PKA phosphorylation of these sites reduces HERG current by 19%–40%. This reduction is due not only to a direct reduction in current, but also to a positive shift in the activation curve by 12–14 mV (Thomas et al. 1999) and to accelerated current deactivation. cAMP itself is also capable of modulating HERG function without PKA phosphorylation (Cui et al. 2000). The carboxyl terminus of HERG is homologous to cyclic nucleotide binding proteins, resulting in cAMP binding that mediates this direct action. This effect of β-adrenergic stimulation to inhibit I_{Kr} may contribute to its ability to delay repolarization and promote EAD-related arrhythmias in LQTS patients, and is counteracted by β-blockade.

4.1.3
The Funny Current, I_f

The hyperpolarization-activated, cyclic nucleotide-gated current, or funny current, is so termed because it has many unusual features. These include activation by hyperpolarization and the ability to carry both Na^+ and K^+ ions, resulting in a reversal potential positive to the resting potential of cardiac cells (DiFrancesco 1993). These properties confer the ability to induce spontaneous diastolic depolarization and pacemaker activity. I_f is most strongly expressed in the SAN, the dominant pacemaker in normal tissue. I_f is encoded by HCN subunit genes, predominantly *HCN1*, *HCN2*, and *HCN4* in the heart (Moroni et al. 2001), with greatest HCN subunit expression in the SAN. I_f gating is regulated by cyclic nucleotides, with β-adrenergic stimulation augmenting I_f via cAMP-mediated increases in current amplitude due primarily to depolarizing activation-curve shifts. A cyclic nucleotide-binding domain is located in the C-terminus of all HCN subunits. This domain inhibits channel gating, but when cAMP is bound, a conformational shift occurs, leaving channel gating unimpeded (Wainger et al. 2001). Increased I_f may underlie ectopic tachycardias, particularly in situations of increased adrenergic drive. β-Adrenergic blockade antagonizes adrenergically induced I_f augmentation and can thereby suppress ectopic tachycardias.

4.1.4
The Inward Rectifier Current, I_{K1}

I_{K1} channels are formed by two transmembrane-domain Kir2.x subunits assembling as tetramers. While effects of β-adrenergic stimulation, cAMP, and PKA have been observed, they are unclear. Currents carried by heterologously expressed Kir2.1 subunits have been found to increase or decrease in different studies (Fakler et al. 1994; Wischmeyer and Karschin 1996). Most experiments on native I_{K1} in cardiomyocytes have shown increases in current as a result of PKA phosphorylation (Koumi et al. 1995b; Koumi et al. 1995a; Tromba and Cohen 1990; Xiao and McArdle 1995).

4.1.5
The Ultra-Rapid Delayed Rectifier Current, I_{Kur}

Kv1.5 α-subunits underlie I_{Kur} in human cardiomyocytes (Feng et al. 1997; Wang et al. 1993). I_{Kur} is selectively expressed in the human atrium, making it an interesting target for the development of AF therapies. Kv1.5 contains both PKA and PKC phosphorylation sites, and is therefore a potentially important adrenergic effector in the heart (Fedida et al. 1993). Indeed, experiments with isoproterenol and the β-blocking agent propranolol have demonstrated that I_{Kur} is affected by PKA phosphorylation (Li et al. 1996a). By increasing I_{Kur} density, β-adrenergic stimulation can shorten APD, thus facilitating the occurrence of AF.

4.1.6
The Transient Outward Current, I_{to}

Expression of I_{to} follows a transmural gradient across the left ventricular wall (Litovsky and Antzelevitch 1988; Wettwer et al. 1994), therefore alterations in its expression can affect APD and lead to arrhythmia. Kv4.3 and Kv1.4 are the principal I_{to} pore-forming subunits, along with KChIP2 as an accessory subunit (Rosati et al. 2003). I_{to} is often downregulated in cardiac disease states (Kaab et al. 1996; Li et al. 2002). Adrenergic effects on I_{to} are complex and the resulting changes unclear (Nakayama and Fozzard 1988). In addition, there is evidence that chronic β-adrenergic blockade may alter cardiac ion-channel function in man, reducing I_{to} and increasing APD (Workman et al. 2003). Such actions would be expected to prevent reentrant arrhythmia to the extent that I_{to} inhibition delays repolarization. However, the precise APD changes caused by I_{to} downregulation are not completely obvious because I_{to} is involved principally in early repolarization, raising the plateau and potentially accelerating later repolarization by activating I_K (Courtemanche et al. 1999).

4.1.7
The Sodium-Calcium Exchanger Current, I_{NCX}

The cardiac sodium-calcium exchanger is encoded by *NCX1* and exchanges three Na^+ ions for one Ca^{2+} ion. NCX carries a net positive charge in the direction of Na^+ movement and is therefore electrogenic. After Ca^{2+} influx via L-type I_{Ca} triggers calcium release from the sarcoplasmic reticulum via the ryanodine receptor, the excess intracellular Ca^{2+} must be removed by the sarcoplasmic reticulum Ca^{2+} pump (SERCA) (70%), and by NCX (30%). A greater expression level or activity of NCX protein will increase the magnitude of the otherwise small I_{NCX} transient inward current and contribute to the formation of DADs. Increased NCX activity occurs in CHF, possibly to compensate for SERCA downregulation, and can lead to DAD promotion (Pogwizd and Bers 2004). β-Blocker treatment of subjects with CHF increases the cardiac levels of SERCA mRNA and protein, while decreasing NCX protein levels (Plank et al. 2003; Yasumura et al. 2003). This reversal of CHF-related remodeling would be expected to prevent DADs related to NCX upregulation.

4.1.8
The cAMP-Activated Chloride Current

The cardiac cAMP-activated chloride current ($I_{Cl.cAMP}$) is a cardiac variant of the cystic fibrosis transmembrane conductance regulator protein, CFTR. Its discovery followed the observation that in some systems, isoproterenol depolarizes resting membrane depolarization to the point that spontaneous automaticity occurs, an effect that depends on transmembrane Cl^- ion movement (Bahinski et al. 1989; Egan et al. 1988; Harvey and Hume 1989; Matsuoka et al. 1990). $I_{Cl.cAMP}$ is more readily demonstrated in the ventricles than the atria (Li et al. 1996b; Warth et al. 1996). Phosphorylation of the channel by PKA is necessary for activation (Hwang et al. 1992). Recent experiments have been unable to provide evidence for $I_{Cl.cAMP}$ in the human myocardium (Li et al. 1996b). This may be the result of downregulation due to cardiovascular disease, suppression of the current by cell isolation, or true absence in the human heart. While $I_{Cl.cAMP}$ could contribute to a variety of arrhythmia mechanisms and is strongly enhanced by β-adrenergic stimulation (Hume et al. 2000), its precise role in arrhythmogenesis in vivo is unclear, largely because of a lack of specific blockers.

4.1.9
The L-Type Calcium Channel, $I_{Ca,L}$

The cardiac AP plateau is maintained predominantly by the inward current, $I_{Ca,L}$. In the SAN, L-type Ca^{2+} channels are also involved in pacemaker function as the main phase 0 depolarizing current. The main $I_{Ca,L}$ pore-forming subunit in the heart is $Ca_v1.2$, or α_{1C}, which is very sensitive to class IV antiarrythmetic agents such as nifedipine, diltiazem, and verapamil. Heart failure, ischemic heart disease, and AF may be associated with decreases in the expression of $I_{Ca,L}$ and an accompanying decrease in $Ca_v1.2$ protein and mRNA expression. This may cause APD and refractory period shortening and contribute to reentrant arrhythmias, particularly in AF (Nattel 2002). β-AR stimulation increases $I_{Ca,L}$ conductance via the cAMP-dependant PKA pathway. Several phosphorylation sites are present on the I_{Ca} β subunit, and an important PKA phosphorylation site exists at serine 1928 on the intracellular C-terminus of the α-subunit. Phosphorylation increases channel activity, leading to increased cellular Ca^{2+} loading and potentially contributing to cellular Ca^{2+} overload. β-Blocker therapy suppresses sympathetic signaling, tending to prevent Ca^{2+} overload.

4.2
β-Blocker Actions Mediated by Effects Other Than on Ion Channels

The β-blocker actions on ion channels discussed above produce a variety of important effects on experimental and clinical arrhythmias. In addition, effects due to actions on targets other than ion channels may have important consequences for arrhythmias.

4.2.1
Role of Anti-ischemic Actions

β-Blockade reduces the SAN rate by decreasing both $I_{Ca,L}$ (the principal phase 0 current in SAN) and I_f. This heart-rate reducing action appears to contribute importantly to mortality reduction by β-blockers in post-myocardial infarction patients, possibly because of anti-ischemic effects (Kjekshus 1986). Based on their lack of direct action on determinants of automaticity in atria and ventricles, β-blockers have little direct effect on atrial and ventricular ectopic beat frequencies. However, β-blockers may be quite effective in preventing ventricular tachyarrhythmias caused by acute ischemia in experimental models (Khan et al. 1972) and are the most effective drugs available for preventing arrhythmic sudden death in patients with active coronary artery disease (Nattel and Waters 1990; Reiter 2002). These properties are much more likely due to anti-ischemic than direct electrophysiological actions.

4.2.2
Role in Remodeling

Neurohumoral stimulation plays a major role in the myocardial deterioration associated with CHF (Katz 2003). A variety of cardiac ion channels is remodeled by β-adrenergic stimulation (Zhang et al. 2002). Circulating norepinephrine concentrations are an important predictor of arrhythmic death in CHF patients, and β-blockers are effective in preventing sudden death in the CHF population (Reiter 2002). Abnormal Ca^{2+} handling, likely central to the arrhythmic diathesis in CHF patients, is normalized by chronic exposure to a β-blocker (Plank et al. 2003).

5
Types of Arrhythmia Treated by β-Blockers

The major factor mediating the salutary effect of β-adrenergic blockers in cardiac arrhythmias is counteraction of the arrhythmogenic actions of catecholamine that facilitate (1) triggered activity due to intracellular Ca^{2+} overload-induced delayed afterdepolarizations, (2) automaticity in the conduction system and abnormal automaticity in diseased myocardium, (3) reentry due to increased heterogeneities of depolarization and repolarization in diseased myocardium, and (4) repolarization impairments caused by abnormalities in repolarizing K^+-currents. Therefore, β-blockers are useful in the treatment and prevention of various disorders of rhythms, as discussed below.

5.1
Prophylactic Use of β-Blockers in Myocardial Infarction

Randomized, controlled clinical trials have demonstrated that β-adrenergic blockade decreases not only the incidence of ventricular fibrillation (VF) within the first few days of acute myocardial infarction (ISIS Collaborative Group 1988; Ryden et al. 1983), but also late sudden arrhythmic death mortality up to 1–3 years after infarction primarily (Anonymous 1981; Anonymous 1982). In pooled data from 18,000 patients treated over long-term post-infarct periods with several different β-blockers, sudden death was reduced 32%–50% (Yusuf et al. 1985). Moreover, a recent report showed that in pooled data from two post-myocardial infarction trials (Cairns et al. 1997; Julian et al. 1997), total mortality rate reduction was greater when β-blockers were administered along with the broad-spectrum antiarrhythmic amiodarone compared with amiodarone alone (Boutitie et al. 1999). This result indicates that amiodarone, which has non-competitive β-antagonist properties, does not replace β-blockers, and it underlines the significance of the use of β-blockers.

5.2
Prophylactic Use of β-Blockers in Congestive Heart Failure

There have been four large randomized, controlled trials of β-blockers in patients with CHF, demonstrating reductions in mortality and sudden death, compared to placebo controls (Anonymous 1999a,b; Packer et al. 1996, 2001). Pooled results from three clinical trials show that the reduction in sudden death is equal to or greater than the reduction in all-cause death (37%, 35%, respectively) and the reduction rate of death due to progression of CHF is not statistically significant (Cleophas and Zwinderman 2001). These findings indicate that a major benefit of β-blockers in CHF is the prevention of sudden arrhythmic death (Cleophas and Zwinderman 2001). Such benefits may be due to the prevention of proarrhythmic effects of β-adrenergic stimulation due to changes in ion-channel function, as discussed above, as well as to the prevention of deleterious β-adrenergic effects to promote ventricular remodeling.

5.3
β-Blockers in Patients with Other Structural Heart Diseases and Ventricular Arrhythmias

Patients who survive life-threatening ventricular tachyarrhythmias, such as sustained monomorphic ventricular tachycardia (VT), polymorphic VT or VF, are at high risk for recurrent arrhythmias. When these tachyarrhythmias occur in the setting of structural heart disease, they can usually be provoked by programmed electrical stimulation. In most patients, β-blockers have little effect in preventing inducibility of the arrhythmia or in terminating VT.

The anti-fibrillatory mechanisms by which β-blockers reduce sudden death in ischemic heart disease and CHF are not understood completely. However, in experimental and clinical studies, β-blockers increase VF threshold and reduce dispersion of repolarization in the ischemic myocardium (Reiter and Reiffel 1998). Moreover, β-blockers attenuate ventricular remodeling (Eichhorn and Bristow 1996; St John and Ferrari 2002), indicating the role of modification of development of the substrate for lethal ventricular arrhythmias.

Other structural heart diseases in which β-blockers are considered for the treatment of ventricular tachyarrhythmias are dilated cardiomyopathy (DCM) and hypertrophic cardiomyopathy (HCM). Sudden, unexpected death can be the first presentation of these diseases and there is a close relationship between the occurrence of ventricular tachyarrhythmias and sudden death. The Metoprolol in Dilated Cardiomyopathy (MDC) trial (Waagstein et al. 1993) showed a 34% decrease in mortality and need for heart transplantation. VT occurs in patients with arrhythmogenic right ventricular dysplasia, which may be very difficult to control medically. Although implantable cardioverter-defibrillators are the intervention of choice in such individuals, ventricular tachyarrhythmias tend to occur in a setting of enhanced sympathetic drive and β-blockers are believed to be of value.

5.4
Long QT Syndrome

Congenital LQTS is characterized by prolonged ventricular repolarization and increased susceptibility to TdP leading to sudden cardiac death, with EADs likely playing a central role in arrhythmogenesis (Ackerman and Clapham 1997). Several LQTS-related genes are involved in the molecular pathogenesis (Curran et al. 1995; Keating and Sanguinetti 2001). Recent genotype–phenotype correlation studies have demonstrated genotype-specific differences in response to catecholamines, triggers for cardiac events, and responses to β-blockers as therapeutic agents (Moss et al. 2000; Schwartz et al. 2001; Shimizu et al. 2003). LQT1 patients (with a mutation in the I_{Ks} α-subunit $KvLQT1$) have a greater QT prolongation response to the adrenergic agonist epinephrine than LQT2 patients (with a mutation in the I_{Kr} α-subunit gene $HERG$; Shimizu et al. 2003). This difference is likely due to the important role of I_{Ks} in offsetting adrenergic enhancement of $I_{Ca,L}$. Cardiac events occur during exercise in LQT1 patients, whereas LQT2 patients experience episodes during emotion or at rest, and LQT3 patients are at greatest risk at rest or while asleep (Schwartz et al. 2001; Wilde and Roden 2000). The recurrence rate of cardiac events in LQT1 patients during β-blocker treatment is lower than for LQT2 and LQT3 patients (Schwartz et al. 2001). Moreover, the incidence of cardiac arrest or sudden death among LQT1 patients treated with β-blockers is very low when compared to previous studies (Schwartz et al 2001). Therefore, β-blockers are particularly recommended for LQT1 patients, but may also be useful for other patients with LQTS, possibly because of inhibitory effects of β-adrenergic stimulation on I_{Kr}.

5.5
Catecholaminergic Polymorphic Ventricular Tachycardia

This is a rare arrhythmogenic disorder characterized by exercise-induced bidirectional or polymorphic VT. This disorder may cause sudden death and has been linked to mutations in cardiac ryanodine receptor genes, which are responsible for sarcoplasmic reticulum Ca^{2+} release upon systolic Ca^{2+} entry through L-type Ca^{2+} channels (Priori et al. 2001). The resulting ryanodine receptor dysfunction promotes DAD formation (Viatchenko-Karpinski et al. 2004), and increased Ca^{2+} entry through $I_{Ca,L}$ under β-adrenergic stimulation likely triggers DADs and tachyarrhythmias in such patients. In one case report, intravenous propranolol terminated VT immediately and long-term nadolol therapy effectively prevented further arrhythmias (De Rosa et al. 2004), but a recent study demonstrated that β-blockers completely controlled catecholaminergic VT in only 41% of cases, and 22% died during follow up (Sumitomo et al. 2003).

5.6
Idiopathic Ventricular Tachycardia

Several discrete forms of VT without structural heart disease have been identified. The most common type is adenosine-sensitive monomorphic VT originating from the right ventricular outflow tract with a left bundle branch block ECG pattern and an inferior axis. This tachyarrhythmia is typically catecholamine sensitive and responds to β-blockade. However, these adenosine-sensitive outflow tachycardias are now commonly cured by radiofrequency catheter ablation, and therefore long-term use of β-blockers is uncommon. Verapamil-sensitive reentrant VT originates in the region of the left posterior fascicle and has a characteristic right bundle branch block and leftward axis morphology. β-Blockers are not effective for this arrhythmia. Some forms of VT appear to be induced by exercise, presumably at least in part because of adrenergic dependence, and may respond well to β-blocker therapy (Woelfel et al. 1984).

5.7
Supraventricular Tachycardias

Reentry involving the AV node can be suppressed by β-blockade to the extent that background adrenergic $I_{Ca,L}$ enhancement is necessary to sustain conduction in the reentry circuit. Although β-blockers were once used fairly widely for this type of arrhythmia, they have been largely supplanted by more effective drugs (direct inhibitors of $I_{Ca,L}$ such as verapamil and purinergic agonists such as adenosine) for acute termination and by radiofrequency ablation for prevention of recurrence. Atrial tachycardias (ATs) are categorized as either focal or macroreentrant. Focal ATs are caused by automatic, triggered, or microreentrant mechanisms (Chen et al. 1994). β-Blockers may have some value for the automatic or triggered forms. However, because of the great efficacy of radiofrequency ablation, this is usually the treatment of choice for recurrent arrhythmias. Macroreentrant AT is not affected by β-blockade, because of the limited role of β-adrenergic tone in maintaining conduction in the reentrant circuit, which is usually determined by Na^+-channel availability and the refractory period of atrial tissue. Similar considerations apply for atrial flutter, which is caused by a form of atrial macroreentry.

5.8
Atrial Fibrillation (AF)

AF is characterized by irregular and chaotic atrial fibrillatory waves at a rate of 350 to 600 beats per minute (bpm) and the ventricular response is irregular, typically at a rate of 120–160 bpm. The ventricular response is determined by the filtering action of the AV node. Many of the clinical manifestations are determined by the ventricular response, and if the ventricular response is kept physiological with the use of drugs that affect AV nodal function patients may be kept asymptomatic. The mechanisms of AF are complex and may include a variety of types of reentry, as well as rapid activity from ectopic foci, particularly in the pulmonary veins (Nattel 2002). Two general approaches are available for AF therapy: (1) stopping AF and maintaining sinus rhythm ("rhythm control" strategy) and (2) allowing the patient to remain in AF but controlling the ventricular response ("rate control" strategy) and preventing thromboembolic complications with anticoagulation. Although sinus rhythm maintenance is the most attractive approach, it is often difficult to achieve and controlled trials have shown that the control of ventricular rate may achieve as good or better clinical results (Nattel 2003). β-Blockers have some efficacy in preventing AF (Kuhlkamp et al. 2000). They may be particularly useful in preventing AF in the elderly (Psaty et al. 1997). β-Blockers are particularly effective in preventing AF in patients undergoing cardiac surgery. AF occurs in about 30% of patients after open heart surgery. Postoperative AF prolongs significantly the duration of hospitalization and increases hospital cost (Reddy 2001). In a meta-analysis of randomized trials of pharmacological interventions for prevention of AF, β-blockers significantly reduced the incidence of postoperative AF (Crystal et al. 2002). However, despite preventing AF occurrence, β-blockers have not been shown to significantly reduce length of hospital stay or hospital costs (Connolly et al. 2003).

Recently, the important role of pulmonary vein (PV) focal activity in AF was demonstrated (Haissaguerre et al. 1998). Ablation of arrhythmogenic PV foci or PV isolation can cure AF in a significant proportion of patients (Haissaguerre et al. 1998; Pappone et al. 2000). Chen et al. evaluated the effects of various anti-arrhythmic drugs on ectopic activity arising from the pulmonary veins and found that propranolol reduces the density of such ectopy (Chen et al. 1999). PV isolation seems very effective in patients with paroxysmal AF occurring during states associated with increased adrenergic activity (so-called adrenergic PAF; Oral et al. 2004). Thus, increased sympathetic activity may play an important role in ectopic impulse formation initiating AF. In addition, an anti-ischemic action may be involved in the efficacy of β-blockers for AF, in view of the ability of acute myocardial ischemia to promote AF maintenance (Sinno et al. 2003). Overall, however, the efficacy of β-blockade in preventing AF is relatively low.

Recent randomized controlled trials have demonstrated that there are no differences in symptoms, morbidity or quality-of-life between rhythm versus rate control strategies for AF therapy (Van Gelder et al. 2002; Wyse et al. 2002). However, rate control has advantages of less serious and common adverse effects—because the drugs used are more innocuous—and a potentially reduced risk of stroke because of the wider use of anticoagulation therapy. There has therefore been increased emphasis on therapy aimed, not at preventing AF, but at keeping the ventricular rate as physiological as possible. By reducing the effect of adrenergic tone to promote AV nodal conduction, β-blockers are valuable drugs for ventricular rate control. They have advantages over alternatives like digoxin in that rate is controlled during exercise as well as rest, and are in wide use for this indication (Nattel et al. 2002).

6
Pharmacokinetic and Pharmacological Properties of β-Blockers Relative to Choice of Agent

A variety of properties differentiate the various drugs available for therapeutic use as β-blockers (for review, see Shand 1983). The available agents differ in their selectivity for β_1 versus β_2-AR blockade, with atenolol and metoprolol being among the more β_1-selective agents available. β_1-Selectivity may help to avoid adverse effects (such as bronchospasm) in at-risk patients; however, selectivity is never absolute and caution must still be used. Lipophilic agents are more readily able to cross the blood–brain barrier, potentially more likely to produce central nerve system (CNS) adverse effects but possibly having greater beneficial actions related to inhibition of CNS β-adrenergic neurotransmission. Lipophilic agents also tend to be eliminated more rapidly by hepatic biotransformation and to have shorter half-lives. Some β-blockers, such as propranolol and sotalol, may have direct membrane actions on cardiac ion channels that are independent of β-blockade. In the case of sotalol, this results in class III antiarrhythmic action due to K^+ channel inhibition, with attendant additional antiarrhythmic effects, but also attendant risks of causing TdP arrhythmias. Finally, some agents, like practolol and acebutolol, are partial agonists with intrinsic sympathomimetic activity (ISA). ISA may be used to advantage when the objective is β-blockade only in situations of enhanced adrenergic tone and not at rest (e.g., patients with adverse effects from β-blockade at rest). In practice, this may be difficult to exploit, because ISA may not be sufficient to prevent effective β-blockade at rest, on one hand, and may negate beneficial effects resulting from resting β-blockade, on the other.

It remains unclear whether all β-blockers have comparable antiarrhythmic efficacy. Clearly, sotalol has additional antiarrhythmic actions due to its class III properties. However, there may be differences in efficacy for certain indications among β-blockers without membrane action. Perhaps because slowing resting heart rate may be very important for mortality prevention by β-blockers in post-myocardial infarction patients (Hjalmarson et al. 1990; Kjekshus 1986), drugs with ISA appear to be relatively ineffective in reducing mortality in post-MI patients (Freemantle et al. 1999). The drugs that have been shown consistently effective in preventing sudden death rate in coronary artery disease patients (timolol, propranolol, and metoprolol) have no ISA and are all lipophilic, whereas there is much less evidence for benefit from the hydrophilic β-blocker atenolol (ISIS Collaborative Group 1986). Thus, a component of the β-blocker-induced reduction of sudden death in coronary-disease patients may be mediated via CNS effects. In a meta-analysis of 71 secondary and primary prevention trials after MI, β_1-selectivity, lipophilicity, absence of membrane stabilizing properties, and absence of ISA appeared to be associated with a greater risk reduction for ischemic sudden death compared with β-blockers without these properties (Soriano et al. 1997).

Among the β-blockers shown to benefit patients with CHF, metoprolol and bisoprolol are relatively β_1 selective, and carvedilol is a nonselective $\beta_1/\beta_2/\alpha_1$ blocking agent. All of these are lipophilic, suggesting a possible role for CNS effects. The recently reported Carvedilol Or Metoprolol European Trial (COMET) represents an attempt to study the relative merits of carvedilol versus metoprolol (Poole-Wilson et al. 2003). The COMET investigators concluded that carvedilol extended survival compared with intermediate-release metoprolol. This difference may be because carvedilol has actions beyond β-blockade, such as vasodilating properties (related to α-blockade) and antioxidant actions. In patients with CHF, vasodilating β-blockers have a greater effect in reducing overall mortality than non-vasodilating agents, particularly in patients with non-ischemic heart disease (Bonet et al. 2000). However, questions about the interpretation of these findings remain, in view of the fact that the COMET trial did not use the dose or formulation of metoprolol that was shown to prolong life in a previous placebo-controlled trial (Goldstein and Hjalmarson 1999). Further studies are needed to define the role of specific β-blocker properties on outcomes in CHF patients.

7
Conclusions

β-Blocking agents have traditionally been viewed as weak antiarrhythmic drugs because of their limited effect on ectopic beat frequency and recurrent tachyarrhythmia incidence. However, they have proved to be the most useful pharmaceutical agents in preventing sudden death in patients with ischemic heart disease, CHF, and congenital LQTS. Because of the wide role of β-adrenergic stimulation in modulating the function of a broad range of cardiac ion channels and in determining the natural history of diseases like CHF and ischemic heart disease, β-blockers are an important group of compounds for the prevention of cardiac arrhythmias. Furthermore, compared to Na^+ and K^+ channel blocking drugs, β-blockers are relatively free of proarrhythmic risk and are therefore much safer to use in clinical practice. With further insights into the role of the adrenergic nervous system and the mechanisms of G protein-coupled receptor signal transduction and function, the clinical use of β-blocking drugs is likely to expand and become more effective.

References

Abbott GW, Sesti F, Splawski I, Buck ME, Lehmann MH, Timothy KW, Keating MT, Goldstein SA (1999) MiRP1 forms IKr potassium channels with HERG and is associated with cardiac arrhythmia. Cell 97:175–187

Ackerman MJ, Clapham DE (1997) Ion channels—basic science and clinical disease. N Engl J Med 336:1575–1586

Ahlquist RP (1948) A study of the adrenotropic receptors. Am J Physiol 153:586–600

Anonymous (1981) Timolol-induced reduction in mortality and reinfarction in patients surviving acute myocardial infarction. N Engl J Med 304:801–807

Anonymous (1982) A randomized trial of propranolol in patients with acute myocardial infarction. I. Mortality results. JAMA 247:1707–1714

Anonymous (1999a) Effect of metoprolol CR/XL in chronic heart failure: Metoprolol CR/XL Randomised Intervention Trial in Congestive Heart Failure (MERIT-HF). Lancet 353:2001–2007

Anonymous (1999b) The Cardiac Insufficiency Bisoprolol Study II (CIBIS-II): a randomised trial. Lancet 353:9–13

Antzelevitch C (2003) Molecular genetics of arrhythmias and cardiovascular conditions associated with arrhythmias. Pacing Clin Electrophysiol 26:2194–2208

Arch JR, Kaumann AJ (1993) Beta 3 and atypical beta-adrenoceptors. Med Res Rev 13:663–729

Bahinski A, Nairn AC, Greengard P, Gadsby DC (1989) Chloride conductance regulated by cyclic AMP-dependent protein kinase in cardiac myocytes. Nature 340:718–721

Barcroft H, Talbot JF (1968) Oliver and Schafer's discovery of the cardiovascular action of suprarenal extract. Postgrad Med J 44:6–8

Barhanin J, Lesage F, Guillemare E, Fink M, Lazdunski M, Romey G (1996) K(V)LQT1 and lsK (minK) proteins associate to form the I(Ks) cardiac potassium current. Nature 384:78–80

Belardinelli L, Isenberg G (1983) Actions of adenosine and isoproterenol on isolated mammalian ventricular myocytes. Circ Res 53:287–297

Beuckelmann DJ, Nabauer M, Erdmann E (1993) Alterations of K+ currents in isolated human ventricular myocytes from patients with terminal heart failure. Circ Res 73:379–385

Bigger JT Jr (1985) Digitalis toxicity. J Clin Pharmacol 25:514–521

Bohm M, Diet F, Feiler G, Kemkes B, Erdmann E (1988) Alpha-adrenoceptors and alpha-adrenoceptor-mediated positive inotropic effects in failing human myocardium. J Cardiovasc Pharmacol 12:357–364

Bonet S, Agusti A, Arnau JM, Vidal X, Diogene E, Galve E, Laporte JR (2000) Beta-adrenergic blocking agents in heart failure: benefits of vasodilating and non-vasodilating agents according to patients' characteristics: a meta-analysis of clinical trials. Arch Intern Med 160:621–627

Boutitie F, Boissel JP, Connolly SJ, Camm AJ, Cairns JA, Julian DG, Gent M, Janse MJ, Dorian P, Frangin G (1999) Amiodarone interaction with beta-blockers: analysis of the merged EMIAT (European Myocardial Infarct Amiodarone Trial) and CAMIAT (Canadian Amiodarone Myocardial Infarction Trial) databases. The EMIAT and CAMIAT Investigators. Circulation 99:2268–2275

Bristow MR, Minobe WA, Raynolds MV, Port JD, Rasmussen R, Ray PE, Feldman AM (1993) Reduced beta 1 receptor messenger RNA abundance in the failing human heart. J Clin Invest 92:2737–2745

Brodde OE (1991) Beta 1- and beta 2-adrenoceptors in the human heart: properties, function, and alterations in chronic heart failure. Pharmacol Rev 43:203–242

Brodde OE, Vogelsang M, Broede A, Michel-Reher M, Beisenbusch-Schafer E, Hakim K, Zerkowski HR (1998) Diminished responsiveness of Gs-coupled receptors in severely failing human hearts: no difference in dilated versus ischemic cardiomyopathy. J Cardiovasc Pharmacol 31:585–594

Burnell RH, Maxwell GM (1971) The cardiovascular effects of terbutaline. Eur J Pharmacol 15:383–385

Bylund DB, Eikenberg DC, Hieble JP, Langer SZ, Lefkowitz RJ, Minneman KP, Molinoff PB, Ruffolo RR Jr, Trendelenburg U (1994) International Union of Pharmacology nomenclature of adrenoceptors. Pharmacol Rev 46:121–136

Cairns JA, Connolly SJ, Roberts R, Gent M (1997) Randomised trial of outcome after myocardial infarction in patients with frequent or repetitive ventricular premature depolarisations: CAMIAT. Canadian Amiodarone Myocardial Infarction Arrhythmia Trial Investigators. Lancet 349:675–682

Cameron JS, Myerburg RJ, Wong SS, Gaide MS, Epstein K, Alvarez TR, Gelband H, Guse PA, Bassett AL (1983) Electrophysiologic consequences of chronic experimentally induced left ventricular pressure overload. J Am Coll Cardiol 2:481–487

CAST Investigators (1989) Preliminary report: effect of encainide and flecainide on mortality in a randomized trial of arrhythmia suppression after myocardial infarction. The Cardiac Arrhythmia Suppression Trial (CAST) Investigators. N Engl J Med 321:406–412

Chen SA, Chiang CE, Yang CJ, Cheng CC, Wu TJ, Wang SP, Chiang BN, Chang MS (1994) Sustained atrial tachycardia in adult patients. Electrophysiological characteristics, pharmacological response, possible mechanisms, and effects of radiofrequency ablation. Circulation 90:1262–1278

Chen SA, Hsieh MH, Tai CT, Tsai CF, Prakash VS, Yu WC, Hsu TL, Ding YA, Chang MS (1999) Initiation of atrial fibrillation by ectopic beats originating from the pulmonary veins: electrophysiological characteristics, pharmacological responses, and effects of radiofrequency ablation. Circulation 100:1879–1886

Cleophas TJ, Zwinderman AH (2001) Beta-blockers and heart failure: meta-analysis of mortality trials. Int J Clin Pharmacol Ther 39:383–388

Connolly SJ, Cybulsky I, Lamy A, Roberts RS, O'Brien B, Carroll S, Crystal E, Thorpe KE, Gent M (2003) Double-blind, placebo-controlled, randomized trial of prophylactic metoprolol for reduction of hospital length of stay after heart surgery: the beta-Blocker Length Of Stay (BLOS) study. Am Heart J 145:226–232

Courtemanche M, Ramirez RJ, Nattel S (1999) Ionic targets for drug therapy and atrial fibrillation-induced electrical remodeling: insights from a mathematical model. Cardiovasc Res 42:477–489

Cranefield PF (1977) Action potentials, afterpotentials, and arrhythmias. Circ Res 41:415–423

Cranefield PF, Wit AL, Hoffman BF (1973) Genesis of cardiac arrhythmias. Circulation 47:190–204

Crystal E, Connolly SJ, Sleik K, Ginger TJ, Yusuf S (2002) Interventions on prevention of postoperative atrial fibrillation in patients undergoing heart surgery: a meta-analysis. Circulation 106:75–80

Cui J, Melman Y, Palma E, Fishman GI, McDonald TV (2000) Cyclic AMP regulates the HERG K(+) channel by dual pathways. Curr Biol 10:671–674

Curran ME, Splawski I, Timothy KW, Vincent GM, Green ED, Keating MT (1995) A molecular basis for cardiac arrhythmia: HERG mutations cause long QT syndrome. Cell 80:795–803

De Ferrari GM, Viola MC, D'Amato E, Antolini R, Forti S (1995) Distinct patterns of calcium transients during early and delayed afterdepolarizations induced by isoproterenol in ventricular myocytes. Circulation 91:2510–2515

De Rosa G, Delogu AB, Piastra M, Chiaretti A, Bloise R, Priori SG (2004) Catecholaminergic polymorphic ventricular tachycardia: successful emergency treatment with intravenous propranolol. Pediatr Emerg Care 20:175–177

DiFrancesco D (1993) Pacemaker mechanisms in cardiac tissue. Annu Rev Physiol 55:455–472

Dixon RA, Kobilka BK, Strader DJ, Benovic JL, Dohlman HG, Frielle T, Bolanowski MA, Bennett CD, Rands E, Diehl RE (1986) Cloning of the gene and cDNA for mammalian beta-adrenergic receptor and homology with rhodopsin. Nature 321:75–79

Dresel PE (1960) Blockade of some cardiac actions of adrenaline by dichloroisoproterenol. Can J Med Sci 38:375–381

Dzimiri N (1999) Regulation of beta-adrenoceptor signaling in cardiac function and disease. Pharmacol Rev 51:465–501

Egan TM, Noble D, Noble SJ, Powell T, Twist VW, Yamaoka K (1988) On the mechanism of isoprenaline- and forskolin-induced depolarization of single guinea-pig ventricular myocytes. J Physiol 400:299–320

Eichhorn EJ, Bristow MR (1996) Medical therapy can improve the biological properties of the chronically failing heart. A new era in the treatment of heart failure. Circulation 94:2285–2296

Eichhorn EJ, Bristow MR (2001) The Carvedilol Prospective Randomized Cumulative Survival (COPERNICUS) trial. Curr Control Trials Cardiovasc Med 2:20–23

Engelhardt S, Bohm M, Erdmann E, Lohse MJ (1996) Analysis of beta-adrenergic receptor mRNA levels in human ventricular biopsy specimens by quantitative polymerase chain reactions: progressive reduction of beta 1-adrenergic receptor mRNA in heart failure. J Am Coll Cardiol 27:146–154

Fabiato A, Fabiato F (1975) Contractions induced by a calcium-triggered release of calcium from the sarcoplasmic reticulum of single skinned cardiac cells. J Physiol 249:469–495

Fakler B, Brandle U, Glowatzki E, Zenner HP, Ruppersberg JP (1994) Kir2.1 inward rectifier K+ channels are regulated independently by protein kinases and ATP hydrolysis. Neuron 13:1413–1420

Fedida D, Wible B, Wang Z, Fermini B, Faust F, Nattel S, Brown AM (1993) Identity of a novel delayed rectifier current from human heart with a cloned K+ channel current. Circ Res 73:210–216

Feng J, Wible B, Li GR, Wang Z, Nattel S (1997) Antisense oligodeoxynucleotides directed against Kv1.5 mRNA specifically inhibit ultrarapid delayed rectifier K+ current in cultured adult human atrial myocytes. Circ Res 80:572–579

Freemantle N, Cleland J, Young P, Mason J, Harrison J (1999) beta Blockade after myocardial infarction: systematic review and meta regression analysis. BMJ 318:1730–1737

Gaspo R, Bosch RF, Talajic M, Nattel S (1997) Functional mechanisms underlying tachycardia-induced sustained atrial fibrillation in a chronic dog model. Circulation 96:4027–4035

Gelband H, Bassett AL (1973) Depressed transmembrane potentials during experimentally induced ventricular failure in cats. Circ Res 32:625–634

Goldstein S, Hjalmarson A (1999) The mortality effect of metoprolol CR/XL in patients with heart failure: results of the MERIT-HF Trial. Clin Cardiol 22 Suppl 5:V30–V35

Haissaguerre M, Jais P, Shah DC, Takahashi A, Hocini M, Quiniou G, Garrigue S, Le Mouroux A, Le Metayer P, Clementy J (1998) Spontaneous initiation of atrial fibrillation by ectopic beats originating in the pulmonary veins. N Engl J Med 339:659–666

Hakim K, Fischer M, Gunnicker M, Poenicke K, Zerkowski HR, Brodde OE (1997) Functional role of beta2-adrenoceptors in the transplanted human heart. J Cardiovasc Pharmacol 30:811–816

Han W, Chartier D, Li D, Nattel S (2001a) Ionic remodeling of cardiac Purkinje cells by congestive heart failure. Circulation 104:2095–2100

Han W, Wang Z, Nattel S (2001b) Slow delayed rectifier current and repolarization in canine cardiac Purkinje cells. Am J Physiol Heart Circ Physiol 280:H1075–H1080

Harvey RD, Hume JR (1989) Autonomic regulation of a chloride current in heart. Science 244:983–985

Hauswirth O, Noble D, Tsien RW (1968) Adrenaline: mechanism of action on the pacemaker potential in cardiac Purkinje fibers. Science 162:916–917

Hjalmarson A, Gilpin EA, Kjekshus J, Schieman G, Nicod P, Henning H, Ross J Jr (1990) Influence of heart rate on mortality after acute myocardial infarction. Am J Cardiol 65:547–553

Hohnloser SH (1997) Proarrhythmia with class III antiarrhythmic drugs: types, risks, and management. Am J Cardiol 80:82G–89G

Hume JR, Duan D, Collier ML, Yamazaki J, Horowitz B (2000) Anion transport in heart. Physiol Rev 80:31–81

Hwang TC, Horie M, Nairn AC, Gadsby DC (1992) Role of GTP-binding proteins in the regulation of mammalian cardiac chloride conductance. J Gen Physiol 99:465–489

Ihl-Vahl R, Eschenhagen T, Kubler W, Marquetant R, Nose M, Schmitz W, Scholz H, Strasser RH (1996) Differential regulation of mRNA specific for beta 1- and beta 2-adrenergic receptors in human failing hearts. Evaluation of the absolute cardiac mRNA levels by two independent methods. J Mol Cell Cardiol 28:1–10

ISIS Collaborative Group (1986) Randomised trial of intravenous atenolol among 16 027 cases of suspected acute myocardial infarction: ISIS-1. First International Study of Infarct Survival Collaborative Group. Lancet 2:57–66

ISIS Collaborative Group (1988) Mechanisms for the early mortality reduction produced by beta-blockade started early in acute myocardial infarction: ISIS-1. ISIS-1 (First International Study of Infarct Survival) Collaborative Group. Lancet 1:921–923

Janse MJ, Wit AL (1989) Electrophysiological mechanisms of ventricular arrhythmias resulting from myocardial ischemia and infarction. Physiol Rev 69:1049–1169

January CT, Riddle JM (1989) Early afterdepolarizations: mechanism of induction and block. A role for L-type Ca2+ current. Circ Res 64:977–990

Julian DG, Camm AJ, Frangin G, Janse MJ, Munoz A, Schwartz PJ, Simon P (1997) Randomised trial of effect of amiodarone on mortality in patients with left-ventricular dysfunction after recent myocardial infarction: EMIAT. European Myocardial Infarct Amiodarone Trial Investigators. Lancet 349:667–674

Kaab S, Nuss HB, Chiamvimonvat N, O'Rourke B, Pak PH, Kass DA, Marban E, Tomaselli GF (1996) Ionic mechanism of action potential prolongation in ventricular myocytes from dogs with pacing-induced heart failure. Circ Res 78:262–273

Katz AM (2003) Pathophysiology of heart failure: identifying targets for pharmacotherapy. Med Clin North Am 87:303–316

Katzung BG, Morgenstern JA (1977) Effects of extracellular potassium on ventricular automaticity and evidence for a pacemaker current in mammalian ventricular myocardium. Circ Res 40:105–111

Kaumann AJ (1996) (−)-CGP 12177-induced increase of human atrial contraction through a putative third beta-adrenoceptor. Br J Pharmacol 117:93–98

Kaumann AJ, Hall JA, Murray KJ, Wells FC, Brown MJ (1989) A comparison of the effects of adrenaline and noradrenaline on human heart: the role of beta 1- and beta 2-adrenoceptors in the stimulation of adenylate cyclase and contractile force. Eur Heart J 10 Suppl B:29–37

Keating MT, Sanguinetti MC (2001) Molecular and cellular mechanisms of cardiac arrhythmias. Cell 104:569–580

Kelman GR, Palmer KN, Cross MR (1969) Cardiovascular effects of AH.3365 (salbutamol). Nature 221:1251

Khan MI, Hamilton JT, Manning GW (1972) Protective effect of beta adrenoceptor blockade in experimental coronary occlusion in conscious dogs. Am J Cardiol 30:832–837

Kjekshus JK (1986) Importance of heart rate in determining beta-blocker efficacy in acute and long-term acute myocardial infarction intervention trials. Am J Cardiol 57:43F–49F

Kleiman RB, Houser SR (1989) Outward currents in normal and hypertrophied feline ventricular myocytes. Am J Physiol 256:H1450–H1461

Koumi S, Backer CL, Arentzen CE, Sato R (1995a) beta-Adrenergic modulation of the inwardly rectifying potassium channel in isolated human ventricular myocytes. Alteration in channel response to beta-adrenergic stimulation in failing human hearts. J Clin Invest 96:2870–2881

Koumi S, Wasserstrom JA, Ten Eick RE (1995b) beta-Adrenergic and cholinergic modulation of the inwardly rectifying K+ current in guinea-pig ventricular myocytes. J Physiol 486:647–659

Kuhlkamp V, Schirdewan A, Stangl K, Homberg M, Ploch M, Beck OA (2000) Use of metoprolol CR/XL to maintain sinus rhythm after conversion from persistent atrial fibrillation: a randomized, double-blind, placebo-controlled study. J Am Coll Cardiol 36:139–146

Lakatta EG (1992) Functional implications of spontaneous sarcoplasmic reticulum Ca2+ release in the heart. Cardiovasc Res 26:193–214

Lazzara R, Scherlag BJ (1988) Generation of arrhythmias in myocardial ischemia and infarction. Am J Cardiol 61:20A–26A

Lee CO, Kang DH, Sokol JH, Lee KS (1980) Relation between intracellular Na ion activity and tension of sheep cardiac Purkinje fibers exposed to dihydro-ouabain. Biophys J 29:315–330

Leenen FH, Davies RA, Fourney A (1998) Catecholamines and heart function in heart transplant patients: effects of beta1- versus nonselective beta-blockade. Clin Pharmacol Ther 64:522–535

Li D, Melnyk P, Feng J, Wang Z, Petrecca K, Shrier A, Nattel S (2000) Effects of experimental heart failure on atrial cellular and ionic electrophysiology. Circulation 101:2631–2638

Li GR, Feng J, Wang Z, Fermini B, Nattel S (1996a) Adrenergic modulation of ultrarapid delayed rectifier K+ current in human atrial myocytes. Circ Res 78:903–915

Li GR, Feng J, Wang Z, Nattel S (1996b) Transmembrane chloride currents in human atrial myocytes. Am J Physiol 270:C500–C507

Li GR, Lau CP, Ducharme A, Tardif JC, Nattel S (2002) Transmural action potential and ionic current remodeling in ventricles of failing canine hearts. Am J Physiol Heart Circ Physiol 283:H1031–H1041

Litovsky SH, Antzelevitch C (1988) Transient outward current prominent in canine ventricular epicardium but not endocardium. Circ Res 62:116–126

Liu DW, Antzelevitch C (1995) Characteristics of the delayed rectifier current (IKr and IKs) in canine ventricular epicardial, midmyocardial, and endocardial myocytes. A weaker IKs contributes to the longer action potential of the M cell. Circ Res 76:351–365

Luo CH, Rudy Y (1994) A dynamic model of the cardiac ventricular action potential. II. Afterdepolarizations, triggered activity, and potentiation. Circ Res 74:1097–1113

Malfatto G, Rosen TS, Rosen MR (1988) The response to overdrive pacing of triggered atrial and ventricular arrhythmias in the canine heart. Circulation 77:1139–1148

Marx SO, Kurokawa J, Reiken S, Motoike H, D'Armiento J, Marks AR, Kass RS (2002) Requirement of a macromolecular signaling complex for beta adrenergic receptor modulation of the KCNQ1-KCNE1 potassium channel. Science 295:496–499

Matsuoka S, Ehara T, Noma A (1990) Chloride-sensitive nature of the adrenaline-induced current in guinea-pig cardiac myocytes. J Physiol 425:579–598

McDevitt DG (1989) In vivo studies on the function of cardiac beta-adrenoceptors in man. Eur Heart J 10 Suppl B:22–28

Moroni A, Gorza L, Beltrame M, Gravante B, Vaccari T, Bianchi ME, Altomare C, Longhi R, Heurteaux C, Vitadello M, Malgaroli A, DiFrancesco D (2001) Hyperpolarization-activated cyclic nucleotide-gated channel 1 is a molecular determinant of the cardiac pacemaker current I(f). J Biol Chem 276:29233–29241

Moss AJ, Zareba W, Hall WJ, Schwartz PJ, Crampton RS, Benhorin J, Vincent GM, Locati EH, Priori SG, Napolitano C, Medina A, Zhang L, Robinson JL, Timothy K, Towbin JA, Andrews ML (2000) Effectiveness and limitations of beta-blocker therapy in congenital long-QT syndrome. Circulation 101:616–623

Motomura S, Reinhard-Zerkowski H, Daul A, Brodde OE (1990) On the physiologic role of beta-2 adrenoceptors in the human heart: in vitro and in vivo studies. Am Heart J 119:608–619

Nakayama T, Fozzard HA (1988) Adrenergic modulation of the transient outward current in isolated canine Purkinje cells. Circ Res 62:162–172

Nattel S (2000) Acquired delayed rectifier channelopathies: how heart disease and antiarrhythmic drugs mimic potentially-lethal congenital cardiac disorders. Cardiovasc Res 48:188–190

Nattel S (2002) New ideas about atrial fibrillation 50 years on. Nature 415:219–226

Nattel S (2003) Rhythm versus rate control for atrial fibrillation management: what recent randomized clinical trials allow us to affirm. CMAJ 168:572–573

Nattel S, Li D (2000) Ionic remodeling in the heart: pathophysiological significance and new therapeutic opportunities for atrial fibrillation. Circ Res 87:440–447

Nattel S, Quantz MA (1988) Pharmacological response of quinidine induced early afterdepolarisations in canine cardiac Purkinje fibres: insights into underlying ionic mechanisms. Cardiovasc Res 22:808–817

Nattel S, Waters D (1990) What is an antiarrhythmic drug? From clinical trials to fundamental concepts. Am J Cardiol 66:96–99

Nattel S, Khairy P, Roy D, Thibault B, Guerra P, Talajic M, Dubuc M (2002) New approaches to atrial fibrillation management: a critical review of a rapidly evolving field. Drugs 62:2377–2397

Nordin C, Siri F, Aronson RS (1989) Electrophysiologic characteristics of single myocytes isolated from hypertrophied guinea-pig hearts. J Mol Cell Cardiol 21:729–739

Nuttall A, Snow HM (1982) The cardiovascular effects of ICI 118,587: A beta 1-adrenoceptor partial agonist. Br J Pharmacol 77:381–388

Oral H, Chugh A, Scharf C, Hall B, Cheung P, Veerareddy S, Daneshvar GF, Pelosi F Jr, Morady F (2004) Pulmonary vein isolation for vagotonic, adrenergic, and random episodes of paroxysmal atrial fibrillation. J Cardiovasc Electrophysiol 15:402–406

Packer M, Bristow MR, Cohn JN, Colucci WS, Fowler MB, Gilbert EM, Shusterman NH (1996) The effect of carvedilol on morbidity and mortality in patients with chronic heart failure. U.S. Carvedilol Heart Failure Study Group. N Engl J Med 334:1349–1355

Packer M, Coats AJ, Fowler MB, Katus HA, Krum H, Mohacsi P, Rouleau JL, Tendera M, Castaigne A, Roecker EB, Schultz MK, DeMets DL (2001) Effect of carvedilol on survival in severe chronic heart failure. N Engl J Med 344:1651–1658

Pappone C, Rosanio S, Oreto G, Tocchi M, Gugliotta F, Vicedomini G, Salvati A, Dicandia C, Mazzone P, Santinelli V, Gulletta S, Chierchia S (2000) Circumferential radiofrequency ablation of pulmonary vein ostia: a new anatomic approach for curing atrial fibrillation. Circulation 102:2619–2628

Plank DM, Yatani A, Ritsu H, Witt S, Glascock B, Lalli MJ, Periasamy M, Fiset C, Benkusky N, Valdivia HH, Sussman MA (2003) Calcium dynamics in the failing heart: restoration by beta-adrenergic receptor blockade. Am J Physiol Heart Circ Physiol 285:H305–H315

Pogwizd SM, Bers DM (2004) Cellular basis of triggered arrhythmias in heart failure. Trends Cardiovasc Med 14:61–66

Poole-Wilson PA, Swedberg K, Cleland JG, Di Lenarda A, Hanrath P, Komajda M, Lubsen J, Lutiger B, Metra M, Remme WJ, Torp-Pedersen C, Scherhag A, Skene A (2003) Comparison of carvedilol and metoprolol on clinical outcomes in patients with chronic heart failure in the Carvedilol Or Metoprolol European Trial (COMET): randomised controlled trial. Lancet 362:7–13

Priori SG, Napolitano C (2004) Genetics of cardiac arrhythmias and sudden cardiac death. Ann N Y Acad Sci 1015:96–110

Priori SG, Napolitano C, Tiso N, Memmi M, Vignati G, Bloise R, Sorrentino V, Danieli GA (2001) Mutations in the cardiac ryanodine receptor gene (hRyR2) underlie catecholaminergic polymorphic ventricular tachycardia. Circulation 103:196–200

Psaty BM, Manolio TA, Kuller LH, Kronmal RA, Cushman M, Fried LP, White R, Furberg CD, Rautaharju PM (1997) Incidence of and risk factors for atrial fibrillation in older adults. Circulation 96:2455–2461

Reddy P (2001) Does prophylaxis against atrial fibrillation after cardiac surgery reduce length of stay or hospital costs? Pharmacotherapy 21:338–344

Reiter MJ (2002) Beta-adrenergic blocking drugs as antifibrillatory agents. Curr Cardiol Rep 4:426–433

Reiter MJ, Reiffel JA (1998) Importance of beta blockade in the therapy of serious ventricular arrhythmias. Am J Cardiol 82:9I–19I

Rodefeld MD, Beau SL, Schuessler RB, Boineau JP, Saffitz JE (1996) Beta-adrenergic and muscarinic cholinergic receptor densities in the human sinoatrial node: identification of a high beta 2-adrenergic receptor density. J Cardiovasc Electrophysiol 7:1039–1049

Roden DM (2004) Drug-induced prolongation of the QT interval. N Engl J Med 350:1013–1022

Rosati B, Grau F, Rodriguez S, Li H, Nerbonne JM, McKinnon D (2003) Concordant expression of KChIP2 mRNA, protein and transient outward current throughout the canine ventricle. J Physiol 548:815–822

Rossner KL, Sachs HG (1978) Electrophysiological study of Syrian hamster hereditary cardiomyopathy. Cardiovasc Res 12:436–443

Ryden L, Ariniego R, Arnman K, Herlitz J, Hjalmarson A, Holmberg S, Reyes C, Smedgard P, Svedberg K, Vedin A, Waagstein F, Waldenstrom A, Wilhelmsson C, Wedel H, Yamamoto M (1983) A double-blind trial of metoprolol in acute myocardial infarction. Effects on ventricular tachyarrhythmias. N Engl J Med 308:614–618

Sanguinetti MC, Curran ME, Zou A, Shen J, Spector PS, Atkinson DL, Keating MT (1996) Coassembly of K(V)LQT1 and minK (IsK) proteins to form cardiac I(Ks) potassium channel. Nature 384:80–83

Schafers RF, Poller U, Ponicke K, Geissler M, Daul AE, Michel MC, Brodde OE (1997) Influence of adrenoceptor and muscarinic receptor blockade on the cardiovascular effects of exogenous noradrenaline and of endogenous noradrenaline released by infused tyramine. Naunyn Schmiedebergs Arch Pharmacol 355:239–249

Schwartz PJ, Priori SG, Spazzolini C, Moss AJ, Vincent GM, Napolitano C, Denjoy I, Guicheney P, Breithardt G, Keating MT, Towbin JA, Beggs AH, Brink P, Wilde AA, Toivonen L, Zareba W, Robinson JL, Timothy KW, Corfield V, Wattanasirichaigoon D, Corbett C, Haverkamp W, Schulze-Bahr E, Lehmann MH, Schwartz K, Coumel P, Bloise R (2001) Genotype-phenotype correlation in the long-QT syndrome: gene-specific triggers for life-threatening arrhythmias. Circulation 103:89–95

Shand DG (1983) Clinical pharmacology of the beta-blocking drugs: implications for the postinfarction patient. Circulation 67:I2–I5

Shimizu W, Noda T, Takaki H, Kurita T, Nagaya N, Satomi K, Suyama K, Aihara N, Kamakura S, Sunagawa K, Echigo S, Nakamura K, Ohe T, Towbin JA, Napolitano C, Priori SG (2003) Epinephrine unmasks latent mutation carriers with LQT1 form of congenital long-QT syndrome. J Am Coll Cardiol 41:633–642

Sicouri S, Antzelevitch C (1995) Electrophysiologic characteristics of M cells in the canine left ventricular free wall. J Cardiovasc Electrophysiol 6:591–603

Singh BN, Vaughan Williams EM (1970) A third class of anti-arrhythmic action. Effects on atrial and ventricular intracellular potentials, and other pharmacological actions on cardiac muscle, of MJ 1999 and AH 3474. Br J Pharmacol 39:675–687

Sinno H, Derakhchan K, Libersan D, Merhi Y, Leung TK, Nattel S (2003) Atrial ischemia promotes atrial fibrillation in dogs. Circulation 107:1930–1936

Sneader W (2001) The discovery and synthesis of epinephrine. Drug News Perspect 14:491–494

Soriano JB, Hoes AW, Meems L, Grobbee DE (1997) Increased survival with beta-blockers: importance of ancillary properties. Prog Cardiovasc Dis 39:445–456

St John SM, Ferrari VA (2002) Prevention of left ventricular remodeling after myocardial infarction. Curr Treat Options Cardiovasc Med 4:97–108

Steinfath M, Danielsen W, von der LH, Mende U, Meyer W, Neumann J, Nose M, Reich T, Schmitz W, Scholz H (1992) Reduced alpha 1- and beta 2-adrenoceptor-mediated positive inotropic effects in human end-stage heart failure. Br J Pharmacol 105:463–469

Stengl M, Volders PG, Thomsen MB, Spatjens RL, Sipido KR, Vos MA (2003) Accumulation of slowly activating delayed rectifier potassium current (IKs) in canine ventricular myocytes. J Physiol 551:777–786

Sumitomo N, Harada K, Nagashima M, Yasuda T, Nakamura Y, Aragaki Y, Saito A, Kurosaki K, Jouo K, Koujiro M, Konishi S, Matsuoka S, Oono T, Hayakawa S, Miura M, Ushinohama H, Shibata T, Niimura I (2003) Catecholaminergic polymorphic ventricular tachycardia: electrocardiographic characteristics and optimal therapeutic strategies to prevent sudden death. Heart 89:66–70

Thomas D, Zhang W, Karle CA, Kathofer S, Schols W, Kubler W, Kiehn J (1999) Deletion of protein kinase A phosphorylation sites in the HERG potassium channel inhibits activation shift by protein kinase A. J Biol Chem 274:27457–27462

Tromba C, Cohen IS (1990) A novel action of isoproterenol to inactivate a cardiac K+ current is not blocked by beta and alpha adrenergic blockers. Biophys J 58:791–795

Ullman A, Svedmyr N (1988) Salmeterol, a new long acting inhaled beta 2 adrenoceptor agonist: comparison with salbutamol in adult asthmatic patients. Thorax 43:674–678

Ungerer M, Bohm M, Elce JS, Erdmann E, Lohse MJ (1993) Altered expression of beta-adrenergic receptor kinase and beta 1-adrenergic receptors in the failing human heart. Circulation 87:454–463

Van Gelder IC, Hagens VE, Bosker HA, Kingma JH, Kamp O, Kingma T, Said SA, Darmanata JI, Timmermans AJ, Tijssen JG, Crijns HJ (2002) A comparison of rate control and rhythm control in patients with recurrent persistent atrial fibrillation. N Engl J Med 347:1834–1840

Viatchenko-Karpinski S, Terentyev D, Gyorke I, Terentyeva R, Volpe P, Priori SG, Napolitano C, Nori A, Williams SC, Gyorke S (2004) Abnormal calcium signaling and sudden cardiac death associated with mutation of calsequestrin. Circ Res 94:471–477

Volders PG, Sipido KR, Carmeliet E, Spatjens RL, Wellens HJ, Vos MA (1999) Repolarizing K+ currents ITO1 and IKs are larger in right than left canine ventricular midmyocardium. Circulation 99:206–210

Volders PG, Stengl M, van Opstal JM, Gerlach U, Spatjens RL, Beekman JD, Sipido KR, Vos MA (2003) Probing the contribution of IKs to canine ventricular repolarization: key role for beta-adrenergic receptor stimulation. Circulation 107:2753–2760

Vos MA, Gorgels AP, Leunissen JD, Wellens HJ (1990) Flunarizine allows differentiation between mechanisms of arrhythmias in the intact heart. Circulation 81:343–349

Waagstein F, Bristow MR, Swedberg K, Camerini F, Fowler MB, Silver MA, Gilbert EM, Johnson MR, Goss FG, Hjalmarson A (1993) Beneficial effects of metoprolol in idiopathic dilated cardiomyopathy. Metoprolol in Dilated Cardiomyopathy (MDC) Trial Study Group. Lancet 342:1441–1446

Wainger BJ, DeGennaro M, Santoro B, Siegelbaum SA, Tibbs GR (2001) Molecular mechanism of cAMP modulation of HCN pacemaker channels. Nature 411:805–810

Waldo AL, Camm AJ, deRuyter H, Friedman PL, MacNeil DJ, Pauls JF, Pitt B, Pratt CM, Schwartz PJ, Veltri EP (1996) Effect of d-sotalol on mortality in patients with left ventricular dysfunction after recent and remote myocardial infarction. The SWORD Investigators. Survival With Oral d-Sotalol. Lancet 348:7–12

Walsh KB, Kass RS (1991) Distinct voltage-dependent regulation of a heart-delayed IK by protein kinases A and C. Am J Physiol 261:C1081–C1090

Wang Z, Fermini B, Nattel S (1993) Sustained depolarization-induced outward current in human atrial myocytes. Evidence for a novel delayed rectifier K+ current similar to Kv1 5 cloned channel currents. Circ Res 73:1061–1076

Warth JD, Collier ML, Hart P, Geary Y, Gelband CH, Chapman T, Horowitz B, Hume JR (1996) CFTR chloride channels in human and simian heart. Cardiovasc Res 31:615–624

Wettwer E, Amos GJ, Posival H, Ravens U (1994) Transient outward current in human ventricular myocytes of subepicardial and subendocardial origin. Circ Res 75:473–482

Wiederhold KF, Nilius B (1986) Increased sensitivity of ventricular myocardium to intracellular calcium-overload in Syrian cardiomyopathic hamster. Biomed Biochim Acta 45:1333–1337

Wijffels MC, Kirchhof CJ, Dorland R, Allessie MA (1995) Atrial fibrillation begets atrial fibrillation. A study in awake chronically instrumented goats. Circulation 92:1954–1968

Wilde AA, Roden DM (2000) Predicting the long-QT genotype from clinical data: from sense to science. Circulation 102:2796–2798

Williams RS, Bishop T (1981) Selectivity of dobutamine for adrenergic receptor subtypes: in vitro analysis by radioligand binding. J Clin Invest 67:1703–1711

Wischmeyer E, Karschin A (1996) Receptor stimulation causes slow inhibition of IRK1 inwardly rectifying K+ channels by direct protein kinase A-mediated phosphorylation. Proc Natl Acad Sci U S A 93:5819–5823

Wit AL, Cranefield PF (1976) Triggered activity in cardiac muscle fibers of the simian mitral valve. Circ Res 38:85–98

Wit AL, Cranefield PF (1977) Triggered and automatic activity in the canine coronary sinus. Circ Res 41:434–445

Woelfel A, Foster JR, Simpson RJ Jr, Gettes LS (1984) Reproducibility and treatment of exercise-induced ventricular tachycardia. Am J Cardiol 53:751–756

Workman AJ, Kane KA, Russell JA, Norrie J, Rankin AC (2003) Chronic beta-adrenoceptor blockade and human atrial cell electrophysiology: evidence of pharmacological remodelling. Cardiovasc Res 58:518–525

Wyse DG, Waldo AL, DiMarco JP, Domanski MJ, Rosenberg Y, Schron EB, Kellen JC, Greene HL, Mickel MC, Dalquist JE, Corley SD (2002) A comparison of rate control and rhythm control in patients with atrial fibrillation. N Engl J Med 347:1825–1833

Xiao YF, McArdle JJ (1995) Activation of protein kinase A partially reverses the effects of 2,3-butanedione monoxime on the transient outward K+ current of rat ventricular myocytes. Life Sci 57:335–343

Yasumura Y, Takemura K, Sakamoto A, Kitakaze M, Miyatake K (2003) Changes in myocardial gene expression associated with beta-blocker therapy in patients with chronic heart failure. J Card Fail 9:469–474

Yue L, Melnyk P, Gaspo R, Wang Z, Nattel S (1999) Molecular mechanisms underlying ionic remodeling in a dog model of atrial fibrillation. Circ Res 84:776–784

Yusuf S, Peto R, Lewis J, Collins R, Sleight P (1985) Beta blockade during and after myocardial infarction: an overview of the randomized trials. Prog Cardiovasc Dis 27:335–371

Zhang LM, Wang Z, Nattel S (2002) Effects of sustained beta-adrenergic stimulation on ionic currents of cultured adult guinea pig cardiomyocytes. Am J Physiol Heart Circ Physiol 282:H880–H889

Zipes DP, Arbel E, Knope RF, Moe GK (1974) Accelerated cardiac escape rhythms caused by ouabain intoxication. Am J Cardiol 33:248–253

HEP (2006) 171:267–286
© Springer-Verlag Berlin Heidelberg 2006

Experimental Therapy of Genetic Arrhythmias: Disease-Specific Pharmacology

S. G. Priori (✉) · C. Napolitano · M. Cerrone

Molecular Cardiology Laboratories, IRCCS Fondazione Salvatore Maugeri, Via Ferrata 8, 27100 Pavia, Italy
spriori@fsm.it

Abstract The integration between molecular biology and clinical practice requires the achievement of fundamental steps to link basic science to diagnosis and management of patients. In the last decade, the study of genetic bases of human diseases has achieved several milestones, and it is now possible to apply the knowledge that stems from the identification of the genetic substrate of diseases to clinical practice. The first step along the process of linking molecular biology to clinical medicine is the identification of the genetic bases of inherited diseases. After this important goal is achieved, it becomes possible to extend research to understand the functional impairments of mutant protein(s) and to link them to clinical manifestations (genotype–phenotype correlation). In genetically heterogeneous diseases, it may be possible to identify locus-specific risk stratification and management algorithms. Finally, the most ambitious step in the study of genetic disease is to discover a novel pharmacological therapy targeted at correcting the inborn defect (locus-specific therapy) or even to "cure" the DNA abnormality by replacing the defective gene with gene therapy. At present, this curative goal has been successful only for very few diseases. In the field of inherited arrhythmogenic diseases, several genes have been discovered, and genetics is now emerging as a source of information contributing not only to a better diagnosis but also to risk stratification and management of patients. The functional characterization of mutant proteins has opened new perspectives about the possibility of performing gene-specific or mutation-specific therapy. In this chapter, we will briefly summarize the genetic bases of inherited arrhythmogenic conditions and we will point out how the information derived from molecular genetics has influenced the "optimal use of traditional therapies" and has paved the way to the development of gene-specific therapy.

Keywords Cardiac arrhythmias · Sudden death · Genetic ·
Genotype-phenotype correlation · Gene specific therapy

1
Long QT Syndrome

1.1
Clinical Presentation and Molecular Bases

The long QT syndrome (LQTS) is an inherited disease characterized by an abnormally prolonged ventricular repolarization that creates a vulnerable substrate for the development of life-threatening arrhythmias (Schwartz et al. 2000b). Two phenotypic variants have been described: the autosomal-dominant Romano-Ward Syndrome (RW) and the autosomal-recessive Jervell–Lange-Nielsen syndrome (JLN); in the latter, the cardiac phenotype is associated with sensorineural deafness.

In the early 1990s, Keating and colleagues published the first evidence that LQTS is caused by mutations in genes encoding subunits of cardiac ion channels (Curran et al. 1995; Wang et al. 1995, 1996). Other reports followed and eight different genes responsible for abnormally prolonged ventricular repolarization have now been identified (Abbott et al. 1999; Curran et al. 1995; Mohler et al. 2003; Sanguinetti et al. 1996; Splawski et al. 2004; Tristani-Firouzi et al. 2002; Wang et al. 1995, 1996).

The RW syndrome includes six genetic subtypes: LQT1 caused by mutations in the KCNQ1 gene (Wang et al. 1996), encoding the α-subunit of the I_{Ks} channel. LQT2 caused by mutations in the KCNH2 gene, encoding the α-subunit of the I_{Kr} channel (Curran et al. 1995). LQT3 is caused by mutations in the SCN5A gene that encodes for the cardiac sodium channel (Wang et al. 1995). LQT4 is caused by mutations in the ankyrin B gene (Mohler et al. 2003) that encodes a chaperon-like protein that regulates the localization of ion channels in the membrane of the cardiac myocytes. This latter gene has been reported only recently and only a few families have been genotyped worldwide. LQT5, due to mutations in *KCNE1*, the β-subunit of the I_{Ks} channel (Sanguinetti et al. 1996), and LQT6, due to mutations in *KCNE2*, the β-subunit of the I_{Kr} channel (Abbott et al. 1999), are relatively uncommon and account for less than 5% of genotyped patients. Indeed, more than 90% of genotyped patients belong to the LQT1, LQT2, and LQT3 variants of LQTS.

Overall, the molecular screening of the open reading frame of these genes allows successful genotyping in 60%–70% of genetically affected individuals, suggesting that additional LQTS-related genes have yet to be discovered.

The JLN syndrome is a recessive disease allelic to LQT1 and LQT5, in which homozygous or compound heterozygous mutations recapitulate the phenotype: prolonged Q-T interval, cardiac arrhythmias, and deafness (Neyroud et al. 1997; Schulze-Bahr et al. 1997).

Functional characterization of several mutants identified in LQTS has helped defining a tight link between QT prolongation and DNA abnormalities by showing that mutations of the α- or β-subunits of the I_{Ks} or the I_{Kr} potassium channels found in LQT1, LQT2, LQT5, and LQT6 patients impair the repolarization process by reducing outward currents conducted through these channels. Interestingly, multiple mechanisms of functional failure have been reported. In some instances, the defective protein loses the ability to tetramerize, causing haploinsufficiency, i.e. a reduction of 50% of the potassium current, while in other cases defective proteins co-assemble with wildtype subunits exerting a dominant-negative effect. Furthermore, mutations could impair the transport of proteins from the Golgi apparatus to the cell membrane (trafficking). At variance with all the other LQTS loci, mutations identified in LQT3 patients cause a gain of function, with an increased inward I_{Na} current (Bennett et al. 1995).

1.2
Traditional Therapies and the Need for Locus-Specific Treatments

1.2.1
Antiadrenergic Therapy and ICD

β-Blockers are the standard treatment for LQTS. The efficacy of these drugs has been documented since the early studies (Schwartz 1985), and it has been confirmed in the large group of patients enrolled in the International LQTS Registry (Moss et al. 2000) and in our LQTS Italian Registry (Priori et al. 2004).

The increasing availability of a large series of patients genotyped as LQT1, LQT2, and LQT3 has demonstrated that the genetic substrate influences the response to therapy (Priori et al. 2004). In the largest study completed so far to investigate the genotype-specific efficacy of β-blockers, we showed that LQT1 patients have a better response than LQT2 and LQT3 (Priori et al. 2004). The latter two groups of patients remain at risk of symptoms and cardiac arrest despite therapy with a relative risk of 2.8 and 4.0, respectively (Priori et al. 2004). The incomplete efficacy of β-blockers in LQT2 and LQT3 provides a strong rationale to search for locus-specific treatments that may allow better prevention of arrhythmic events.

1.2.2
Locus-Specific Modulation of Transmembrane Currents

The hints to develop a locus-specific therapy for LQT2 came from the experimental evidence demonstrating that the conductance of I_{Kr} is increased in the presence of high extracellular K^+ (Sanguinetti and Jurkiewicz 1991). The hypothesis was therefore put forth that an increase in $[K^+]^o$ would enhance I_{Kr} to compensate for the presence of the mutation. Accordingly, Compton et al. (1996) were the first to propose and to show that exogenous administration of potassium supplements (reaching levels ≥ 1.5 mEq/l above baseline) effectively shortens Q-T interval among LQT2 patients (Compton et al. 1996). A controlled clinical trial is ongoing that is expected to provide further information on the value of this locus-specific treatment not only on QT duration but also on the occurrence of cardiac events.

Since mutations associated to LQT3 phenotype lead to a gain of function of the sodium channel, the use of sodium channels blockers appeared as a reasonable locus-specific approach to treat these patients. Based upon experimental evidence obtained in our lab (Priori et al. 1996; Fig. 1), we provided the initial evidence that mexiletine effectively shortens the Q-T interval in LQT3 patients

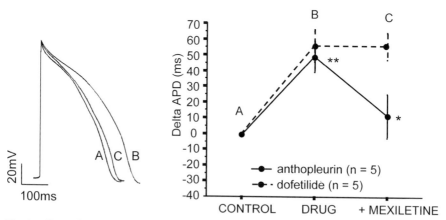

Fig. 1 Effect of mexiletine in an LQT3. *Left panel:* action potential duration (APD) in response to mexiletine in an LQT3 model in isolated guinea pig cardiac myocytes. Superimposed action potentials recorded at baseline (*A*), and during exposure to anthopleurin (*B*), and anthopleurin plus mexiletine (*C*). *Right panel:* summary of experiments in a pharmacological model of LQT2 (mimicked by dofetilide—*dotted line*) and LQT3 (mimicked by anthopleurin—*continuous line*) show a significant APD shortening upon mexiletine (*C*) exposure only in the LQT3 model; . **$p<0.001$ vs control; *$p<0.001$ vs dofetilide. (Modified from Priori et al. 1996)

(Schwartz et al. 1995). Shortly after, we also demonstrated the short-term efficacy of mexiletine to prevent lethal events (Schwartz et al. 2000a; Fig. 2). However, as of today there are no long-term prospective data demonstrating that mexiletine improves survival in LQT3 patients. Interestingly, we (Schwartz et al. 2000a) and others (Kehl et al. 2004) have successfully used mexiletine in LQT3 newborns patients with functional AV block in whom the QT shortening obtained with this drug has been able to restore 1:1 AV conduction.

Recent observations have introduced the concept that the efficacy of mexiletine in preventing cardiac events may be predicted by the biophysical properties of the mutations (Rivolta et al. 2004). In this study, we showed that in

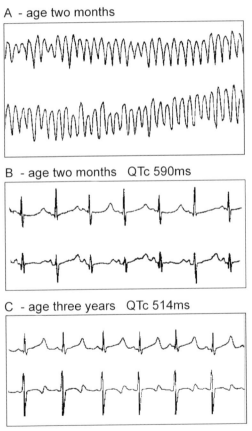

Fig. 2a–c Clinical use of mexiletine in a LQT3 patient. Electrocardiograms at the time of admission to the hospital (**a** and **b**), and during mexiletine treatment (**c**). At hospital admission, the 44-day-old infant had ventricular fibrillation (**a**). After the restoration of sinus rhythm, the corrected Q-T interval was found to be markedly prolonged (648 ms, **b**). Oral mexiletine was administered and the child's corrected Q-T interval, albeit still prolonged, was significantly reduced (510 ms, **c**). (Modified from Schwartz et al. 2000a)

vitro testing of mexiletine on two mutants (*P1332L, Y1795C*) allowed a reliable "prediction" of the response to the drug of the mutation carriers.

A further development in the use of sodium channel blockers in LQT3 was provided by Benhorin et al. (2000) and Windle et al. (2001) who investigated the response to flecainide in two LQT3 families, harboring the D1790G and the *ΔKPQ* mutations. They demonstrated that, similarly to mexiletine, flecainide may shorten Q-T interval duration. In vitro experiments on *D1790G* showed a unique pharmacological response, consisting in a selective block by flecainide but not by lidocaine of the heterologously expressed mutant channels during repetitive stimulations (use-dependent block; Abriel et al. 2000). In general, flecainide should be used with caution in unselected LQT3 patients. Indeed, when we tested flecainide in 13 unselected LQT3 patients (Priori et al. 2000c), we observed ST-segment elevation, resembling a Brugada syndrome (BrS) ECG in 6 of them (Fig. 3). These data are not surprising as it is known that prolongation of Q-T interval and ST segment elevation in right precordial leads may co-exist in some patients who are usually described as having "overlapping phenotypes" (Bezzina et al. 1999; Grant et al. 2002). Since ST-segment elevation in right precordial leads is regarded as a marker

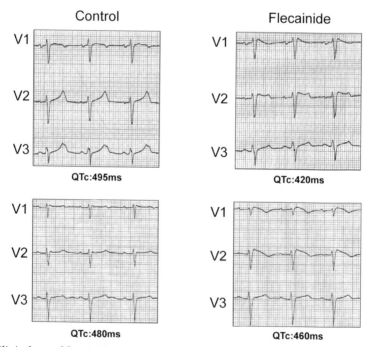

Fig. 3 Clinical use of flecainide in LQT3. Examples of ST segment elevation observed in two LQT3 patients upon intravenous administration of 2 mg/kg of flecainide. (Modified from Priori et al. 2000c)

of electrical instability, the use of flecainide in LQT3 should be used only in selected patients when it is demonstrated that they do not develop ST segment elevation.

In summary, it appears evident that sodium channel blockade is a promising approach for the treatment of LQT3 patients. However, the experimental data (Abriel et al. 2000; Rivolta et al. 2004) strongly suggest that not all I_{Na} blockers may be equally effective and that mutation-specific, more than locus-specific, treatments will be required in order to achieve the best possible clinical efficacy.

1.2.3
Rescue of Defective Proteins

Experimental investigations (Zhou et al. 1998) have led to the appreciation that defective intracellular processing of mutant channels is an important pathophysiological mechanism in LQTS. While Zhou and co-workers focused on defective trafficking of KCNH2 (HERG) mutants, other authors suggested that also KCNQ1 mutants may be retained intracellularly and fail to reach the plasmalemma (Bianchi et al. 1999; Gouas et al. 2004). Based on this evidence, several groups have initiated a very challenging set of investigations aimed at restoring protein trafficking. Preliminary experiments on the N470D KCNH2 mutation demonstrated that if cells are cultured at lower temperature (27 °C instead of 37 °C) or in the presence of compounds such E4031 (a class III antiarrhythmic agent), astemizole, or cisapride, the trafficking into the plasma membrane is restored (Zhou et al. 1999). Unfortunately, rescue of trafficking occurred only at drug concentrations that also block the channel (Zhou et al. 1999). Therefore, these drugs are unable to normalize the duration of repolarization. Other drugs have been tested in an attempt to separate the effect on trafficking from I_{Kr} blockade. When cultured with fexofenadine, a metabolite of terfenadine, with a weak I_{Kr} blocking effect, two different KCNH2 mutants recovered their function at a concentration that could not block the channel (Rajamani et al. 2002). Along the same line, thapsigargin, an inhibitor of sarcoendoplasmic reticulum calcium-ATPase (SERCA) transporter, has been shown to rescue other trafficking-defective KCNH2 mutants (Delisle et al. 2003). Overall, the most recent findings demonstrate that it is possible to dissociate the blocking activity on the channel from the ability to restore normal trafficking; as a consequence, this therapeutic strategy is now regarded as being closer to the bedside.

In summary, gene-specific therapy of LQTS is still in a preliminary phase. The most robust evidence of efficacy concerns the use of mexiletine for LQT3 patients; at present, however, mexiletine should still be regarded only as an adjunctive treatment to β-blockade or to the implantable cardioverter defibrillator to prevent cardiac events in high-risk LQT3 patients.

2
Brugada Syndrome

2.1
Clinical Aspects

BrS is a primary electrical disease associated with ventricular arrhythmias and sudden cardiac death (SCD) in young individuals with a typical ECG pattern of ST-segment elevation in the right precordial leads with or without right bundle branch block (Brugada and Brugada 1992).

SCD usually occurs during sleep and in the early morning hours (Matsuo et al. 1999). The initiating mechanism of arrhythmias is a short-coupled extrasystolic beat triggering polymorphic VT that degenerates into ventricular fibrillation (VF). Even if some pediatric cases have been reported (Priori et al. 2000b), most commonly the onset of symptoms is in the third to fourth decade of life (Suzuki et al. 2000), and males are at higher risk of arrhythmic events than females (Brugada et al. 2000; Priori et al. 2002b; Wilde et al. 2002).

2.2
Genetic Bases and Pathophysiology

BrS is transmitted as an autosomal-dominant trait. In 1998, Chen et al. (1998) reported BrS patients and families harboring mutations in the SCN5A gene, thus suggesting that one form of BrS is allelic to LQT3. Unfortunately, no other gene has been linked to BrS, and *SCN5A* accounts only for 20% of clinically affected patients; consequently, the genetic substrate can be identified only in a minority of clinically affected individuals.

More than 50 BrS *SCN5A* mutations have been reported so far, and functional expression studies showed a spectrum of biophysical abnormalities all leading to a loss of function (Priori et al. 2003a): (1) a failure of the channel to express (haploinsufficiency); (2) a shift of voltage- and time-dependent channel activation, inactivation, or re-activation; (3) entry of the I_{Na} into an intermediate, slowly recovering, state of inactivation; (4) accelerated inactivation.

In order to recapitulate the consequences of the different mutations in a hypothesis that accounts for the distinguishing electrocardiographic pattern of the syndrome, Antzelevitch proposed that the electrophysiological mechanism of BrS is an outward shift of net transmembrane current at the end of phase 1 of the action potential in the right ventricular epicardium (Antzelevitch 2001). This effect may be related to the differential transmural and left-to-right expression level of I_{To}, which is highest in epicardial cells and particularly those of the right ventricle. Accelerated inactivation or reduction of I_{Na} in BrS may leave I_{To} unopposed during phase 1 of the action potential. The I_{To} -mediated spike-and-dome morphology in the right ventricular epicardium, but not in the endocardium, could generate a prominent J-point, and the typical BrS

ECG (Antzelevitch 2001). This hypothesis is so far the only one put forward that accounts for the BrS phenotype, and it has inspired the development of gene-specific therapies.

2.3
Traditional Therapies and the Need for Locus-Specific Treatments

2.3.1
Therapy: ICD Indications and Efficacy

At present, there is no effective pharmacological treatment to prevent arrhythmic events in BrS. Therefore, the implantable cardioverter–defibrillator (ICD) is the only option for high-risk individuals. There is agreement on the use of ICD for secondary prevention (Priori et al. 2001) and in all high-risk individuals, i.e., those with a history of syncope and a spontaneously abnormal ECG (Priori et al. 2001; Priori et al. 2002b). The management of asymptomatic patients and of those in whom diagnosis is possible only upon provocative test with sodium channel blockers, is still debated, as no conclusive evidence exists for risk stratification of these subjects. The risk of experiencing a cardiac event in a lifetime is estimated around 8% (Priori et al. 2001). The predictive value of inducibility [induction of a VF during programmed electrical stimulation (PES)], in asymptomatic patients is still under debate and further data are needed before the use of a prophylactic ICD in these subjects is recommended (Brugada et al. 2003a,b; Eckardt et al. 2002, 2005; Gasparini et al. 2002; Priori et al. 2000a, 2002b). In this scenario, it is clear that the identification of a pharmacological therapy that could counteract the electrophysiological abnormalities and reduce the risk of arrhythmias would be extremely welcome.

2.3.2
Novel Therapies Based on Pathophysiology

Based on Antzelevitch's hypothesis to account for the phenotype of the disease, most studies have tried to identify strategies for blocking the I_{To} current. Since selective I_{To} blockers are not available, a variety of drugs with less specific blocking properties has been used in experimental models and in preliminary clinical trials. In vitro studies suggested that quinidine normalizes the I_{To}-induced ST-segment elevation (Yan and Antzelevitch 1999) and preliminary clinical evidence shows that this drug may prevent arrhythmia inducibility at PES (Belhassen et al. 1999). These results have been recently confirmed by Hermida et al. (2004) in a larger cohort of asymptomatic patients in whom quinidine prevented VT/VF inducibility in 76% of treated patients. Further data from the Belhassen's group (Belhassen et al. 2004) confirmed the long-term efficacy of quinidine in preventing VF induction at PES and the occurrence of spontaneous arrhythmias. However, a 36% incidence of side effects leading to drug discontinuation was recorded, limiting the success rate of this treatment. Cilostazol, an oral phosphodiesterase III inhibitor marketed as an antiplatelet agent, can normalize the ST-segment in BrS patients. It increases I_{Ca} by inhibiting phosphodiesterase activity in ventricular myocytes and it decreases I_{To} by accelerating heart rate; this latter effect is secondary to the increase of I_{Ca} in the sinus node. Cilostazol has been used in only one patient (Tsuchiya et al. 2002), and further investigations are needed to confirm its efficacy in BrS. Finally, tedisamil, an anti-arrhythmic agent that blocks I_{To} and other potassium currents, has been proposed as an alternative to quinidine for in BrS patients (Freestone and Lip 2004), but at present this drug is still in the pre-marketing evaluation phase.

2.3.3
Mutation-Specific Therapy

Mutations leading to impairment of protein trafficking have been identified in *SCN5A* and associated with BrS (Valdivia et al. 2002, 2004) Interestingly, mexiletine was able to restore I_{Na} by rescuing the proper localization of the protein (Valdivia et al. 2004). Whether mexiletine could achieve the same result in vivo is still unknown, but these data provide interesting insight for a mutation-specific treatment for BrS patients.

3
Short QT Syndrome

3.1
Clinical Features and Genetic Bases

The short QT syndrome (SQTS) has been described as an autosomal-dominant disease characterized by an abbreviated repolarization that fails to show dynamic changes during heart rate variations (Gussak et al. 2000). In analogy with other inherited diseases, SQTS has been reported as a cause of syncope, SCD, and atrial arrhythmias in a structurally intact heart (Gaita et al. 2003). Electrophysiological study in SQTS patients showed short refractory periods in ventricles and atria, and a high rate of inducibility of ventricular arrhythmias (Gaita et al. 2003).

Three genes have been associated to SQTS: *KNCQ1* (Bellocq et al. 2004), *KCNH2* (Brugada et al. 2004) and *KCNJ2* (Priori et al. 2005). Since a loss-of-function mutation in these two genes causes LQTS, as expected SQTS mutations produce a "gain of function," thus accelerating the ventricular repolarization rate (shortening of action potential) in cardiac myocytes.

3.2
Therapy Based on Pathophysiology

Preliminary studies evaluated the effect of antiarrhythmic drugs targeted to counteract the specific effects of mutations identified in SQTS patients (Gaita et al. 2004).

Carriers of gain of function *KCNH2* mutations were initially treated with I_{Kr} blocking agents, such as sotalol and ibutilide (Gaita et al. 2004). Unexpectedly, both drugs did not increase significantly the Q-T interval. A possible explanation for this result was provided by Brugada et al. (2004), who showed that in vitro, sotalol blocks I_{Kr} conducted by wildtype channels (as expected) but it fails to suppress the current in mutant channels. These authors identified, in the lack of rectification of the SQTS mutant channel, the likely cause of such reduced affinity for class III antiarrhythmic agents.

Gaita et al. (2004) showed that quinidine normalizes Q-T interval duration and abolishes inducibility at PES. Quinidine has several electrophysiological properties: It blocks fast-inward sodium current (I_{Na}), I_{Kr}, I_{Ks}, the inward rectifier (I_{K1}), the I_{To}, and the ATP-sensitive potassium currents (I_{KATP}), and all these effects concur to prolong the cardiac action potential. Furthermore, quinidine has a greater affinity for the open state of the I_{Kr} channel, possibly accounting for its higher efficacy in STQS as compared to class III anti-arrhythmic agents.

Long-term follow up in larger populations of SQTS patients is needed to define if treatment with quinidine can affect mortality and whether this drug represents a treatment for all SQTS patients or it is specifically indicated for those with mutations in the KCNH2 gene.

4
Catecholaminergic Polymorphic Ventricular Tachycardia

4.1
Clinical Features and Molecular Bases

Coumel and colleagues (Coumel et al. 1978) in 1978 and Leenhardt and colleagues (Leenhardt et al. 1995) in 1995 initially described the catecholaminergic polymorphic ventricular tachycardia (CPVT) as a peculiar clinical entity that could lead to stress-induced syncope or SCD in young people with an autosomal-dominant pattern of transmission in some cases. The characterizing pattern of arrhythmias in CPVT patients is the so-called bi-directional tachycardia, a ventricular arrhythmia presenting a 180° alternans of the QRS axis on a beat-to-beat basis. More recent observations have pointed to the fact that CPVT patients may also show irregular polymorphic VT without a "stable" QRS vector alternans. At variance with the other inherited arrhythmogenic syndromes, the baseline ECG is unremarkable. Therefore, diagnosis is uniquely based on the demonstration of ventricular tachycardia elicited during exercise or emotional stress. Symptoms typically manifest during childhood with most of SCD occurring before age 20 (Leenhardt et al. 1995; Priori et al. 2002a).

In 1999, Swan et al. (1999) mapped a CPVT locus to chromosome 1q42-q43, and in 2001 our group (Priori et al. 2001b) identified the cardiac ryanodine receptor gene *RyR2* as the gene involved in the pathogenesis of CPVT. This evidence was subsequently confirmed (Laitinen et al. 2001).

The ryanodine receptor is localized across the membrane of the sarcoplasmic reticulum (SR) and it releases Ca^{2+} from SR in response to the calcium entry through the L-type channels, during phase 2 of the cardiac action potential.

All *RyR2* mutations reported so far are missense (single amino acid substitutions) located in functionally important regions of the protein: the transmembrane domain, the Ca^{2+} binding sites, and the FKBP12.6 (calstabin 2) binding domain. Experimental data suggest that CPVT mutations destabilize the protein with consequent Ca^{2+} overload during repolarization and electric diastole, thus facilitating the occurrence of delayed afterdepolarizations (DADs) (Marks et al. 2002). Functional characterization of the mutants (Jiang et al. 2002; Wehrens et al. 2003) confirmed that they all produce abnormal Ca^{2+} release in response to adrenergic stimulation.

In 2001, Lahat et al. (2001) described the autosomal recessive variant of CPVT and linked it to a mutation in the CASQ2 gene on chromosome 1p11-p13 that encodes for calsequestrin. Autosomal-dominant and autosomal-recessive CPVT have very similar clinical presentation.

Calsequestrin is another protein involved in Ca^{2+} homeostasis by serving the major Ca^{2+} buffering protein into the SR cisternae. Therefore, *CASQ2* has a direct role in the modulation of the excitation–contraction coupling. Functional studies (Viatchenko-Karpinski et al. 2004) showed that mutant CASQ2 proteins displays altered Ca^{2+} binding and reduced buffering properties. Furthermore, the interplay between calsequestrin and ryanodine receptor may also be affected, resulting in uncoordinated release. Thus, the mechanism of *CASQ2* and *RyR2* mutations share the presence of abnormal intracellular Ca^{2+} handling, possibly leading to DADs and triggered activity upon adrenergic stimulation.

4.2
Therapy

Since its initial description (Coumel et al. 1978; Leenhardt et al. 1995), antiadrenergic treatment appeared as the most effective and appropriate therapy in CPVT to limit the detrimental consequences of adrenergic activation on heart rhythm. β-Blockers achieve a satisfactory control of arrhythmias, and they are effective both for prophylaxis of arrhythmias and for the suppression of incessant arrhythmias during the acute phase (De Rosa et al. 2004; Leenhardt et al. 1995; Priori et al. 2001b, 2002a). However, not all patients have a satisfactory response to these drugs. Leenhardt et al. (1995) described SCD and syncope in 10% of patients (2 out of 21). In our series, 30% of patients on β-blockers required an ICD and about 50% of them received appropriate shocks during a 2-year follow-up (Cerrone et al. 2004; Priori et al. 2002a). Other antiarrhythmic drugs have been used anecdotally but with unsatisfactory results (Sumitomo et al. 2003).

4.2.1
Stabilization of RyR2 Channel: A Novel Direction for Therapy

Based on the consideration that several *RyR2* mutations occur in the FKBP12.6 binding domain, the role of calstabin 2 (also known as FKBP12.6) in the pathophysiology of CPVT has been thoroughly investigated. Calstabin 2 is a RyR2 regulatory subunit that stabilizes the channel in the closed state, thus preventing abnormal diastolic Ca^{2+} release. Several missense mutations found in CPVT appear to decrease the calstabin 2 affinity for RyR2, eventually leading to leaky channels that contribute to the occurrence of DADs and triggered arrhythmias (Lehnart et al. 2004a).

Abnormal Ca^{2+} release and decreased affinity for binding of calstabin 2 are also involved in the pathogenesis of ventricular dysfunction and possibly SCD during heart failure (Lehnart et al. 2004a; Yano et al. 2003). The experimental agent JTV519 is a 1,4-benzothiazepine derivative that was shown to inhibit progression of heart failure in a canine model, probably by increasing the binding of calstabin 2 to RyR2 (Yano et al. 2003).

Based on these observations, JTV519 has been tested in experimental models of abnormal Ca^{2+} handling and CPVT, to investigate whether it could prevent ventricular arrhythmias. In knock-out calstabin 2 (−/−) mice, JTV519 did not prevent the occurrence of ventricular arrhythmias, while it was effective in calstabin 2 haploinsufficient (−/+) mice, suggesting that the anti-arrhythmic potential of the drug resides in the recover of the binding of calstabin to RyR2 channels (Wehrens et al. 2004). Further experiments on a RyR2 mutant (Lehnart et al. 2004b) showed that JTV519 restores the normal activity of the channel. These findings suggest that RyR stabilization in the closed state, by recovering calstabin 2 affinity, could represent a novel therapeutic strategy to prevent the occurrence of life-threatening arrhythmias in affected patients.

5
Conclusions

In the past decade, molecular biology has allowed to elucidate the genetic background of several inherited diseases predisposing to cardiac arrhythmias

Table 1 Proposed gene-specific therapies in inherited arrhythmogenic diseases (see text for details)

Disease	Gene	Effect	Treatment
LQT2	KCNH2	Reduced I_{Kr}	Increase I_{Kr}: potassium supplements
			Rescue of trafficking: fexofenadine, thapsigargin
LQT3	SCN5A	Increase of I_{Na}	Block of I_{Na}: mexiletine
BrS	SCN5A	Reduction of I_{Na}	Block of I_{To}: quinidine, tedisamil, cilostazol
			Rescue of trafficking-defective mutants: mexiletine
SQTS	KNCH2	Increased I_{Kr}	Block of I_{Kr}: quinidine
CPVT (autosomal dominant)	RYR2	Intracellular Ca^{2+} overload	Recover of FKBP12.6 binding: JTV519

and SCD. Functional characterization of mutant proteins has provided fascinating insights about the electrophysiological derangements that account for the phenotypes. Now that genetics has already entered clinical cardiology, playing a role for diagnosis and for novel risk-stratification strategies, fundamental research has already set its next goal, and several groups are turning their attention toward the development of locus-specific therapies. Preliminary experimental studies have been successful and a few clinical pilot projects are indicating how to direct further research (Table 1). The availability of new models used in basic research, such as expression of mutant proteins in cardiac myocytes and the use of transgenic animals, opens very promising perspectives for the development of novel treatments tailored to the correction of genetically determined electrophysiological abnormalities.

References

Abbott GW, Sesti F, Splawski I, Buck ME, Lehmann MH, Timothy KW, Keating MT, Goldstein SA (1999) MiRP1 forms IKr potassium channels with HERG and is associated with cardiac arrhythmia. Cell 97:175–187

Abriel H, Wehrens XH, Benhorin J, Kerem B, Kass RS (2000) Molecular pharmacology of the sodium channel mutation D1790G linked to the long-QT syndrome. Circulation 102:921–925

Antzelevitch C (2001) The Brugada syndrome: ionic basis and arrhythmia mechanisms. J Cardiovasc Electrophysiol 12:268–272

Belhassen B, Viskin S, Fish R, Glick A, Setbon I, Eldar M (1999) Effects of electrophysiologic-guided therapy with Class IA antiarrhythmic drugs on the long-term outcome of patients with idiopathic ventricular fibrillation with or without the Brugada syndrome. J Cardiovasc Electrophysiol 10:1301–1312

Belhassen B, Glick A, Viskin S (2004) Efficacy of quinidine in high-risk patients with Brugada syndrome. Circulation 110:1731–1737

Bellocq C, van Ginneken AC, Bezzina CR, Alders M, Escande D, Mannens MM, Baro I, Wilde AA (2004) Mutation in the KCNQ1 gene leading to the short QT-interval syndrome. Circulation 109:2394–2397

Benhorin J, Taub R, Goldmit M, Kerem B, Kass RS, Windman I, Medina A (2000) Effects of flecainide in patients with new SCN5A mutation: mutation-specific therapy for long-QT syndrome? Circulation 101:1698–1706

Bennett PB, Yazawa K, Makita N, George AL Jr (1995) Molecular mechanism for an inherited cardiac arrhythmia. Nature 376:683–685

Bezzina C, Veldkamp MW, van Den Berg MP, Postma AV, Rook MB, Viersma JW, van Langen IM, Tan-Sindhunata G, Bink-Boelkens MT, Der Hout AH, Mannens MM, Wilde AA (1999) A single Na(+) channel mutation causing both long-QT and Brugada syndromes. Circ Res 85:1206–1213

Bianchi L, Shen Z, Dennis AT, Priori SG, Napolitano C, Ronchetti E, Bryskin R, Schwartz PJ, Brown AM (1999) Cellular dysfunction of LQT5-minK mutants: abnormalities of IKs, IKr and trafficking in long QT syndrome. Hum Mol Genet 8:1499–1507

Brugada J, Brugada P, Brugada R (2000) Sudden death (VI). The Brugada syndrome and right myocardiopathies as a cause of sudden death. The differences and similarities. Rev Esp Cardiol 53:275–285

Brugada J, Brugada R, Brugada P (2003a) Determinants of sudden cardiac death in individuals with the electrocardiographic pattern of Brugada syndrome and no previous cardiac arrest. Circulation 108:3092–3096

Brugada P, Brugada J (1992) Right bundle branch block, persistent ST segment elevation and sudden cardiac death: a distinct clinical and electrocardiographic syndrome. A multicenter report. J Am Coll Cardiol 20:1391–1396

Brugada P, Brugada R, Mont L, Rivero M, Geelen P, Brugada J (2003b) Natural history of Brugada syndrome: the prognostic value of programmed electrical stimulation of the heart. J Cardiovasc Electrophysiol 14:455–457

Brugada R, Hong K, Dumaine R, Cordeiro J, Gaita F, Borggrefe M, Menendez TM, Brugada J, Pollevick GD, Wolpert C, Burashnikov E, Matsuo K, Wu YS, Guerchicoff A, Bianchi F, Giustetto C, Schimpf R, Brugada P, Antzelevitch C (2004) Sudden death associated with short-QT syndrome linked to mutations in HERG. Circulation 109:30–35

Cerrone M, Colombi B, Bloise R, Memmi M, Molcalvo C, Potenza D, Drago F, Napolitano C, Bradley DJ, Priori SG (2004) Clinical and molecular characterization of a large cohort of patients affected with catecholaminergic polymorphic ventricular tachycardia. Circulation 110:552 (suppl II) (abstr)

Chen Q, Kirsch GE, Zhang D, Brugada R, Brugada J, Brugada P, Potenza D, Moya A, Borggrefe M, Breithardt G, Ortiz-Lopez R, Wang Z, Antzelevitch C, O'Brien RE, Schulze-Bahr E, Keating MT, Towbin JA, Wang Q (1998) Genetic basis and molecular mechanism for idiopathic ventricular fibrillation. Nature 392:293–296

Compton SJ, Lux RL, Ramsey MR, Strelich KR, Sanguinetti MC, Green LS, Keating MT, Mason JW (1996) Genetically defined therapy of inherited long-QT syndrome. Correction of abnormal repolarization by potassium. Circulation 94:1018–1022

Coumel P, Fidelle J, Lucet V, Attuel P, Bouvrain Y (1978) Catecholaminergic-induced severe ventricular arrhythmias with Adams-Stokes syndrome in children: report of four cases. Br Heart J 40:28–37

Curran ME, Splawski I, Timothy KW, Vincent GM, Green ED, Keating MT (1995) A molecular basis for cardiac arrhythmia: HERG mutations cause long QT syndrome. Cell 80:795–803

De Rosa G, Delogu AB, Piastra M, Chiaretti A, Bloise R, Priori SG (2004) Catecholaminergic polymorphic ventricular tachycardia: successful emergency treatment with intravenous propranolol. Pediatr Emerg Care 20:175–177

Delisle BP, Anderson CL, Balijepalli RC, Anson BD, Kamp TJ, January CT (2003) Thapsigargin selectively rescues the trafficking defective LQT2 channels G601S and F805C. J Biol Chem 278:35749–35754

Eckardt L, Kirchhof P, Schulze-Bahr E, Rolf S, Ribbing M, Loh P, Bruns HJ, Witte A, Milberg P, Borggrefe M, Breithardt G, Wichter T, Haverkamp W (2002) Electrophysiologic investigation in Brugada syndrome; yield of programmed ventricular stimulation at two ventricular sites with up to three premature beats. Eur Heart J 23:1394–1401

Eckardt L, Probst V, Smits JPP, Bahr ES, Wolpert C, Schimpf R, Wichter T, Boisseau P, Heinecke A, Breithardt G, Borggrefe M, LeMarec H, Bocker D, Wilde AAM (2005) Long-term prognosis of individuals with right precordial ST-segment-elevation Brugada syndrome. Circulation 111:257–263

Freestone B, Lip GY (2004) Tedisamil: a new novel antiarrhythmic. Expert Opin Investig Drugs 13:151–160

Gaita F, Giustetto C, Bianchi F, Wolpert C, Schimpf R, Riccardi R, Grossi S, Richiardi E, Borggrefe M (2003) Short QT Syndrome: a familial cause of sudden death. Circulation 108:965–970

Gaita F, Giustetto C, Bianchi F, Schimpf R, Haissaguerre M, Calo L, Brugada R, Antzelevitch C, Borggrefe M, Wolpert C (2004) Short QT syndrome: pharmacological treatment. J Am Coll Cardiol 43:1494–1499

Gasparini M, Priori SG, Mantica M, Coltorti F, Napolitano C, Galimberti P, Bloise R, Ceriotti C (2002) Programmed electrical stimulation in Brugada syndrome: how reproducible are the results? J Cardiovasc Electrophysiol 13:880–887

Gouas L, Bellocq C, Berthet M, Potet F, Demolombe S, Forhan A, Lescasse R, Simon F, Balkau B, Denjoy I, Hainque B, Baro I, Guicheney P (2004) New KCNQ1 mutations leading to haploinsufficiency in a general population; defective trafficking of a KvLQT1 mutant. Cardiovasc Res 63:60–68

Grant AO, Carboni MP, Neplioueva V, Starmer CF, Memmi M, Napolitano C, Priori S (2002) Long QT syndrome, Brugada syndrome, and conduction system disease are linked to a single sodium channel mutation. J Clin Invest 110:1201–1209

Gussak I, Brugada P, Brugada J, Wright RS, Kopecky SL, Chaitman BR, Bjerregaard P (2000) Idiopathic short QT interval: a new clinical syndrome? Cardiology 94:99–102

Hermida JS, Denjoy I, Clerc J, Extramiana F, Jarry G, Milliez P, Guicheney P, Di Fusco S, Rey JL, Cauchemez B, Leenhardt A (2004) Hydroquinidine therapy in Brugada syndrome. J Am Coll Cardiol 43:1853–1860

Jiang D, Xiao B, Zhang L, Chen SR (2002) Enhanced basal activity of a cardiac Ca2+ release channel (ryanodine receptor) mutant associated with ventricular tachycardia and sudden death. Circ Res 91:218–225

Kehl HG, Haverkamp W, Rellensmann G, Yelbuz TM, Krasemann T, Vogt J, Schulze-Bahr E (2004) Images in cardiovascular medicine. Life-threatening neonatal arrhythmia: successful treatment and confirmation of clinically suspected extreme long QT-syndrome-3. Circulation 109:e205–e206

Lahat H, Pras E, Olender T, Avidan N, Ben Asher E, Man O, Levy-Nissenbaum E, Khoury A, Lorber A, Goldman B, Lancet D, Eldar M (2001) A missense mutation in a highly conserved region of CASQ2 is associated with autosomal recessive catecholamine-induced polymorphic ventricular tachycardia in Bedouin families from Israel. Am J Hum Genet 69:1378–1384

Laitinen PJ, Brown KM, Piippo K, Swan H, Devaney JM, Brahmbhatt B, Donarum EA, Marino M, Tiso N, Viitasalo M, Toivonen L, Stephan DA, Kontula K (2001) Mutations of the cardiac ryanodine receptor (RyR2) gene in familial polymorphic ventricular tachycardia. Circulation 103:485–490

Leenhardt A, Lucet V, Denjoy I, Grau F, Ngoc DD, Coumel P (1995) Catecholaminergic polymorphic ventricular tachycardia in children. A 7-year follow-up of 21 patients. Circulation 91:1512–1519

Lehnart SE, Wehrens XH, Marks AR (2004a) Calstabin deficiency, ryanodine receptors, and sudden cardiac death. Biochem Biophys Res Commun 322:1267–1279

Lehnart SE, Wehrens XHT, Laitinen PJ, Reiken SR, Deng SX, Cheng Z, Landry DW, Kontula K, Swan H, Marks AR (2004b) Sudden death in familial polymorphic ventricular tachycardia associated with calcium release channel (ryanodine receptor) leak. Circulation 109:3208–3214

Marks AR, Priori S, Memmi M, Kontula K, Laitinen PJ (2002) Involvement of the cardiac ryanodine receptor/calcium release channel in catecholaminergic polymorphic ventricular tachycardia. J Cell Physiol 190:1–6

Matsuo K, Kurita T, Inagaki M, Kakishita M, Aihara N, Shimizu W, Taguchi A, Suyama K, Kamakura S, Shimomura K (1999) The circadian pattern of the development of ventricular fibrillation in patients with Brugada syndrome. Eur Heart J 20:465–470

Mohler PJ, Schott JJ, Gramolini AO, Dilly KW, Guatimosim S, duBell WH, Song LS, Haurogne K, Kyndt F, Ali ME, Rogers TB, Lederer WJ, Escande D, Le Marec H, Bennett V (2003) Ankyrin-B mutation causes type 4 long-QT cardiac arrhythmia and sudden cardiac death. Nature 421:634–639

Moss AJ, Zareba W, Hall WJ, Schwartz PJ, Crampton RS, Benhorin J, Vincent GM, Locati EH, Priori SG, Napolitano C, Medina A, Zhang L, Robinson JL, Timothy K, Towbin JA, Andrews ML (2000) Effectiveness and limitations of beta-blocker therapy in congenital long-QT syndrome. Circulation 101:616–623

Neyroud N, Tesson F, Denjoy I, Leibovici M, Donger C, Barhanin J, Faure S, Gary F, Coumel P, Petit C, Schwartz K, Guicheney P (1997) A novel mutation in the potassium channel gene KVLQT1 causes the Jervell and Lange-Nielsen cardioauditory syndrome. Nat Genet 15:186–189

Priori SG, Napolitano C, Cantu F, Brown AM, Schwartz PJ (1996) Differential response to Na+ channel blockade, beta-adrenergic stimulation, and rapid pacing in a cellular model mimicking the SCN5A and HERG defects present in the long-QT syndrome. Circ Res 78:1009–1015

Priori SG, Napolitano C, Gasparini M, Pappone C, Della BP, Brignole M, Giordano U, Giovannini T, Menozzi C, Bloise R, Crotti L, Terreni L, Schwartz PJ (2000a) Clinical and genetic heterogeneity of right bundle branch block and ST-segment elevation syndrome: A prospective evaluation of 52 families. Circulation 102:2509–2515

Priori SG, Napolitano C, Giordano U, Collisani G, Memmi M (2000b) Brugada syndrome and sudden cardiac death in children. Lancet 355:808–809

Priori SG, Napolitano C, Schwartz PJ, Bloise R, Crotti L, Ronchetti E (2000c) The elusive link between LQT3 and Brugada syndrome: the role of flecainide challenge. Circulation 102:945–947

Priori SG, Aliot E, Blomstrom-Lundqvist C, Bossaert L, Breithardt G, Brugada P, Camm AJ, Cappato R, Cobbe SM, Di Mario C, Maron BJ, McKenna WJ, Pedersen AK, Ravens U, Schwartz PJ, Truz-Gluza M, Vardas P, Wellens HJJ, Zipes DP (2001a) Task force on sudden cardiac death of the European Society of Cardiology. Eur Heart J 22:1374–1450

Priori SG, Napolitano C, Tiso N, Memmi M, Vignati G, Bloise R, Sorrentino V, Danieli GA (2001b) Mutations in the cardiac ryanodine receptor gene (hRyR2) underlie catecholaminergic polymorphic ventricular tachycardia. Circulation 103:196–200

Priori SG, Napolitano C, Memmi M, Colombi B, Drago F, Gasparini M, DeSimone L, Coltorti F, Bloise R, Keegan R, Cruz Filho FE, Vignati G, Benatar A, DeLogu A (2002a) Clinical and molecular characterization of patients with catecholaminergic polymorphic ventricular tachycardia. Circulation 106:69–74

Priori SG, Napolitano C, Gasparini M, Pappone C, Bella PD, Giordano U, Bloise R, Giustetto C, De Nardis R, Grillo M, Ronchetti E, Faggiano G, Nastoli J (2002b) Natural history of Brugada syndrome. Insights for risk stratification and management. Circulation 105:1342–1347

Priori SG, Rivolta I, Napolitano C (2003a) Genetics of long QT, Brugada and other Channellopathies. In: Zipes DP, Jalife J (eds) Cardiac electrophysiology, 4th edn. Elsevier, Philadelphia, pp 462–470

Priori SG, Schwartz PJ, Napolitano C, Bloise R, Ronchetti E, Grillo M, Vicentini A, Spazzolini C, Nastoli J, Bottelli G, Folli R, Cappelletti D (2003b) Risk stratification in the long-QT syndrome. N Engl J Med 348:1866–1874

Priori SG, Napolitano C, Schwartz PJ, Grillo M, Bloise R, Ronchetti E, Moncalvo C, Tulipani C, Veia A, Bottelli G, Nastoli J (2004) Association of long QT syndrome loci and cardiac events among patients treated with beta-blockers. JAMA 292:1341–1344

Priori SG, Pandit SV, Rivolta I, Berenfeld O, Ronchetti E, Dhamoon A, Napolitano C, Anumonow J, di Barletta MR, Gudapakkam S, Bosi G, Stramba-Badiale M, Jalife J (2005) A novel form of short QT syndrome (SQT3) is caused by a mutation in the KCNJ2 gene. Circ Res. 96:800–807

Rajamani S, Anderson CL, Anson BD, January CT (2002) Pharmacological rescue of human K(+) channel long-QT2 mutations: human ether-a-go-go-related gene rescue without block. Circulation 105:2830–2835

Rivolta I, Giarda E, Nastoli J, Ronchetti E, Napolitano C, Priori SG (2004) In vitro characterization of the electrophysiological effects of mexiletine on SCN5A mutants predicts clinical response in LQT3 patients. Circulation 110(17):III-230, 26-10 (abstr)

Sanguinetti MC, Jurkiewicz NK (1991) Delayed rectifier outward K+ current is composed of two currents in guinea pig atrial cells. Am J Physiol 260:H393–H399

Sanguinetti MC, Curran ME, Zou A, Shen J, Spector PS, Atkinson DL, Keating MT (1996) Coassembly of K(V)LQT1 and minK (IsK) proteins to form cardiac I(Ks) potassium channel. Nature 384:80–83

Schulze-Bahr E, Wang Q, Wedekind H, Haverkamp W, Chen Q, Sun Y, Rubie C, Hordt M, Towbin JA, Borggrefe M, Assmann G, Qu X, Somberg JC, Breithardt G, Oberti C, Funke H (1997) KCNE1 mutations cause Jervell and Lange-Nielsen syndrome. Nat Genet 17:267–268

Schwartz PJ (1985) Idiopathic long QT syndrome: progress and questions. Am Heart J 109:399–411

Schwartz PJ, Priori SG, Locati EH, Napolitano C, Cantu F, Towbin JA, Keating MT, Hammoude H, Brown AM, Chen LS (1995) Long QT syndrome patients with mutations of the SCN5A and HERG genes have differential responses to Na+ channel blockade and to increases in heart rate. Implications for gene-specific therapy. Circulation 92:3381–3386

Schwartz PJ, Priori SG, Dumaine R, Napolitano C, Antzelevitch C, Stramba-Badiale M, Richard TA, Berti MR, Bloise R (2000a) A molecular link between the sudden infant death syndrome and the long-QT syndrome. N Engl J Med 343:262–267

Schwartz PJ, Priori SG, Napolitano C (2000b) The long QT syndrome. In: Zipes DP, Jalife J (eds) Cardiac electrophysiology from cell to bedside, 3rd edn. WB Saunders Co, Philadelphia, pp 597–615

Splawski I, Timothy KW, Sharpe LM, Decher N, Kumar P, Bloise R, Napolitano C, Schwartz PJ, Joseph RM, Condouris K, Tager-Flusberg H, Priori SG, Sanguinetti MC, Keating MT (2004) Ca(V)1.2 calcium channel dysfunction causes a multisystem disorder including arrhythmia and autism. Cell 119:19–31

Sumitomo N, Harada K, Nagashima M, Yasuda T, Nakamura Y, Aragaki Y, Saito A, Kurosaki K, Jouo K, Koujiro M, Konishi S, Matsuoka S, Oono T, Hayakawa S, Miura M, Ushinohama H, Shibata T, Niimura I (2003) Catecholaminergic polymorphic ventricular tachycardia: electrocardiographic characteristics and optimal therapeutic strategies to prevent sudden death. Heart 89:66–70

Suzuki H, Torigoe K, Numata O, Yazaki S (2000) Infant case with a malignant form of Brugada syndrome. J Cardiovasc Electrophysiol 11:1277–1280

Swan H, Piippo K, Viitasalo M, Heikkila P, Paavonen T, Kainulainen K, Kere J, Keto P, Kontula K, Toivonen L (1999) Arrhythmic disorder mapped to chromosome 1q42-q43 causes malignant polymorphic ventricular tachycardia in structurally normal hearts. J Am Coll Cardiol 34:2035–2042

Tristani-Firouzi M, Jensen JL, Donaldson MR, Sansone V, Meola G, Hahn A, Bendahhou S, Kwiecinski H, Fidzianska A, Plaster N, Fu YH, Ptacek LJ, Tawil R (2002) Functional and clinical characterization of KCNJ2 mutations associated with LQT7 (Andersen syndrome). J Clin Invest 110:381–388

Tsuchiya T, Ashikaga K, Honda T, Arita M (2002) Prevention of ventricular fibrillation by cilostazol, an oral phosphodiesterase inhibitor, in a patient with Brugada syndrome. J Cardiovasc Electrophysiol 13:698–701

Valdivia CR, Ackerman MJ, Tester DJ, Wada T, McCormack J, Ye B, Makielski JC (2002) A novel SCN5A arrhythmia mutation, M1766L, with expression defect rescued by mexiletine. Cardiovasc Res 55:279–289

Valdivia CR, Tester DJ, Rok BA, Porter CB, Munger TM, Jahangir A, Makielski JC, Ackerman MJ (2004) A trafficking defective, Brugada syndrome-causing SCN5A mutation rescued by drugs. Cardiovasc Res 62:53–62

Viatchenko-Karpinski S, Terentyev D, Gyorke I, Terentyeva R, Volpe P, Priori SG, Napolitano C, Nori A, Williams SC, Gyorke S (2004) Abnormal calcium signaling and sudden cardiac death associated with mutation of calsequestrin. Circ Res 94:471–477

Wang Q, Shen J, Splawski I, Atkinson D, Li Z, Robinson JL, Moss AJ, Towbin JA, Keating MT (1995) SCN5A mutations associated with an inherited cardiac arrhythmia, long QT syndrome. Cell 80:805–811

Wang Q, Curran ME, Splawski I, Burn TC, Millholland JM, VanRaay TJ, Shen J, Timothy KW, Vincent GM, de Jager T, Schwartz PJ, Toubin JA, Moss AJ, Atkinson DL, Landes GM, Connors TD, Keating MT (1996) Positional cloning of a novel potassium channel gene: KVLQT1 mutations cause cardiac arrhythmias. Nat Genet 12:17–23

Wehrens XH, Lehnart SE, Huang F, Vest JA, Reiken SR, Mohler PJ, Sun J, Guatimosim S, Song LS, Rosemblit N, D'Armiento JM, Napolitano C, Memmi M, Priori SG, Lederer WJ, Marks AR (2003) FKBP12.6 deficiency and defective calcium release channel (ryanodine receptor) function linked to exercise-induced sudden cardiac death. Cell 113:829–840

Wehrens XH, Lehnart SE, Reiken SR, Deng SX, Vest JA, Cervantes D, Coromilas J, Landry DW, Marks AR (2004) Protection from cardiac arrhythmia through ryanodine receptor-stabilizing protein calstabin2. Science 304:292–296

Wilde AA, Antzelevitch C, Borggrefe M, Brugada J, Brugada R, Brugada P, Corrado D, Hauer RN, Kass RS, Nademanee K, Priori SG, Towbin JA (2002) Proposed diagnostic criteria for the Brugada syndrome: consensus report. Circulation 106:2514–2519

Windle JR, Geletka RC, Moss AJ, Zareba W, Atkins DL (2001) Normalization of ventricular repolarization with flecainide in long QT syndrome patients with SCN5A:DeltaKPQ mutation. Ann Noninvasive Electrocardiol 6:153–158

Yan GX, Antzelevitch C (1999) Cellular basis for the Brugada syndrome and other mechanisms of arrhythmogenesis associated with ST-segment elevation. Circulation 100:1660–1666

Yano M, Kobayashi S, Kohno M, Doi M, Tokuhisa T, Okuda S, Suetsugu M, Hisaoka T, Obayashi M, Ohkusa T, Kohno M, Matsuzaki M (2003) FKBP12.6-mediated stabilization of calcium-release channel (ryanodine receptor) as a novel therapeutic strategy against heart failure. Circulation 107:477–484

Zhou Z, Gong Q, Epstein ML, January CT (1998) HERG channel dysfunction in human long QT syndrome. Intracellular transport and functional defects. J Biol Chem 273:21061–21066

Zhou Z, Gong Q, January CT (1999) Correction of defective protein trafficking of a mutant HERG potassium channel in human long QT syndrome. Pharmacological and temperature effects. J Biol Chem 274:31123–31126

HEP (2006) 171:287–304

Mutation-Specific Pharmacology of the Long QT Syndrome

R. S. Kass[1] (✉) · A. J. Moss[2]

[1] Department of Pharmacology, Columbia University College of Physicians and Surgeons,
New York NY, 10032, USA
RSK20@Columbia.edu

[2] Heart Research Follow-up Program, Department of Medicine,
University of Rochester School of Medicine and Dentistry, Rochester NY, 14642, USA

Abstract The congenital long QT syndrome is a rare disease in which inherited mutations of genes coding for ion channel subunits, or channel interacting proteins, delay repolarization of the human ventricle and predispose mutation carriers to the risk of serious or fatal arrhythmias. Though a rare disorder, the long QT syndrome has provided invaluable insight from studies that have bridged clinical and pre-clinical (basic science) medicine. In this brief review, we summarize some of the key clinical and genetic characteristics of this disease and highlight novel findings about ion channel structure, function, and the causal relationship between channel dysfunction and human disease, that have come from investigations of this disorder.

Keywords Na$^+$ channel blocker · Lidocaine · Flecainide · Local anesthetic · Mutation ·
Channelopathies · Polymorphism · Structural determinants · Antiarrhythmic ·
Proarrhythmic · VGSC · TTX · Tonic block · Use-dependent block · Na$_V$1.5 · Na$_V$1.1 ·
SCN5A · SCN1A · Pharmacokinetics · Pharmacodynamics · Recovery from block ·
Singh–Vaughan Williams · Sicilian Gambit · CAST · CYP · Cytochrome enzymes ·
Long QT syndrome · Brugada syndrome · Conduction disorders · Isoform specificity ·
Molecular determinants

Abbreviations

LQTS	Long QT syndrome
QTc	Heart rate-corrected QT
RWS	Romano–Ward syndrome
β-ARs	β-Adrenergic receptors

1
Background

The common form of long QT syndrome (LQTS), Romano–Ward syndrome
(RWS), is a heterogeneous, autosomal-dominant genetic disease caused by mutations of genes coding for ion channels expressed in the heart. These channels
regulate cardiac rhythm by controlling electrical activity of the cardiac cycle.
Dysfunction in channels expressed in ventricular (and presumably Purkinje
fiber) cells delays cellular repolarization, causing the disease phenotype: prolonged QT intervals of the ECG. This channelopathy is clinically manifest by
syncope and sudden death from ventricular arrhythmias, notably torsades de
pointes (TdP) (Moss et al. 1991). Clinically, LQTS is identified by abnormal Q-
T interval prolongation on the ECG. The QT prolongation reflects prolonged
cellular action potentials and may arise from either a decrease in repolarizing
cardiac membrane currents or an increase in depolarizing cardiac currents.
These altered currents must occur late in the cardiac cycle to account for the
prolonged Q-T interval. Most commonly, QT prolongation is produced by delayed repolarization due to reductions in either the rapidly or slowly activating
delayed repolarizing cardiac potassium (K$^+$) currents, I_{Kr} or I_{Ks} (Sanguinetti
and Spector 1997). Less commonly, QT prolongation results from prolonged
depolarization due to a small persistent inward "leak" in cardiac sodium (Na$^+$)
current I_{Na} (Bennett et al. 1995). Most recently, mutations in genes coding for
important cardiac calcium channels, the so-called L-type calcium channels,
have also been shown to dramatically prolong the Q-T interval and cause LQTS
(Splawski et al. 2004).

Patients with LQTS are usually identified by QT prolongation on the ECG
during clinical evaluation of unexplained syncope, as part of a family study
when one family member has been identified with the syndrome, or in the
investigation of patients with congenital neural deafness. The first family with
LQTS was reported in 1957 by Jervell and Lange-Nielsen and was thought to be

an autosomal recessive disorder (Jervell and Lange-Nielsen 1957), but in 1997 it was shown to result from a double-dominant, homozygous mutation involving the KvLQT1 gene (Splawski et al. 1997a), now called the KCNQ1 gene. The more common autosomal dominant RWS was described in 1963–1964, and over 300 different mutations involving seven different genes (*LQT1–7*) have now been reported (Splawski et al. 2000). Most of the clinical information currently available regarding LQTS relates to the RWS. There is considerable variability in the clinical presentation of LQTS due to the different genotypes, different mutations, variable penetrance of the mutations, and possible genetic and environmental modifying factors. Clinical criteria have been developed to determine the probability of having LQTS, and genotype screening of suspect LQTS individuals and of family members from known LQTS families has progressively increased the number of subjects with genetically confirmed LQTS. The genes associated with LQTS have been numerically ordered by the chronology of their discovery (LQT1, LQT2, LQT3, ... LQT7), with 95% of the known mutations located in the first three of the seven identified LQTS genes. Current prophylactic and preventive therapy for LQTS to reduce the incidence of syncope and sudden death has involved left cervico-thoracic sympathetic ganglionectomy, β-blockers, pacemakers, implanted defibrillators, and gene/mutation-specific pharmacologic therapy (Moss 2003).

2
Arrhythmia Risk Factors Are Mutation/Gene-Specific

The discovery that distinct LQTS variants are associated with genes coding for different ion channel subunits has had a major impact on the diagnosis and analysis of LQTS patients. Critical evaluation of clinical data has revealed that there are distinct risk factors associated with the different LQTS genotypes, and that these must be taken into account during patient evaluation and diagnosis. The greatest difference in risk factors becomes apparent in comparing LQT3 syndrome patients (*SCN5A* mutations) and patients with LQT1 syndrome (*KCNQ1* mutations) or LQT2 syndrome (*hERG* mutations). The potential for understanding a mechanistic basis for arrhythmia risk was realized soon after the first genetic information relating mutations in genes coding for distinct ion channels became available, (Priori et al. 1997) but is still the focus of extensive clinical and basic investigation. In one such study, which focused on patients with KCNQ1 (LQT1), hERG (LQT2), and SCN5A (LQT3) mutations, a clear difference in arrhythmia risk emerged, and this difference appeared in a gene-specific manner. In the case of SCN5A mutation carriers (LQT3), risk of cardiac events was greatest during rest, (bradycardia) when sympathetic nerve activity is expected to be low. In contrast, cardiac events in LQT2 syndrome patients were associated with arousal and/or conditions in which patients were startled, whereas LQT1 syndrome patients were found to be at greatest risk of experiencing cardiac events during exercise or conditions associated with elevated sympathetic nerve activity (Schwartz et al. 2001).

Additional evidence has continued to support the view that under conditions in which sympathetic nerve activity is likely to be high, such as during periods of exercise, patients harboring LQT1 mutations (Ackerman et al. 1999; Paavonen et al. 2001; Takenaka et al. 2003) are likely to experience dysfunctional regulation in cardiac electrical activity and hence an increased arrhythmia risk. The contrast between the role of adrenergic input and/or heart rate in the arrhythmia risk of LQT1 and LQT3 patients is clear and has raised the possibility of distinct therapeutic strategies in the management of patients with these LQTS variants. In fact β-blocker therapy has been shown to be most effective in preventing recurrence of cardiac events and lowering the death rate in LQT1 and LQT2 syndrome patients but is much less effective in the treatment of LQT3 syndrome patients (Moss et al. 2000; Priori 2004). β-Blocking drugs have minimal effects on the QTc interval but are associated with a significant reduction in cardiac events in LQTS patients, probably because these drugs modulate the stimulation of β-adrenergic receptors (β-ARs) and hence the regulation of downstream signaling targets during periods of elevated sympathetic nerve activity. Clinical data for genotyped patients continues to provide strong support for the hypothesis that the effectiveness of β-blocking drugs depends critically on the genetic basis of the disease with recent data providing evidence that there is still a high rate of cardiac events in LQT2 and LQT3 patients treated with β-blocking drugs (Priori et al. 2004). Consequently, even β-blockers do not provide absolute protection against fatal cardiac arrhythmias.

3
Mutation-Specific Pharmacology: Role of the Sodium Channel

The SCN5A gene encodes the α-subunit of the major cardiac voltage-gated sodium channel (George et al. 1995). Voltage-gated Na^+ channels are integral membrane proteins (Catterall 1995, 1996) that not only underlie excitation in excitable cells, but determine the vulnerability of the heart to dysfunctional rhythm by controlling the number of channels available to conduct inward Na^+ movement (Rivolta et al. 2001). Na^+ channels open in response to membrane depolarization, allowing a rapid selective influx of Na^+ which serves to further depolarize excitable cells and initiate multiple cellular signals (Catterall 2000). Within milliseconds of opening, Na^+ channels enter a non-conducting inactivated state (Stuhmer et al. 1989; Patton et al. 1992; West et al. 1992; McPhee et al. 1994, 1995, 1998; Kellenberger et al. 1997a,b). Channel inactivation is necessary to limit the duration of excitable cell depolarization. Therefore disruption of inactivation by inherited mutations, which delays cellular repolarization, is associated with a diverse range of human diseases including myotonias

(Yang et al. 1994), epilepsy and seizure disorders (Kearney et al. 2001; Lossin et al. 2002), autism (Weiss et al. 2003), and sudden cardiac death (Keating and Sanguinetti 2001; Kass and Moss 2003).

The Na$^+$ channel α-subunit, which forms the ion-conducting pore and contains channel gating components, consists of four homologous domains (I to IV; Sato et al. 2001). Each domain contains six α-helical transmembrane repeats (S1–S6), for which mutagenesis studies have revealed key functional roles (Catterall 2000). Voltage-dependent inactivation of Na$^+$ channels is a consequence of voltage-dependent activation (Aldrich et al. 1983), and inactivation is characterized by at least two distinguishable kinetic components: an initial rapid component (fast inactivation) and a slower component (slow inactivation). Within milliseconds of opening, Na$^+$ channels enter a non-conducting inactivated state as the inactivation gate, the cytoplasmic loop linking domains III and IV of the α-subunit, occludes the open pore (Stuhmer et al. 1989; Patton et al. 1992; West et al. 1992; McPhee et al. 1994, 1995, 1998; Kellenberger et al. 1996). Fast Na$^+$ channel inactivation is due to rapid block of the inner mouth of the channel pore by the cytoplasmic linker between domains III and IV that occurs within milliseconds of membrane depolarization (Vassilev et al. 1988; Stuhmer et al. 1989; Vassilev et al. 1989; West et al. 1992). Nuclear magnetic resonance (NMR) analysis of this inactivation linker (gate) in solution has revealed a rigid helical structure that is positioned such that it can block the pore, providing a structural explanation of the functional studies (Rohl et al. 1999) and a biological mechanism of inhibiting channel conduction.

The residues that form a hydrophobic triplet (IFM) in the III–IV linker are involved in inactivation gating (West et al. 1992). The IFM motif has been suggested to function as a 'latch' that holds the inactivation gate shut. Cysteine scanning of the residues I1485, F1486, and M1487 in the human cardiac Na$^+$ channel revealed that these amino acids contribute to stabilizing the fast-inactivation particle (Deschenes et al. 1999) in analogy to the brain Na$^+$ channel (Stuhmer et al. 1989; Sheets et al. 2000).

4
Na$^+$ Channel Block by Local Anesthetics Is Linked to Channel Inactivation

Blockade of voltage-dependent Na$^+$ channels has long been recognized as a potential therapeutic approach to the management of many cardiac arrhythmias, but with considerable risk of toxic side effects (Rosen et al. 1975). The discovery that mutant forms of Na$^+$ channels linked to inherited human cardiac arrhythmias might make distinct targets for Na$^+$ channel blocking drugs (An et al. 1996; Wang et al. 1997; Dumaine et al. 1996; Dumaine and Kirsch 1998; Nagatomo et al. 2000; Viswanathan et al. 2001) has stimulated reinvestigation of the molecular determinants of Na$^+$ channel blockade in the heart.

Voltage-dependent block of Na^+ channels by local anesthetics and related drugs has been well described within the framework of the modulated receptor hypothesis, which proposes that allosteric changes in a drug receptor occur when changes in voltage induce changes in channel conformation states (Hille 1977; Hondeghem and Katzung 1977). Extensive mutagenesis experiments have been performed with several different drugs in many sodium channel isoforms in an effort to define the molecular determinants of drug binding. While a clear consensus has not been reached regarding precisely where drug binds and there is certainly variability in drugs, isoforms, and how the data are interpreted, the current evidence strongly suggests that most drugs tested bind in the pore of the channel on the intracellular side of the selectivity filter.

Furthermore, mutagenesis studies by several groups find specific amino acid residues that contribute to drug binding on the S6 segment of domains I, III, and IV. The most dramatic effects on drug binding can be attributed primarily to two aromatic residues on DIV S6, a phenylalanine at position 11760 (F1760) and a tyrosine at position 1767 (Y1767) using $Na_V1.5$ numbering, that are conserved among sodium channel isoforms (Ragsdale et al. 1994 1996; Li et al. 1999; Weiser et al. 1999).

5
LQT-3 Mutations: A Common Phenotype Caused by a Range of Mutation-Induced Channel Function

Different LQT3 mutations can result in distinct functional changes in the activity of the sodium channel, but with similar degrees of QT prolongation and cardiac arrhythmias (Wang et al. 1995a,b). For example, the nine-base-pair deletion with loss of three amino acids (Δ-KPQ) in the linker between the third and fourth domains of the α-alpha unit of the sodium channel and three missense mutations in this gene (N1325S, R1623Q, and R1644H) all promote sustained and inappropriate sodium entry into the myocardial cell during the plateau phase of the action potential, resulting in prolonged ventricular repolarization and the LQTS phenotype. This mutation occurs in the cytoplasmic peptide that links two domains of the channel: domain III and domain IV (see Sect. 9), and, not surprisingly, alters the stability of inactivated, or nonconducting channels. In contrast, the functional consequences of the D1790G missense mutation on sodium channel gating are quite different. This mutation does not promote sustained inward sodium current, but rather causes a negative shift in steady-state inactivation with a similar LQTS phenotype (Abriel et al. 2000b). Despite these functional differences in channel activity, the phenotypical effect of the mutations is the same: QT prolongation. Interestingly, in vitro studies have shown that the D1790G mutation alters the response of the sodium channel to adrenergic stimulation, a finding that may have implications for triggers of this unique mutation (Tateyama et al. 2003). It should be noted that other mutations in the SCN5A gene can result in the Brugada syndrome and conduction system disorders without QT prolongation. At least one mutation (1795insD) has been shown to have a dual effect with inappropriate sodium entry at slow heart rates (LQTS ECG pattern) and reduced sodium entry at fast heart rates (Brugada ECG pattern; Veldkamp et al. 2000).

Mutation-specific pharmacologic therapy has been reported in two specific *SCN5A* mutations associated with LQTS. In 1995, Schwartz et al. reported that a single oral dose of the sodium-channel blocker mexiletine administered to seven LQT3 patients with the ΔKPQ deletion produced significant shortening of the QTc interval within 4 h (Schwartz et al. 1995). Similar QTc shortening in LQT3 patients with the ΔKPQ deletion has been reported with lidocaine and tocainide (Rosero et al. 1997). Preliminary clinical experience with flecainide revealed normalization of the QTc interval with low doses of this drug in patients with the ΔKPQ deletion (Windle et al. 2001). In 2000, Benhorin et al. reported the effectiveness of open-label oral flecainide in shortening the QTc in eight asymptomatic subjects with the D1790G mutation (Benhorin et al. 2000).

In the *SCN5A*-ΔKPQ deletion mutation, flecainide has high affinity for the sodium-channel protein and provides almost complete correction of the impaired inactivation (Nagatomo et al. 2000). A recent randomized, double-blind, placebo-controlled clinical trial in six male LQT3 subjects having the ΔKPQ deletion, with four 6-month alternating periods of low-dose flecainide (1.5 to 3.0 mg/kg/day) and placebo therapy (A.J. Moss, unpublished data). The average QTc values during placebo and flecainide therapies were 534 ms and 503 ms, respectively, with a change in QTc from baseline during 6-month flecainide therapy of -29 ms (95% confidence interval, -37 ms to -21 ms; $p<0.001$) at a mean flecainide blood level of 0.11 ± 0.05 µg/ml. At this low flecainide blood level, there were minimal prolongations in P-R and QRS duration and no major adverse cardiac effects.

The *SCN5A*-D1790G mutation changes the sodium channel's interaction with flecainide. This mutation confers a high sensitivity to use-dependent block by flecainide, due in large part to the marked slowing of the repriming of the mutant channels in the presence of the drug (Abriel et al. 2000a). Flecainide tonic block is not affected by the D1790G mutation. These flecainide affects are different from those occurring with the ΔKPQ mutant channels, and may underlie the distinct efficacy of this drug in treating LQT3 patients harboring the D1790G mutation (Liu et al. 2002, 2003).

These flecainide findings in patients with the ΔKPQ and D1790G mutations provide encouraging evidence in support of mutation-specific pharmacologic therapy for two specific forms of the LQT3 disorder. Larger clinical trials with flecainide in patients with these two mutations are needed before this therapy can be recommended as safe and effective for patients with these genetic disorders.

6
Clinical Relevance of Mutations Within Different Regions of the Ion Channel: Structure/Function

The hERG gene encodes the ion channel involved in the rapid component of the delayed rectifier repolarization current (I_{Kr}), and mutations in this gene are responsible for the LQT2 form of LQTS (Nagatomo et al. 2000). Mutations in *hERG* are associated with diminution in the repolarizing I_{Kr} current with resultant prolongation of ventricular repolarization and lengthening of the Q-T interval. During the 1990s, it was appreciated that several drugs such as terfenadine and cisapride caused QT prolongation by reducing I_{Kr} current through the pore region of the hERG channel (Sanguinetti et al. 1996b). These findings raised the question whether mutations in the pore region of the hERG channel would be associated with a more virulent form of LQT2 than mutations in the non-pore region.

In a report from the International LQTS Registry, 44 different *hERG* mutations were identified in 201 subjects, with 14 mutations in 13 locations in the pore region (amino acid residues 550 through 650); (Moss et al. 2002; Fig. 1). Of the subjects, 35 had mutations in the pore region and 166 in non-pore regions. Using birth as the time origin with follow-up through age 40, subjects with pore mutations had more severe clinical manifestations of the genetic disor-

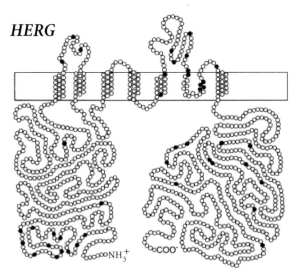

Fig. 1 Schematic representation of hERG potassium channel α-subunit involving the N-terminal portion (NH2), 6 membrane-spanning segments with the pore region extending from segments S5 to S6, and the C-terminus portion (C00⁻). Mutation locations are indicated by *black dots*. Fourteen different mutations were located in 13 locations within the pore region. (Reprinted with permission from Moss et al. 2002)

der and experienced a higher frequency of arrhythmia-related cardiac events at an earlier age than did subjects with non-pore mutations. The cumulative probability of a first cardiac event before β-blockers were, initiated in subjects with pore mutations and non-pore mutations in the hERG channel are shown in Fig. 2, with a hazard ratio in the range of 11 ($p<0.0001$) at an adjusted QTc of 0.50 s. This study involved a limited number of different *hERG* mutations and only a small number of subjects with each mutation. Missense mutations made up 94% of the pore mutations, and thus it was not possible to evaluate risk by the mutation type within the pore region.

These findings indicate that mutations in different regions of the hERG potassium channel can be associated with different levels of risk for cardiac arrhythmias in LQT2. An important question is whether similar region-related risk phenomena exist in the other LQTS channels. Two studies evaluated the clinical risk of mutations located in different regions of the KCNQ1 (LQT1) gene and reported contradictory findings. Zareba et al. found no significant differences in clinical presentation, ECG parameters, and cardiac events among 294 LQT1 patients with *KCNQ1* mutations located in the pre-pore region including N-terminus (1–278), the pore region (279–354), and the post-pore

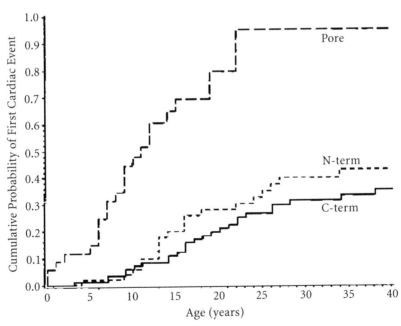

Fig. 2 Kaplan–Meier cumulative probability of first cardiac events from birth through age 40 years for subjects with mutations in pore ($n = 34$), N-terminus ($n = 54$), and C-terminus ($n = 91$) regions of the hERG channel. The curves are significantly different ($p < 0.0001$, log-rank), with the difference caused mainly by the high first-event rate in subjects with pore mutations. (Reprinted with permission from Moss et al. 2002)

region including C-terminus (>354) (Zareba et al. 2003). In contrast, Shimizu et al. studied 66 LQT1 patients and found that mutations in the transmembrane portion of *KCNQ1* were associated with a higher risk of LQTS-related cardiac events and had greater sensitivity to sympathetic stimulation than mutations located in the C-terminal region (Shimizu et al. 2004). These different findings in the two LQT1 studies may reflect, in part, population-related genetic heterogeneity, since the Zareba population was almost entirely Caucasian and the subjects in the Shimizu study were Japanese. Much larger homogeneous populations need to be studied to resolve this issue.

7
Basic Electrophysiology Revealed Through LQTS Studies

Though a rare congenital disorder, LQTS has provided a wealth of information about fundamental mechanisms underlying human cardiac electrophysiology that has come about because of true collaborative interactions between clinical and basic scientists. Our understanding of the mechanisms that control the critical plateau and repolarization phases of the human ventricular action potential has been raised to new levels through these studies which impact on the manner in which both potassium and sodium channels regulate this critical period of electrical activity.

8
Identification of Cardiac Delayed Rectifier Channels

It had been known since 1969 that potassium currents with unique kinetic and voltage-dependent properties were important to the cardiac action potential plateau (Noble and Tsien 1968; Noble and Tsien 1969). Because of the unique voltage-dependence, these currents were referred to as delayed rectifiers. In a pivotal study, Sanguinetti and Jurkiewicz used a pharmacological analysis to demonstrate two distinct components of the delayed rectifier potassium current in heart: I_{Kr} and I_{KS} (Sanguinetti and Jurkiewicz 1990). The I_{KS} component had previously been shown to be under control of the sympathetic nervous system, providing an increase in repolarization currents in the face of β-AR agonists in cellular models (Kass and Wiegers 1982), but the molecular identity and the relevance to human electrophysiology were not only not clear, but controversial. The clear clinical importance and the genetic basis of these potassium currents were revealed through LQTS investigations.

The first report linking potassium channel dysfunction to LQTS revealed the molecular identity of one of the delayed rectifier channels and confirmed the pharmacological evidence for independent channels underlying these currents (Sanguinetti et al. 1995). This report revealed that hERG encodes the α (pore forming) subunit of the I_{Kr} channel and that the rectifying properties of this channel, identified previously by pharmacological dissection, were indigenous to the channel protein. Not only did this work provide the first clear evidence for a role of this channel in the congenital LQTS but also laid the baseline for future studies which would show that it is the hERG channel that underlies almost all cases of acquired LQTS (Sanguinetti et al. 1996a).

In 1996 it was discovered that LQTS variant 1 (LQT1) was caused by mutations in a gene (*KvLQT1/KCNQ1*) coding for an unusual potassium channel subunit that could be studied in heterologous expression systems (Wang et al. 1996) and the KvLQT1 gene product was found to be the α (pore forming) subunit of the I_{KS} channel (Barhanin et al. 1996; Sanguinetti et al. 1996b). Furthermore, these studies indicated that a previously reported, but as-yet poorly understood gene (*mink*) formed a key regulatory subunit of this important channel. Mutations in *mink* (later called *KCNE1*) have subsequently been linked to LQT5 (Splawski et al. 1997b). Now the molecular identity of the two cardiac delayed rectifiers had been established.

Clinical studies had provided convincing evidence linking sympathetic nerve activity and arrhythmia susceptibility in LQTS patients, particularly in patients harboring LQT1 mutations. These data and previous basic reports of the robust sensitivity of the slow delayed rectifier component, I_{KS}, to β-AR agonists (Kass and Wiegers 1982), motivated investigation of the molecular links between KCNQ1/KCNE1 channels to β-AR stimulation which revealed, for the first time, that the KCNQ1/KCNE1 channel is part of a macromolecular signaling complex in human heart (Marx et al. 2002). The channel complexes with an adaptor protein (AKAP 9 or yotiao) that in turn directly binds key enzymes in the β-AR signaling cascade [protein kinase A (PKA) and protein phosphatase 1 (PP1)]. Thus, the binding of yotiao to the KCNQ1 carboxy-terminus recruits signaling molecules to the channel to form a micro-signaling environment to control the phosphorylation state of the channel. When the channel is PKA phosphorylated, there is an increase in repolarizing (potassium channel) current, which provides a repolarization reserve to shorten action potentials. This must occur with the concomitant increase in heart rate, which is the fundamental response to sympathetic nerve stimulation, in order to preserve cardiac function during exercise. Mutations either in *KCNQ1* (Marx et al. 2002) or *KCNE1* (Kurokawa et al. 2003) can disrupt this regulation and create heterogeneity in the cellular response to β-AR stimulation, a novel mechanism that may contribute to the triggering of some arrhythmias in LQT1 and LQT5 (Kass et al. 2003). Importantly, disruption of the regulation of only the potassium channel by these mutations disrupts, at the cellular level, the coordinated response of one, but not all, channel/pump proteins that are regulated by PKA. Because many of the target proteins regulate cellular calcium homeostasis, it is entirely possible that the trigger underlying at least some forms of exercise-induced arrhythmias in LQT1 may be due to dysfunction in cellular calcium handling (Kass et al. 2003).

9
The Cardiac Sodium Channel and the Action Potential Plateau Phase

The report that mutations in *SCN5A*, the gene coding for the α-subunit of the major cardiac sodium channel, were associated with LQTS (Wang et al. 1995a) was surprising because this channel is associated most frequently with impulse conduction and hence the QRS but not the QT waveforms of the ECG. Sodium channels are voltage-gated channels that rapidly enter a non-conducting in-activated state during sustained depolarization such as the cardiac action potential plateau. Importantly, the first *SCN5A* mutation, the ΔKPQ mutation, physically disrupted a cytoplasmic peptide linker in the channel protein that, in basic biochemical and biophysical studies, had been shown to be a critical determinant of sodium channel inactivation: the inactivation gate (Stuhmer et al. 1989; Catterall 1995). This peptide links two domains (III and IV) of the channel and physically moves to occlude the channel pore upon depolar-ization. Once again, the combination of basic and clinical investigation has led to a clear understanding of the molecular basis of this key physiological parameter in human heart. Further, the demonstration that small changes in sodium channel inactivation such as those changes that occur in LQT3 muta-tions, can have life-threatening consequences confirms predictions made more than 50 years ago by Silvio Weidmann. Demonstrated that the cardiac action potential plateau was an exquisitely sensitive period of electrical activity that could adapt, with little energy expenditure, to small changes in ionic currents (Weidmann 1952).

Subsequent investigations of LQT3 mutations have revealed that not only is the domain III/IV intracellular linker key to inactivation and maintenance of the action potential plateau (and hence Q-T interval), but the channel carboxy terminal (C-T) domain is essential in this process also, and not only is disrup-tion of the inactivation gate a mechanism by which LQT3 arrhythmias can be generated, but much more subtle changes in channel gating can also underlie these arrhythmias. For example, one LQT3 mutation (the I1768V mutation) speeds the recovery from inactivation in a voltage-dependent manner, and this leads to augmentation of depolarizing current during the repolarization phase of the action potential. The consequence is delayed repolarization, which underlies the clinical phenotype—prolonged QT (Clancy et al. 2003).

10
The Sodium Channel Inactivation Gate as a Molecular Complex

Recent work in which biochemical and functional experiments were combined directly addressed the question of whether or not the C-terminus may have a direct structural role in the control of channel inactivation, and, if so, how the C-T domain affects stabilization of the inactivated Na^+ channel. The conclusion from this work is that the cardiac sodium channel inactivation gate is a molecular complex, providing additional structural insight into the role of the carboxy-terminal domain in regulating channel activity. Experimental data support the view that the III–IV linker interacts directly with the carboxy terminal domain of the channel to stabilize inactivated channels (Motoike et al. 2004).

In these experiments, biochemical evidence was presented for direct physical interaction between the C-T domain of the channel and the III–IV linker inactivation gate. These biochemical data are remarkably consistent with a role of the C-terminal/III–IV linker in stabilization of the inactivated state. Further, using glutamate scaning of the III–IV linker peptide, a region on the linker was suggested to be the motif that coordinates III–IV linker/C-T interactions, and this motif was found to be distinct from the III–IV linker motif previously identified as the region that coordinates binding of the inactivation gate to the inner mouth of the channel pore. These data provided strong evidence that the inactivation gate of the voltage-dependent Na^+ channel is a molecular complex that consists of the III–IV linker and the C-terminal domain of the channel and that this interaction underlies the stabilization of the inactivated state by the C-T domain during prolonged depolarization. Uncoupling of this complex destabilizes inactivation and increases the likelihood of channel re-opening during prolonged depolarization.

11
Summary and Future Directions

Investigation into the molecular basis of inherited cardiac arrhythmias caused by mutations of the α-subunit of the principal cardiac sodium channel (Nav1.5) has led to an appreciation of the role of the carboxy terminal domain of the channel in regulating channel gating. Theoretical and experimental structural analysis of the channel C-T domain provides strong evidence for a highly structured region of the channel and that interactions between the C-T domain and the channel inactivation gate are necessary to control channel activity that directly affects action potential, and hence QT, duration in the heart. This structured region thus provides a novel target against which to develop drugs that have the potential to regulate the activity of this key cardiac ion channel, not by blocking the conduction pore, but by regulating, in an allosteric manner, channel gating. Investigations into the mechanisms underlying the clinical observations that LQT1 patients are at elevated arrhythmia risk during exercise have led to the unraveling of the molecular architecture of a critically important cardiac potassium channel and its interconnection to the sympathetic nervous system.

We have made considerable progress in understanding the importance of ion channel structure to human physiology since the first ion channel was cloned in 1982. We now have a better understanding of the molecular genetics, ion channel structures, and cellular electrophysiology that contribute to the genesis of cardiac arrhythmias. Much of this improved insight has come directly from investigations of LQTS and other inherited arrhythmias and is being translated into more effective and more rational therapy for patients with electrical disorders of the cardiac rhythm. Much remains to be accomplished, and this will be done thorough continued collaboration of basic and clinical scientists in many ways based on the foundations laid by studies of LQTS.

References

Abriel H, Motoike H, Kass RS (2000a) KChAP: a novel chaperone for specific K (+) channels key to repolarization of the cardiac action potential. Focus on "KChAP as a chaperone for specific K (+) channels" [editorial; comment]. Am J Physiol Cell Physiol 278:C863–C864

Abriel H, Wehrens XH, Benhorin J, Kerem B, Kass RS (2000b) Molecular pharmacology of the sodium channel mutation D1790G linked to the long-QT syndrome. Circulation 102:921–925

Ackerman MJ, Tester DJ, Porter CJ (1999) Swimming, a gene-specific arrhythmogenic trigger for inherited long QT syndrome. Mayo Clin Proc 74:1088–1094

Aldrich RW, Corey DP, Stevens CF (1983) A reinterpretation of mammalian sodium channel gating based on single channel recording. Nature 306:436–441

An RH, Bangalore R, Rosero SZ, Kass RS (1996) Lidocaine block of LQT-3 mutant human Na+ channels. Circ Res 79:103–108

Barhanin J, Lesage F, Guillemare E, Fink M, Lazdunski M, Romey G (1996) K (V)LQT1 and lsK (minK) proteins associate to form the I (Ks) cardiac potassium current. Nature 384:78–80

Benhorin J, Taub R, Goldmit M, Kerem B, Kass RS, Windman I, Medina A (2000) Effects of flecainide in patients with new SCN5A mutation: mutation-specific therapy for long-QT syndrome? Circulation 101:1698–1706

Bennett PB, Yazawa K, Makita N, George AL (1995) Molecular mechanism for an inherited cardiac arrhythmia. Nature 376:683–685

Catterall WA (1995) Structure and function of voltage-gated ion channels. Annu Rev Biochem 64:493–531

Catterall WA (1996) Molecular properties of sodium and calcium channels [review]. J Bioenerg Biomembr 28:219–230

Catterall WA (2000) From ionic currents to molecular mechanisms: the structure and function of voltage-gated sodium channels. Neuron 26:13–25

Clancy CE, Tateyama M, Liu H, Wehrens XHT, Kass RS (2003) Non-equilibrium gating in cardiac Na+ channels: an original mechanism of arrhythmia. Circulation 107:2233–2237

Deschenes I, Trottier E, Chahine M (1999) Cysteine scanning analysis of the IFM cluster in the inactivation gate of a human heart sodium channel. Cardiovasc Res 42:521–529

Dumaine R, Kirsch GE (1998) Mechanism of lidocaine block of late current in long Q-T mutant Na+ channels. Am J Physiol 274:H477–H487

Dumaine R, Wang Q, Keating MT, Hartmann HA, Schwartz PJ, Brown AM, Kirsch GE (1996) Multiple mechanisms of Na+ channel linked long-QT syndrome. Circ Res 78:916–924

George AL, Varkony TA, Drabkin HA, Han J, Knops JF, Finley WH, Brow GB, Ward DC, Haas M (1995) Assignment of the human heart tetrodotoxin-resistant voltage-gated Na+ channel alpha-subunit gene (SCN5A) to band 3P21. Cytogenet Cell Genet 68:67–70

Hille B (1977) Local anesthetics: hydrophilic and hydrophobic pathways for the drug-receptor reaction. J Gen Physiol 69:497–515

Hondeghem LM, Katzung BG (1977) Time- and voltage-dependent interactions of antiarrhythmic drugs with cardiac sodium channels [review]. Biochim Biophys Acta 472:373–398

Jervell A, Lange-Nielsen F (1957) Congenital deaf-mutism, functional heart disease with prolongation of the Q-T interval and sudden death. Am Heart J 54:59–68

Kass RS, Moss AJ (2003) Long QT syndrome: novel insights into the mechanisms of cardiac arrhythmias. J Clin Invest 112:810–815

Kass RS, Wiegers SE (1982) The ionic basis of concentration-related effects of noradrenaline on the action potential of calf cardiac Purkinje fibres. J Physiol (Lond) 322:541–558

Kass RS, Kurokawa J, Marx SO, Marks AR (2003) Leucine/isoleucine zipper coordination of ion channel macromolecular signaling complexes in the heart. Roles in inherited arrhythmias. Trends Cardiovasc Med 13:52–56

Kearney JA, Plummer NW, Smith MR, Kapur J, Cummins TR, Waxman SG, Goldin AL, Meisler MH (2001) A gain-of-function mutation in the sodium channel gene Scn2a results in seizures and behavioral abnormalities. Neuroscience 102:307–317

Keating MT, Sanguinetti MC (2001) Molecular and cellular mechanisms of cardiac arrhythmias. Cell 104:569–580

Kellenberger S, Scheuer T, Catterall WA (1996) Movement of the Na+ channel inactivation gate during inactivation. J Biol Chem 271:30971–30979

Kellenberger S, West JW, Catterall WA, Scheuer T (1997a) Molecular analysis of potential hinge residues in the inactivation gate of brain type IIA Na+ channels. J Gen Physiol 109:607–617

Kellenberger S, West JW, Scheuer T, Catterall WA (1997b) Molecular analysis of the putative inactivation particle in the inactivation gate of brain type IIA Na+ channels. J Gen Physiol 109:589–605

Kurokawa J, Chen L, Kass RS (2003) Requirement of subunit expression for cAMP-mediated regulation of a heart potassium channel. Proc Natl Acad Sci U S A 100:2122–2127

Li HL, Galue A, Meadows L, Ragsdale DS (1999) A molecular basis for the different local anesthetic affinities of resting versus open and inactivated states of the sodium channel. Mol Pharmacol 55:134–141

Liu H, Tateyama M, Clancy CE, Abriel H, Kass RS (2002) Channel openings are necessary but not sufficient for use-dependent block of cardiac Na$^{(+)}$ channels by flecainide: evidence from the analysis of disease-linked mutations. J Gen Physiol 120:39–51

Liu H, Atkins J, Kass RS (2003) Common molecular determinants of flecainide and lidocaine block of heart Na$^{(+)}$ channels: evidence from experiments with neutral and quaternary flecainide analogues. J Gen Physiol 121:199–214

Lossin C, Wang DW, Rhodes TH, Vanoye CG, George AL Jr (2002) Molecular basis of an inherited epilepsy. Neuron 34:877–884

Marx SO, Kurokawa J, Reiken S, Motoike H, D'Armiento J, Marks AR, Kass RS (2002) Requirement of a macromolecular signaling complex for beta adrenergic receptor modulation of the KCNQ1-KCNE1 potassium channel. Science 295:496–499

McPhee JC, Ragsdale DS, Scheuer T, Catterall WA (1994) A mutation in segment IVS6 disrupts fast inactivation of sodium channels. Proc Natl Acad Sci U S A 91:12346–12350

McPhee JC, Ragsdale DS, Scheuer T, Catterall WA (1995) A critical role for transmembrane segment IVS6 of the sodium channel alpha subunit in fast inactivation. J Biol Chem 270:12025–12034

McPhee JC, Ragsdale DS, Scheuer T, Catterall WA (1998) A critical role for the S4-S5 intracellular loop in domain IV of the sodium channel alpha-subunit in fast inactivation. J Biol Chem 273:1121–1129

Moss AJ (2003) Long QT Syndrome. JAMA 289:2041–2044

Moss AJ, Schwartz PJ, Crampton RS, Tzivoni D, Locati EH, MacCluer J, Hall WJ, Weitkamp L, Vincent M, Garso A, Robinson JL, Benhorin J, Choi S (1991) The long QT syndrome: prospective longitudinal study of 328 families. Circulation 84:1136–1144

Moss AJ, Zareba W, Hall WJ, Schwartz PJ, Crampton RS, Benhorin J, Vincent GM, Locati EH, Priori SG, Napolitano C, Medina A, Zhang L, Robinson JL, Timothy K, Towbin JA, Andrews ML (2000) Effectiveness and limitations of beta-blocker therapy in congenital long-QT syndrome. Circulation 101:616–623

Moss AJ, Zareba W, Kaufman ES, Gartman E, Peterson DR, Benhorin J, Towbin JA, Keating MT, Priori SG, Schwartz PJ, Vincent GM, Robinson JL, Andrews ML, Feng C, Hall WJ, Medina A, Zhang L, Wang Z (2002) Increased risk of arrhythmic events in long-QT syndrome with mutations in the pore region of the human ether-a-go-go-related gene potassium channel. Circulation 105:794–799

Motoike HK, Liu H, Glaaser IW, Yang AS, Tateyama M, Kass RS (2004) The Na+ channel inactivation gate is a molecular complex: a novel role of the COOH-terminal domain. J Gen Physiol 123:155–165

Nagatomo T, January CT, Makielski JC (2000) Preferential block of late sodium current in the LQT3 DeltaKPQ mutant by the class I (C) antiarrhythmic flecainide. Mol Pharmacol 57:101–107

Noble D, Tsien R (1968) The kinetics and rectifier properties of the slow potassium current in cardiac Purkinje fibres. J Physiol (Lond) 195:185–214

Noble D, Tsien RW (1969) Outward membrane currents activated in the plateau range of potentials in cardiac Purkinje fibres. J Physiol (Lond) 200:205–231

Paavonen KJ, Swan H, Piippo K, Hokkanen L, Laitinen P, Viitasalo M, Toivonen L, Kontula K (2001) Response of the QT interval to mental and physical stress in types LQT1 and LQT2 of the long QT syndrome. Heart 86:39–44

Patton DE, West JW, Catterall WA, Goldin AL (1992) Amino acid residues required for fast Na (+)-channel inactivation: charge neutralizations and deletions in the III–IV linker. Proc Natl Acad Sci U S A 89:10905–10909

Priori SG (2004) From trials to guidelines to clinical practice: the need for improvement. Europace 6:176–178

Priori SG, Napolitano C, Paganini V, Cantu F, Schwartz PJ (1997) Molecular biology of the long QT syndrome: impact on management. Pacing Clin Electrophysiol 20:2052–2057

Priori SG, Napolitano C, Schwartz PJ, Grillo M, Bloise R, Ronchetti E, Moncalvo C, Tulipani C, Veia A, Bottelli G, Nastoli J (2004) Association of long QT syndrome loci and cardiac events among patients treated with beta-blockers. JAMA 292:1341–1344

Ragsdale DS, McPhee JC, Scheuer T, Catterall WA (1994) Molecular determinants of state-dependent block of Na+ channels by local anesthetics. Science 265:1724–1728

Ragsdale DS, McPhee JC, Scheuer T, Catterall WA (1996) Common molecular determinants of local anesthetic, antiarrhythmic, and anticonvulsant block of voltage-gated Na+ channels. Proc Natl Acad Sci U S A 93:9270–9275

Rivolta I, Abriel H, Kass RS (2001) Ion channels as targets for drugs. In: Sperelakis N (ed) Cell physiology sourcebook. Academic Press, New York, pp 643–652

Rohl CA, Boeckman FA, Baker C, Scheuer T, Catterall WA, Klevit RE (1999) Solution structure of the sodium channel inactivation gate. Biochemistry 38:855–861

Rosen MR, Hoffman BF, Wit AL (1975) Electrophysiology and pharmacology of cardiac arrhythmias. V Cardiac antiarrhythmic effects of lidocaine. Am Heart J 89:526–536

Rosero SZ, Zareba W, Robinson JL, (Moss A 1997) Gene-specfic therapy for long QT syndrome: QT shortening with lidocaine and tocainide in patients with mutation of the sodidum channel gene. Ann Noninvasive Electrocardiol 2:274–278

Sanguinetti MC, Jurkiewicz NK (1990) Two components of cardiac delayed rectifier K+ current. Differential sensitivity to block by class III antiarrhythmic agents. J Gen Physiol 96:195–215

Sanguinetti MC, Spector PS (1997) Potassium channelopathies. Neuropharmacology 36:755–762

Sanguinetti MC, Jiang C, Curran ME, Keating MT (1995) A mechanistic link between an inherited and an acquired cardiac arrhythmia: HERG encodes the IKr potassium channel. Cell 81:299–307

Sanguinetti MC, Curran ME, Spector PS, Keating MT (1996a) Spectrum of HERG K channel dysfunction in an inherited cardiac arrhythmia. Proc Natl Acad Sci U S A 93:2208–2212

Sanguinetti MC, Curran ME, Zou A, Shen J, Spector PS, Atkinson DL, Keating MT (1996b) Coassembly of KvLQT1 and minK (ISK) proteins to form cardiac IKS potassium channel. Nature 384:80–83

Sato C, Ueno Y, Asai K, Takahashi K, Sato M, Engel A, Fujiyoshi Y (2001) The voltage-sensitive sodium channel is a bell-shaped molecule with several cavities. Nature 409:1047–1051

Schwartz PJ, Priori SG, Locati EH, Napolitano C, Cantu F, Towbin JA, Keating MT, Hammoude H, Brown AM, Chen LS (1995) Long QT syndrome patients with mutations of the SCN5A and HERG genes have differential responses to NA+ channel blockade and to increases in heart rate: implications for gene-specific therapy. Circulation 92:3381–3386

Schwartz PJ, Priori SG, Spazzolini C, Moss AJ, Vincent GM, Napolitano C, Denjoy I, Guicheney P, Breithardt G, Keating MT, Towbin JA, Beggs AH, Brink P, Wilde AA, Toivonen L, Zareba W, Robinson JL, Timothy KW, Corfield V, Wattanasirichaigoon D, Corbett C, Haverkamp W, Schulze-Bahr E, Lehmann MH, Schwartz K, Coumel P, Bloise R (2001) Genotype-phenotype correlation in the long-QT syndrome: gene-specific triggers for life-threatening arrhythmias. Circulation 103:89–95

Sheets MF, Kyle JW, Hanck DA (2000) The role of the putative inactivation lid in sodium channel gating current immobilization. J Gen Physiol 115:609–620

Shimizu W, Horie M, Ohno S, Takenaka K, Yamaguchi M, Shimizu M, Washizuka T, Aizawa Y, Nakamura K, Ohe T, Aiba T, Miyamoto Y, Yoshimasa Y, Towbin JA, Priori SG, Kamakura S (2004) Mutation site-specific differences in arrhythmic risk and sensitivity to sympathetic stimulation in the LQT1 form of congenital long QT syndrome: multicenter study in Japan. J Am Coll Cardiol 44:117–125

Splawski I, Timothy KW, Vincent GM, Atkinson DL, Keating MT (1997a) Molecular basis of the long-QT syndrome associated with deafness. N Engl J Med 336:1562–1567

Splawski I, Tristani-Firouzi M, Lehmann MH, Sanguinetti MC, Keating MT (1997b) Mutations in the hminK gene cause long QT syndrome and suppress IKs function. Nat Genet 17:338–340

Splawski I, Shen J, Timothy KW, Lehmann MH, Priori S, Robinson JL, Moss AJ, Schwartz PJ, Towbin JA, Vincent GM, Keating MT (2000) Spectrum of mutations in long-QT syndrome genes: KVLQT1, HERG, SCN5A, KCNE1, and KCNE2. Circulation 102:1178–1185

Splawski I, Timothy KW, Sharpe LM, Decher N, Kumar P, Bloise R, Napolitano C, Schwartz PJ, Joseph RM, Condouris K, Tager-Flusberg H, Priori SG, Sanguinetti MC, Keating MT (2004) Ca (V)1.2 calcium channel dysfunction causes a multisystem disorder including arrhythmia and autism. Cell 119:19–31

Stuhmer W, Conti F, Suzuki H, Wang X, Noda M, Yahagi N, Kubo H, Numa S (1989) Structural parts involved in activation and inactivation of the sodium channel. Nature 339:597–603

Takenaka K, Ai T, Shimizu W, Kobori A, Ninomiya T, Otani H, Kubota T, Takaki H, Kamakura S, Horie M (2003) Exercise stress test amplifies genotype-phenotype correlation in the LQT1 and LQT2 forms of the long-QT syndrome. Circulation 107:838–844

Tateyama M, Rivolta I, Clancy CE, Kass RS (2003) Modulation of cardiac sodium channel gating by protein kinase A can be altered by disease-linked mutation. J Biol Chem 278:46718–46726

Vassilev P, Scheuer T, Catterall WA (1989) Inhibition of inactivation of single sodium channels by a site-directed antibody. Proc Natl Acad Sci U S A 86:8147–8151

Vassilev PM, Scheuer T, Catterall WA (1988) Identification of an intracellular peptide segment involved in sodium channel inactivation. Science 241:1658–1661

Veldkamp MW, Viswanathan PC, Bezzina C, Baartscheer A, Wilde AA, Balser JR (2000) Two distinct congenital arrhythmias evoked by a multidysfunctional Na (+) channel. Circ Res 86:E91–E97

Viswanathan PC, Bezzina CR, George AL, Roden JDM, Wilde AA, Balser JR (2001) Gating-dependent mechanisms for flecainide action in SCN5A-linked arrhythmia syndromes. Circulation 104:1200–1205

Wang DW, Yazawa K, Makita N, George AL, Bennett PB (1997) Pharmacological targeting of long QT mutant sodium channels. J Clin Invest 99:1714–1720

Wang Q, Shen J, Li Z, Timothy K, Vincent GM, Priori SG, Schwartz PJ, Keating MT (1995a) Cardiac sodium channel mutations in patients with long QT syndrome, an inherited cardiac arrhythmia. Hum Mol Genet 4:1603–1607

Wang Q, Shen J, Splawski I, Atkinson D, Li Z, Robinson JL, Moss AJ, Towbin JA, Keating MT (1995b) SCN5A mutations associated with an inherited cardiac arrhythmia, long QT syndrome. Cell 80:805–811

Wang Q, Curran ME, Splawski I, Burn TC, Millholland JM, Vanraay TJ, Shen J, Timothy KW, Vincent GM, Dejager T, Schwartz PJ, Towbin JA, Moss AJ, Atkinson DL, Landes GM, Connors TD, Keating MT (1996) Positional cloning of a novel potassium channel gene— KVLQT1 mutations cause cardiac arrhythmias. Nat Genet 12:17–23

Weidmann S (1952) The electrical constants of Purkinje fibres. J Physiol 118:348–360

Weiser T, Qu Y, Catterall WA, Scheuer T (1999) Differential interaction of R-mexiletine with the local anesthetic receptor site on brain and heart sodium channel alpha-subunits. Mol Pharmacol 56:1238–1244

Weiss LA, Escayg A, Kearney JA, Trudeau M, MacDonald BT, Mori M, Reichert J, Buxbaum JD, Meisler MH (2003) Sodium channels SCN1A, SCN2A and SCN3A in familial autism. Mol Psychiatry 8:186–194

West JW, Patton DE, Scheuer T, Wang Y, Goldin AL, Catterall WA (1992) A cluster of hydrophobic amino acid residues required for fast Na (+)-channel inactivation. Proc Natl Acad Sci U S A 89:10910–10914

Windle JR, Geletka RC, Moss AJ, Zareba W, Atkins DL (2001) Normalization of ventricular repolarization with flecainide in long QT syndrome patients with SCN5A:DeltaKPQ mutation. Ann Noninvasive Electrocardiol 6:153–158

Yang N, Ji S, Zhou M, Ptacek LJ, Barchi RL, Horn R, George AL, (Jr1994) Sodium channel mutations in paramyotonia congenita exhibit similar biophysical phenotypes in vitro. Proc Natl Acad Sci U S A 91:12785–12789

Zareba W, Moss AJ, Sheu G, Kaufman ES, Priori S, Vincent GM, Towbin JA, Benhorin J, Schwartz PJ, Napolitano C, Hall WJ, Keating MT, Qi M, Robinson JL, Andrews ML (2003) Location of mutation in the KCNQ1 and phenotypic presentation of long QT syndrome. J Cardiovasc Electrophysiol 14:1149–1153

HEP (2006) 171:305–330
© Springer-Verlag Berlin Heidelberg 2006

Therapy for the Brugada Syndrome

C. Antzelevitch (✉) · J. M. Fish

Masonic Medical Research Laboratory, 2150 Bleecker Street, Utica NY, 13501, USA
ca@mmrl.edu

Abstract The Brugada syndrome is a congenital syndrome of sudden cardiac death first described as a new clinical entity in 1992. Electrocardiographically characterized by a distinct coved-type ST segment elevation in the right precordial leads, the syndrome is associated with a high risk for sudden cardiac death in young and otherwise healthy adults, and less frequently in infants and children. The ECG manifestations of the Brugada syndrome are often dynamic or concealed and may be revealed or modulated by sodium channel blockers. The syndrome may also be unmasked or precipitated by a febrile state, vagotonic agents, α-adrenergic agonists, β-adrenergic blockers, tricyclic or tetracyclic antidepressants, a combination of glucose and insulin, and hypokalemia, as well as by alcohol and cocaine toxicity. An implantable cardioverter–defibrillator (ICD) is the most widely accepted approach to therapy. Pharmacological therapy aimed at rebalancing the currents active during phase 1 of the right ventricular action potential is used to abort electrical storms, as an adjunct to device therapy, and as an alternative to device therapy when use of an ICD is not possible. Isoproterenol and cilostazol boost calcium channel current, and drugs like quinidine inhibit the transient outward current, acting to diminish the action potential notch and thus suppress the substrate and trigger for ventricular tachycardia/fibrillation (VT/VF).

Keywords Brugada syndrome · Phase 2 reentry · ST segment elevation · I_{Na} · I_{to} · Implantable cardioverter–defibrillator (ICD) · VT · *SCN5A* mutations · Sudden death · Bradycardia

1
Clinical Characteristics and Diagnostic Criteria

The Brugada syndrome typically manifests in the third or fourth decade of life (average age of 41 ± 15 years), although patients have been diagnosed with the syndrome at an age as young as 2 days and as old as 84 years. The prevalence of the disease is estimated to be at least 5 per 10,000 inhabitants in Southeast Asia, where the syndrome is endemic (Nademanee et al. 1997). In Japan, a Brugada syndrome ECG (type 1) is observed in 12 per 10,000 inhabitants; type 2 and type 3 ECGs, which are not diagnostic of Brugada syndrome, are much more prevalent, appearing in 58 per 10,000 inhabitants (Miyasaka et al. 2001). The true prevalence of the disease in the general population is difficult to estimate because the ECG pattern is often concealed (Brugada et al. 2003). Sudden unexplained nocturnal death syndrome (SUNDS also known as SUDS) and Brugada syndrome have been shown to be phenotypically, genetically, and functionally the same disorder (Vatta et al. 2002).

Although syncope and sudden death are a consequence of ventricular tachycardia/fibrillation (VT/VF), approximately 20% of Brugada syndrome patients also develop supraventricular arrhythmias (Morita et al. 2002). Atrial fibrillation (AF) is reported in approximately 10%–20% of cases. Atrio-ventricular (AV) nodal reentrant tachycardia (AVNRT) and Wolf–Parkinson–White (WPW) syndrome have been described as well (Eckardt et al. 2001). Prolonged sinus node recovery time and sino-atrial conduction time (Morita et al. 2004) as well as slowed atrial conduction and atrial standstill have been reported in association with the syndrome (Takehara et al. 2004). A recent study reports that ventricular inducibility is positively correlated with a history of atrial arrhythmias (Bordachar et al. 2004). The incidence of atrial arrhythmias is 27% in Brugada syndrome patients with an indication for ICD vs 13% in patients without an indication for ICD, suggesting a more advanced disease process in patients with spontaneous atrial arrhythmias (Bordachar et al. 2004).

The Brugada syndrome is characterized by an ST segment elevation in the right precordial leads. Three types of ST segment elevation are generally recognized (Wilde et al. 2002a,b). Type 1 is diagnostic of Brugada syndrome and is characterized by a coved ST segment elevation exceeding or at 2 mm (0.2 mV) followed by a negative T wave (Fig. 1). Brugada syndrome is definitively diagnosed when a type 1 ST segment elevation is observed in more than one right-precordial lead (V_1–V_3), in the presence or absence of sodium channel blocking agent, and in conjunction with one of the following: documented ventricular fibrillation, polymorphic ventricular tachycardia, a family history of sudden cardiac death (SCD) (<45 years old), coved type ECGs in family members, inducibility of VT with programmed electrical stimulation, syncope, or nocturnal agonal respiration. The electrocardiographic manifestations of the Brugada syndrome, when concealed, can be unmasked by sodium channel blockers, but also during febrile state or with vagotonic agents (Brugada

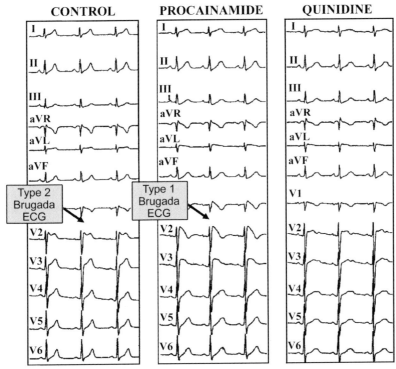

Fig. 1 Twelve-lead electrocardiogram (ECG) tracings in an asymptomatic 26-year-old man with the Brugada syndrome. *Left:* Baseline: type 2 ECG (not diagnostic) displaying a "saddleback-type" ST segment elevation is observed in V₂. *Center:* After intravenous administration of 750 mg procainamide, the type 2 ECG is converted to the diagnostic type 1 ECG consisting of a "coved-type" ST segment elevation. *Right:* A few days after oral administration of quinidine bisulfate (1,500 mg/day, serum quinidine level 2.6 mg/l), ST segment elevation is attenuated, displaying a nonspecific abnormal pattern in the right precordial leads. VF could be induced during control and procainamide infusion, but not after quinidine. (Modified from Belhassen et al. 2002, with permission)

et al. 2000b,c; Miyazaki et al. 1996; Antzelevitch and Brugada 2002). Sodium channel blockers, including flecainide, ajmaline, procainamide, disopyramide, propafenone, and pilsicainide are used to aid in a differential diagnosis when ST segment elevation is not diagnostic under baseline conditions (Brugada et al. 2000c; Shimizu et al. 2000a; Priori et al. 2000).

Type 2 ST segment elevation has a saddleback appearance with an ST segment elevation of ≥2 mm followed by a trough displaying ≥1-mm ST elevation followed by either a positive or biphasic T wave (Fig. 1). Type 3 has either a saddleback or coved appearance with an ST segment elevation of less than 1 mm. Type 2 and type 3 ECG are *not* diagnostic of the Brugada syndrome. These three patterns may be observed spontaneously in serial ECG tracings from the same patient or following the introduction of specific drugs. The diagnosis of Brugada syndrome is also considered positive when a type 2 (saddleback pattern) or type 3 ST segment elevation is observed in more than one right precordial lead under baseline conditions and conversion to the diagnostic type 1 pattern occurs after sodium channel blocker administration (ST segment elevation should be ≥ 2 mm). One or more of the clinical criteria described above need also be present.

Placement of the right precordial leads in a superior position (two intercostal spaces above normal) can increase the sensitivity of the ECG for detecting the Brugada phenotype in some patients, both in the presence and absence of a drug challenge (Shimizu et al. 2000b; Sangwatanaroj et al. 2001).

While most cases of Brugada syndrome display right precordial ST segment elevation, isolated cases of inferior lead (Kalla et al. 2000) or left precordial lead (Horigome et al. 2003) ST segment elevation have been reported in Brugada-like syndromes, in some cases associated with *SCN5A* mutations (Potet et al. 2003).

Minor prolongation of the QT interval may accompany ST segment elevation in the Brugada syndrome (Alings and Wilde 1999; Bezzina et al. 1999; Priori et al. 2000). The QT-interval is prolonged more in the right vs left precordial leads, probably due to a preferential prolongation of action potential duration (APD) in right ventricular (RV) epicardium secondary to accentuation of the action potential notch (Pitzalis et al. 2003). Depolarization abnormalities including prolongation of P wave duration, PR and QRS intervals are frequently observed, particularly in patients linked to *SCN5A* mutations (Smits et al. 2002). PR prolongation likely reflects HV conduction delay (Alings and Wilde 1999a).

2
Genetic Basis

The only gene thus far linked to the Brugada syndrome is *SCN5A*, the gene encoding for the α-subunit of the cardiac sodium channel gene (Chen et al. 1998). *SCN5A* mutations account for 18%–30% of Brugada syndrome cases. Nearly 100 mutations in *SCN5A* have been linked to the syndrome over the past 4 years (see Antzelevitch 2001a; Priori et al. 2002; Balser 2001; Tan et al. 2003 for references; also see http://pc4.fsm.it:81/cardmoc/). Approximately 30 of these mutations have been studied in expression systems and shown to result in loss of function due to: (1) failure of the sodium channel to express; (2) a shift in the voltage- and time-dependence of sodium channel current (I_{Na}) activation, inactivation or reactivation; (3) entry of the sodium channel into an intermediate state of inactivation from which it recovers more slowly; or (4) accelerated inactivation of the sodium channel. Inheritance of the Brugada syndrome is via an autosomal-dominant mode of transmission. A second locus on chromosome 3, close to but apart from the *SCN5A* locus, has recently been linked to the syndrome (Weiss et al. 2002).

3
Cellular and Ionic Basis

The ability of the RV action potential to lose its dome, giving rise to phase 2 reentry and other characteristics of the Brugada syndrome, were identified in the early 1990s and evolved in parallel with the clinical syndrome (Antzelevitch et al. 1991, 2002; Krishnan and Antzelevitch 1991; Krishnan and Antzelevitch 1993).

The ST segment elevation in the Brugada syndrome is thought to be secondary to a rebalancing of the currents active at the end of phase 1, leading to accentuation of the action potential notch in RV epicardium (see Antzelevitch 2001a for references). A transient outward current (I_{to})-mediated spike and dome morphology, or notch, in ventricular epicardium, but not endocardium, generates a voltage gradient responsible for the inscription of the electrocardiographic J wave in larger mammals and in man (Yan and Antzelevitch 1996). ST segment is normally isoelectric because of the absence of transmural voltage gradients at the level of the action potential plateau. Under pathophysiologic conditions, accentuation of the RV notch leads to exaggeration of transmural voltage gradients and thus to accentuation of the J wave, causing an apparent ST segment elevation (Antzelevitch 2001a). The repolarization waves take on a saddleback or coved appearance depending on the timing of repolarization of epicardium relative to endocardium. A delay in epicardial activation and repolarization time leads to progressive inversion of the T wave. The down-sloping ST segment elevation, or accentuated J wave, observed in the experimental wedge models often appears as an R′, suggesting that the appearance of a right bundle branch block (RBBB) morphology in Brugada patients may be due at least in part to early repolarization of RV epicardium, rather than to marked impulse delay or conduction block in the right bundle. Indeed, RBBB criteria are not fully met in many cases of Brugada syndrome (Gussak et al. 1999).

Accentuation of the RV action potential notch can give rise to the typical Brugada ECG without creating an arrhythmogenic substrate (Fig. 2). The arrhythmogenic substrate arises when a further shift in the balance of currents leads to loss of the action potential dome at some epicardial sites but not others. Loss of the action potential dome in epicardium but not endocardium results in the development of a marked transmural dispersion of repolarization and refractoriness, responsible for the development of a vulnerable window. A closely coupled extrasystole can then capture this vulnerable window and induce a reentrant arrhythmia. Loss of the epicardial action potential dome is usually heterogeneous, leading to the development of epicardial dispersion of repolarization. Conduction of the action potential dome from sites at which it

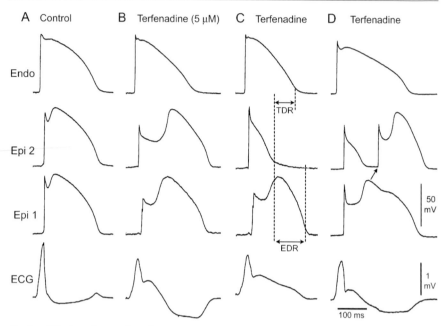

Fig. 2a–d Terfenadine-induced ST segment elevation, T wave inversion, transmural and epicardial dispersion of repolarization, and phase 2 reentry. Each panel shows transmembrane action potentials from one endocardial (*top*) and two epicardial sites together with a transmural ECG recorded from a canine arterially perfused right ventricular wedge preparation. **a** Control (BCL 400 ms). **b** Terfenadine (5 μM) accentuated the epicardial action potential notch creating a transmural voltage gradient that manifests as an ST segment elevation or exaggerated J wave in the ECG. First beat recorded after changing from BCL 800 ms to BCL 400 ms. **c** Continued pacing at BCL 400 ms results in all-or-none repolarization at the end of phase 1 at some epicardial sites but not others, creating a local epicardial dispersion of repolarization (*EDR*) as well as a transmural dispersion of repolarization (*TDR*). **d** Phase 2 reentry occurs when the epicardial action potential dome propagates from a site where it is maintained to regions where it has been lost. (Note: **d** was recorded from a different preparation.) (From Fish and Antzelevitch 2004, with permission)

is maintained to sites at which it is lost causes local re-excitation via a phase 2 reentry mechanism, leading to the development of the very closely coupled extrasystole, which triggers a circus movement reentry in the form of VT/VF (Lukas and Antzelevitch 1996; Yan and Antzelevitch 1999). The phase 2 reentrant beat fuses with the negative T wave of the basic response. Because the extrasystole originates in epicardium, the QRS complex is largely composed of a negative Q wave, which serves to accentuate the inverted T wave, giving the ECG a more symmetrical appearance, a morphology commonly observed in the clinic preceding the onset of polymorphic VT. Support for these hypotheses derives from experiments involving the arterially perfused RV wedge preparation (Yan and Antzelevitch 1999). Further evidence in support of these mechanisms derives from the recent studies of Kurita et al. in which monophasic action potential (MAP) electrodes where positioned on the epicardial and endocardial surfaces of the RV outflow tract (RVOT) in patients with the Brugada syndrome (Kurita et al. 2002; Antzelevitch et al. 2002).

Figure 3 shows the ability of terfenadine-induced phase 2 reentry to generate an extrasystole, couplet, and polymorphic VT/VF. Figure 3d illustrates an example of programmed electrical stimulation-induced VT/VF under similar conditions.

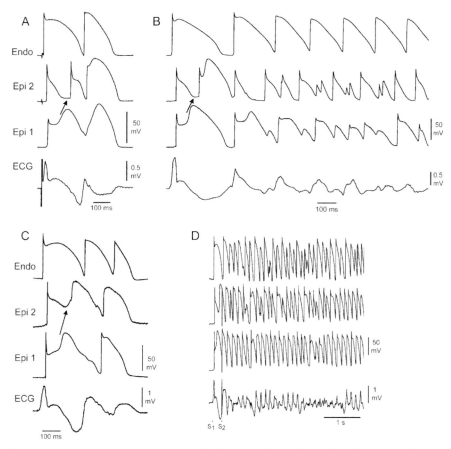

Fig. 3a–d Spontaneous and programmed electrical stimulation-induced polymorphic VT in RV wedge preparations pretreated with terfenadine (5–10 µM). **a** Phase 2 reentry in epicardium gives rise to a closely coupled extrasystole. **b** Phase 2 reentrant extrasystole triggers a brief episode of polymorphic VT. **c** Phase 2 reentrant extrasystole triggers brief reentry. **d** Same impalements and pacing conditions as **c**, however an extra stimulus (S1–S2 = 250 ms) applied to epicardium triggers a polymorphic VT. (From Fish and Antzelevitch 2004, with permission)

Although the genetic mutation is equally distributed between the sexes, the clinical phenotype is 8 to 10 times more prevalent in males than in females. The basis for this sex-related distinction was recently shown to be due to a more prominent I_{to}-mediated action potential notch in the RV epicardium of males vs females (Di Diego et al. 2002). The more prominent I_{to} causes the end of phase 1 of the RV epicardial action potential to repolarize to more negative potentials in tissue and arterially perfused wedge preparations from males, facilitating loss of the action potential dome and the development of phase 2 reentry and polymorphic VT. The gender distinction is not seen in all families; a recent report describes a family without a male predominance of the Brugada phenotype (Hong et al. 2004).

The available information supports the hypothesis that the Brugada syndrome is the result of amplification of heterogeneities intrinsic to the early phases of the action potential among the different transmural cell types. The amplification is secondary to a rebalancing of currents active during phase 1, including a decrease in I_{Na} or I_{Ca} or augmentation of any one of a number of outward currents including I_{Kr}, I_{Ks}, $I_{Cl(Ca)}$, or I_{to} (Fig. 4). ST segment elevation

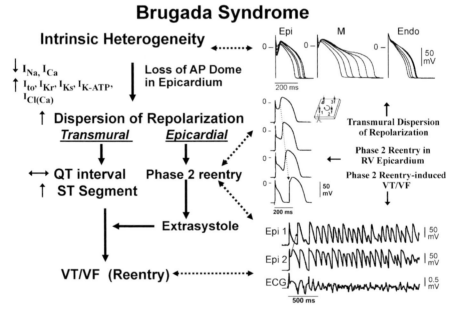

Fig. 4 Proposed mechanism for the Brugada syndrome. A shift in the balance of currents serves to amplify existing heterogeneities by causing loss of the action potential dome at some epicardial, but not endocardial sites. A vulnerable window develops as a result of the dispersion of repolarization and refractoriness within epicardium as well as across the wall. Epicardial dispersion leads to the development of phase 2 reentry, which provides the extrasystole that captures the vulnerable window and initiates VT/VF via a circus movement reentry mechanism. (Modified from Antzelevitch 2001b, with permission)

occurs as a consequence of the accentuation of the action potential notch, eventually leading to loss of the action potential dome in RV epicardium, where I_{to} is most prominent. Loss of the dome gives rise to both a transmural as well as epicardial dispersion of repolarization. The transmural dispersion is responsible for the development of ST segment elevation and the creation of a vulnerable window across the ventricular wall, whereas the epicardial dispersion leads to phase 2 reentry, which provides the extrasystole that captures the vulnerable window, thus precipitating VT/VF. The VT generated is usually polymorphic, resembling a very rapid form of torsade de pointes (TdP) (Fig. 4).

4
Factors That Modulate ECG and Arrhythmic Manifestations of the Brugada Syndrome

ST segment elevation in the Brugada syndrome is often dynamic. The Brugada ECG is often concealed and can be unmasked or modulated by sodium channel blockers, a febrile state, vagotonic agents, α-adrenergic agonists, β-adrenergic blockers, tricyclic or tetracyclic antidepressants, a combination of glucose and insulin, hyperkalemia, hypokalemia, hypercalcemia, and by alcohol and cocaine toxicity (Brugada et al. 2000bc; Miyazaki et al. 1996; Babaliaros and Hurst 2002; Goldgran-Toledano et al. 2002; Tada et al. 2001; Pastor et al. 2001; Ortega-Carnicer et al. 2001; Nogami et al. 2003; Araki et al. 2003). These agents may also induce acquired forms of the Brugada syndrome (Table 1). Until a definitive list of drugs to avoid in the Brugada syndrome is formulated, the list of agents in Table 1 may provide some guidance.

Acute ischemia or myocardial infarction due to vasospasm involving the RVOT mimics ST segment elevation similar to that in Brugada syndrome. This effect is secondary to the depression of I_{Ca} and the activation of I_{K-ATP} during ischemia, and suggests that patients with congenital and possibly acquired forms of Brugada syndrome may be at a higher risk for ischemia-related SCD (Noda et al. 2002).

VF and sudden death in the Brugada syndrome usually occur at rest and at night. Circadian variation of sympatho-vagal balance, hormones, and other metabolic factors likely contribute this circadian pattern. Bradycardia, due to altered symaptho-vagal balance or other factors, may contribute to arrhythmia initiation (Kasanuki et al. 1997; Proclemer et al. 1993; Mizumaki et al. 2004). Abnormal ^{123}I-MIBG uptake in 8 (17%) of the 17 Brugada syndrome patients but none in the control group was demonstrated by Wichter et al. (2002). There was segmental reduction of ^{123}I-MIBG in the inferior and the septal left ventricular wall, indicating presynaptic sympathetic dysfunction. Of note, imaging of the right ventricle, particularly the RVOT, is difficult with this technique, so insufficient information is available concerning sympathetic function in the regions known to harbor the arrhythmogenic substrate. Moreover, it remains

Table 1 Drug-induced Brugada-like ECG patterns

I. Antiarrhythmic drugs
　　1. Na$^+$ channel blockers
　　　Class IC drugs [Flecainide (Krishnan and Josephson 1998; Fujiki et al. 1999; Shimizu
　　　et al. 2000a; Brugada et al. 2000c; Gasparini et al. 2003), Pilsicainide (Takenaka
　　　et al. 1999; Shimizu et al. 2001), Propafenone (Matana et al. 2000)]
　　　Class IA drugs [Ajmaline (Brugada et al. 2000c; Rolf et al. 2003), Procainamide
　　　(Miyazaki et al. 1996; Brugada et al. 2000c), Disopyramide (Miyazaki et al. 1996;
　　　Wilde et al. 2002a), Cibenzoline (Tada et al. 2000)]
　　2. Ca^{2+} channel blockers
　　　Verapamil
II. Antianginal drugs
　　1. Ca^{2+} channel blockers
　　　Nifedipine, diltiazem
　　2. Nitrate
　　　Isosorbide dinitrate, nitroglycerine (Matsuo et al. 1998)
　　3. K$^+$ channel openers
　　　Nicorandil
III. Psychotropic drugs
　　1. Tricyclic antidepressants
　　　Amitriptyline (Bolognesi et al. 1997; Rouleau et al. 2001), Nortriptyline (Tada
　　　et al. 2001), desipramine (Babaliaros and Hurst 2002), clomipramine
　　　(Goldgran-Toledano et al. 2002)
　　2. Tetracyclic antidepressants
　　　Maprotiline (Bolognesi et al. 1997)
　　3. Phenothiazine
　　　Perphenazine (Bolognesi et al. 1997), cyamemazine
　　4. Selective serotonin reuptake inhibitors
　　　Fluoxetine (Rouleau et al. 2001)
IV. Other drugs
　　1. Histaminic H1 receptor antagonists
　　　Dimenhydrinate (Pastor et al. 2001)
　　2. Cocaine intoxication (Ortega-Carnicer et al. 2001; Littmann et al. 2000)
　　3. Alcohol intoxication

Modified from Shimizu (2004) with permission

unclear what role the reduced uptake function plays in the arrhythmogenesis of the Brugada syndrome. If indeed the RVOT is similarly affected, this defect may alter the symaptho-vagal balance in favor of the development of an arrhythmogenic substrate (Litovsky and Antzelevitch 1990; Yan and Antzelevitch 1999).

More recently, Kies and coworkers (Kies et al. 2004) assessed autonomic nervous system function noninvasively in patients with the Brugada syndrome, quantifying myocardial presynaptic and postsynaptic sympathetic function by means of positron emission tomography with the norepinephrine analog 11C-Hydroxyephedrine (11C-HED) and the nonselective β-blocker 11C-CGP 12177 (11C-CGP). Presynaptic sympathetic norepinephrine recycling, assessed by 11C-HED, was found to be globally increased in patients with Brugada syndrome compared with a group of age-matched healthy control subjects, whereas postsynaptic β-adrenoceptor density, assessed by 11C-CGP, was similar in patients and controls. This study provides further evidence in support of an autonomic dysfunction in Brugada syndrome.

Hypokalemia has been implicated as a contributing cause for the high prevalence of SUDS in the northeastern region of Thailand, where potassium deficiency is endemic (Nimmannit et al. 1991; Araki et al. 2003). Serum potassium in the northeastern population is significantly lower than that of the population in Bangkok, which lies in the central part of Thailand, where potassium is abundant in the food. A recent case report highlights the ability of hypokalemia to induce VF in a 60-year-old man who had asymptomatic Brugada syndrome, without a family history of sudden cardiac death (Araki et al. 2003). This patient was initially treated for asthma by steroids, which lowered serum potassium from 3.8 mmol/l on admission to 3.4 and 2.9 mmol/l on the seventh day and eighth day of admission, respectively. Both were associated with unconsciousness. VF was documented during the last episode, which reverted spontaneously to sinus rhythm.

Accelerated inactivation of the sodium channel in *SCN5A* mutations associated with the Brugada syndrome has been shown to be accentuated at higher temperatures (Dumaine et al. 1999), suggesting that a febrile state may unmask the Brugada syndrome by causing loss of function secondary to premature inactivation of I_{Na}. Indeed, numerous case reports have emerged since 1999 demonstrating that febrile illness could reveal the Brugada ECG and precipitate VF (Gonzalez Rebollo et al. 2000; Madle et al. 2002; Saura et al. 2002; Porres et al. 2002; Kum et al. 2002; Antzelevitch and Brugada 2002; Ortega-Carnicer et al. 2003; Dzielinska et al. 2004). Anecdotal reports point to hot baths as a possible precipitating factor. Of note, the northeastern part of Thailand, where the Brugada syndrome is most prevalent, is known for its very hot climate.

5
Approach to Therapy

Table 2 lists the device and pharmacologic therapies evaluated clinically or suggested on the basis of experimental evidence.

Table 2 Device and pharmacologic approach to therapy of the Brugada syndrome

Devices and Ablation
  ICD (Brugada et al. 2000a)
  ? Ablation or Cryosurgery (Haissaguerre et al. 2003)
  ? Pacemaker (van Den Berg et al. 2001)
Pharmacologic Approach to Therapy
  Ineffective
    Amiodarone (Brugada et al. 1998)
    β-Blockers (Brugada et al. 1998)
    Class IC antiarrhythmics
      Flecainide (Shimizu et al. 2000a)
      Propafenone (Matana et al. 2000)
      ? Disopyramide (Chinushi et al. 1997)
    Class IA antiarrhythmics
      Procainamide (Brugada et al. 2000c)
  Effective for treatment of electrical storms
    β-Adrenergic agonists—isoproterenol (Miyazaki et al. 1996; Shimizu et al. 2000b)
    Phosphodiesterase III inhibitors—cilostazol (Tsuchiya et al. 2002)
  Effective general therapy
    Class IA antiarrhythmics
      Quinidine (Belhassen and Viskin 2004; Alings et al. 2001; Belhassen et al. 1999, 2002;
    Yan and Antzelevitch 1999; Hermida et al. 2004; Mok et al. 2004)
  Experimental therapy
  I_{to} blockers—cardioselective and ion channel specific
    Quinidine (Yan and Antzelevitch 1999)
    4-Aminopyridine (Yan and Antzelevitch 1999)
    Tedisamil (Fish et al. 2004b)
    AVE0118 (Fish et al. 2004a)

5.1
Device Therapy

An implantable cardioverter–defibrillator (ICD) is the only proven effective device treatment for the disease (Brugada et al. 1999, 2000a). Recommendations of the Second Brugada Syndrome Consensus Conference (Antzelevitch et al. 2005) for ICD implantation are illustrated in Fig. 5 and summarized as follows:

– Symptomatic patients displaying the type 1 Brugada ECG (either spontaneously or after sodium channel blockade) who present with aborted sudden death should receive an ICD without additional need for electro-

Indications for ICD Implantation in Patients with the Brugada Syndrome

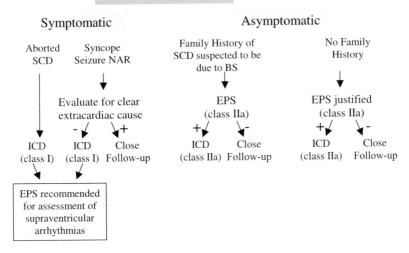

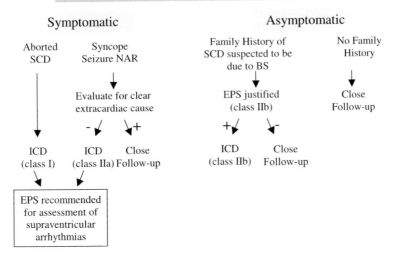

Class I: Clear evidence that procedure or treatment is useful or effective
Class II: Conflicting evidence concerning usefulness or efficacy
Class IIa: Weight of evidence in favor of usefulness or efficacy
Class IIb: Usefulness or efficacy less well established
    BS = Brugada Syndrome; EPS = Electrophysiologic Study; NAR = Nocturnal
    Agonal Respiration; SCD = Sudden Cardiac Death
From (Antzelevitch et al., 2004) with permission

Fig. 5 Indications for ICD implantation in patients with the Brugada syndrome

physiologic study (EPS). Similar patients presenting with related symptoms such as syncope, seizure, or nocturnal agonal respiration should also undergo ICD implantation after non-cardiac causes of these symptoms have been carefully ruled out. EPS is recommended in symptomatic patients only for the assessment of supraventricular arrhythmia.

- Asymptomatic patients displaying a type 1 Brugada ECG (spontaneously or after sodium channel block) should undergo EPS if there is a family history of SCD suspected to be due to Brugada syndrome. EPS may be justified when the family history is negative for SCD if the type 1 ECG occurs spontaneously. If inducible for ventricular arrhythmia, the patient should receive an ICD. Asymptomatic patients who have no family history and who develop a type 1 ECG only after sodium channel blockade should be closely followed-up. As additional data become available, these recommendations will no doubt require further fine-tuning.

The effectiveness of ICD in reverting VF and preventing sudden cardiac death was 100% in a recent multicenter trial in which 258 patients diagnosed with Brugada syndrome received an ICD (Brugada et al. 2004). Appropriate shocks were delivered in 14% 20%, 29%, 38%, and 52% of cases at 1, 2, 3, 4, and 5 years of follow-up, respectively. In the case of initially asymptomatic patients, appropriate ICD discharge was delivered 4%, 6%, 9%, 17%, and 37% at 1, 2, 3, 4, and 5 years of follow-up, respectively.

A recent report highlights the need for therapy other than with ICD. The case involves a patient with the Brugada syndrome who experienced multiple electrical storms, leading to numerous inappropriate ICD discharges. The patient was eventually given a heart transplant (Ayerza et al. 2002).

5.2
Pharmacologic Approach to Therapy

ICD implantation is not an appropriate solution for infants and young children or for patients residing in regions of the world where an ICD is out of reach because of economic factors. Although arrhythmias and sudden cardiac death generally occur during sleep or at rest and have been associated with slow heart rates, a potential therapeutic role for cardiac pacing remains largely unexplored. A recent interesting report by Haissaguerre and coworkers (Haissaguerre et al. 2003) points to focal radiofrequency ablation as a potentially valuable tool in controlling arrhythmogenesis by focal ablation of the ventricular premature beats that trigger VT/VF in the Brugada syndrome. However, data relative to a cryosurgical approach or the use of ablation therapy are very limited at this point in time.

A pharmacologic approach to therapy, based on a rebalancing of currents active during the early phases of the epicardial action potential in the right ventricle so as to reduce the magnitude of the action potential notch and/or restore the action potential dome, has been a focus of basic and clinical research in recent years. Table 2 lists the various pharmacologic agents thus far investigated. Antiarrhythmic agents such as amiodarone and β-blockers have been shown to be ineffective (Brugada et al. 1998). Class IC antiarrhythmic drugs (such as flecainide and propafenone) and class IA agents, such as procainamide, are contraindicated because of their effects to unmask the Brugada syndrome and induce arrhythmogenesis. Disopyramide is a class IA antiarrhythmic that has been demonstrated to normalize ST segment elevation in some Brugada patients but to unmask the syndrome in others (Chinushi et al. 1997).

Because the presence of a prominent transient outward current, I_{to}, is central to the mechanism underlying the Brugada syndrome, the most rational approach to therapy, regardless of the ionic or genetic basis for the disease, is to partially inhibit I_{to}. Cardioselective and I_{to}-specific blockers are not currently available. 4-Aminopyridine (4-AP) is an agent that is ion-channel specific at low concentrations, but is not cardioselective in that it inhibits I_{to} present in the nervous system. Although it is effective in suppressing arrhythmogenesis in wedge models of the Brugada syndrome (Yan and Antzelevitch 1999; Fig. 6), it is unlikely to be of clinical benefit because of neural-mediated and other side effects.

The only agent on the market in the United States with significant I_{to} blocking properties is quinidine. It is for this reason that we suggested several years ago that this agent might be of therapeutic value in the Brugada syndrome (Antzelevitch et al. 1999a). Experimental studies have since shown quinidine to be effective in restoring the epicardial action potential dome, thus normalizing the ST segment and preventing phase 2 reentry and polymorphic VT in experimental models of the Brugada syndrome (Fig. 6; Yan and Antzelevitch 1999). Clinical evidence of the effectiveness of quinidine in normalizing ST segment elevation in patients with the Brugada syndrome has been reported (Figs. 1 and 7; Belhassen et al. 2002; Alings et al. 2001; Belhassen and Viskin 2004).

The effects of quinidine to prevent inducible and spontaneous VF was recently reported by Belhassen and coworkers (Belhassen and Viskin 2004) in a prospective study of 25 Brugada syndrome patients (24 men, 1 woman; 19 to 80 years of age) orally administered 1,483±240 mg quinidine bisulfate. There were 15 symptomatic patients (7 cardiac arrest survivors and 7 with unexplained syncope) and 10 asymptomatic patients. All 25 patients had inducible VF at baseline electrophysiological study. Quinidine prevented VF induction in 22 of the 25 patients (88%). After a follow-up period of 6 months to 22.2 years, all patients were alive. Of 19 patients treated with oral quinidine for 6 to 219 months (56±67 months), none developed arrhythmic events. Administration of quinidine was associated with a 36% incidence of side effects, principally diarrhea, that resolved after drug discontinuation. The authors concluded that quinidine effectively suppresses VF induction as well as spon-

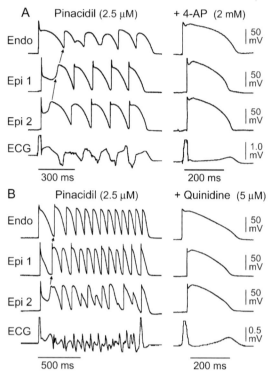

Fig. 6a,b Effects of I_{to} blockers 4-AP and quinidine on pinacidil-induced phase 2 reentry and VT in the arterially perfused RV wedge preparation. In both examples, 2.5 mmol/l pinacidil produced heterogeneous loss of AP dome in epicardium, resulting in ST segment elevation, phase 2 reentry, and VT (*left*); 4-AP (**a**) and quinidine (**b**) restored epicardial AP dome, reduced both transmural and epicardial dispersion of repolarization, normalized the ST segment, and prevented phase 2 reentry and VT in continued presence of pinacidil. (From Yan and Antzelevitch 1999, with permission)

taneous arrhythmias in patients with Brugada syndrome and may be useful as an adjunct to ICD therapy or as an alternative to ICD in cases in which an ICD is refused, unaffordable, or not feasible for any reason. These results are consistent with those reported the same group in prior years (Belhassen et al. 1999, 2002) and more recently by other investigators (Hermida et al. 2004; Mok et al. 2004). The data highlight the need for randomized clinical trials to assess the effectiveness of quinidine, preferably in patients with frequent events who have already received an ICD.

The development of a more cardioselective and I_{to}-specific blocker would be a most welcome addition to the limited therapeutic armamentarium currently available to combat this disease. Another agent being considered for this purpose is the drug tedisamil, currently being evaluated for the treatment of atrial fibrillation. Tedisamil may be more potent than quinidine because it lacks the

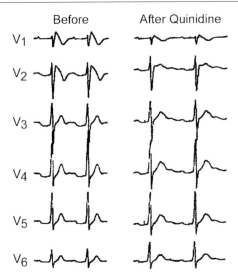

Fig. 7 Precordial leads recorded from a Brugada syndrome patient before and after quinidine (1,500 mg/day). (Modified from Alings et al. 2001, with permission)

inward current blocking actions of quinidine, while potently blocking I_{to}. The effectiveness of tedisamil to suppress phase 2 reentry and VT in a wedge model of the Brugada syndrome is illustrated in Fig. 8 (Fish et al. 2004b).

Quinidine and tedisamil can suppress the substrate and trigger for the Brugada syndrome due to inhibition of I_{to}. Both, however, have the potential to induce an acquired form of the long QT syndrome, secondary to inhibition of the rapidly activating delayed rectifier current, I_{Kr}. Thus, the drugs may substitute one form polymorphic VT for another, particularly under conditions that promote TdP, such as bradycardia and hypokalemia. This effect of quinidine is minimized at high plasma levels because, at these concentrations, quinidine block of I_{Na} counters the effect of I_{Kr} block to increase transmural dispersion of repolarization, the substrate for the development of TdP arrhythmias (Antzelevitch et al. 1999b; Antzelevitch and Shimizu 2002; Belardinelli et al. 2003). Relatively high doses of quinidine (1,000–1,500 mg/day) are recommended in order to effect I_{to} block, but prevent TdP.

Another potential candidate is an agent recently reported to be a relatively selective I_{to} and I_{Kur} blocker, AVE0118 (Fish et al. 2004a). Figure 9 shows the effect of AVE0118 to normalize the ECG and suppress phase 2 reentry in a wedge model of the Brugada syndrome. This drug has the advantage that it does not block I_{Kr}, and therefore does not prolong the QT-interval or have the potential to induce TdP. The disadvantage of this particular drug is that it undergoes first-pass hepatic metabolism and is therefore not effective with oral administration.

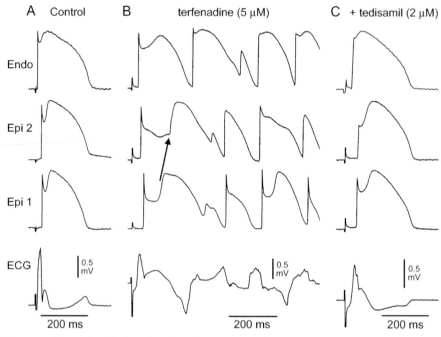

Fig. 8a–c Effects of I_{to} block with tedisamil to suppress phase 2 reentry induced by terfenadine in an arterially perfused canine RV wedge preparation. **a** Control, BCL 800 ms. **b** Terfenadine (5 µM) induces ST segment elevation as a result of heterogeneous loss of the epicardial action potential dome, leading to phase 2 reentry, which triggers an episode of poly VT (BCL = 800 ms). **c** Addition of tedisamil (2 µM) normalizes the ST segment and prevents loss of the epicardial action potential dome and suppresses phase 2 reentry induced polymorphic VT (BCL = 800 ms)

Appropriate clinical trials are needed to establish the effectiveness of all of the above pharmacologic agents as well as the possible role of pacemakers.

Agents that boost the calcium current, such as β-adrenergic agents like isoproterenol, are useful as well (Antzelevitch 2001a; Yan and Antzelevitch 1999; Tsuchiya et al. 2002). Isoproterenol, sometimes in combination with quinidine, has been shown to be effective in normalizing ST segment elevation in patients with the Brugada syndrome and in controlling electrical storms, particularly in children (Alings et al. 2001; Shimizu et al. 2000b; Suzuki et al. 2000; Tanaka et al. 2001; Belhassen et al. 2002; Mok et al. 2004).

A recent addition to the pharmacological armamentarium is the phosphodiesterase III inhibitor, cilostazol (Tsuchiya et al. 2002), which normalizes the ST segment, most likely by augmenting calcium current (I_{Ca}) as well as by reducing I_{to} secondary to an increase in heart rate.

Finally, another potential pharmacologic approach to therapy is to augment I_{Na} active during phase 1 of the epicardial action potential. This theoretical

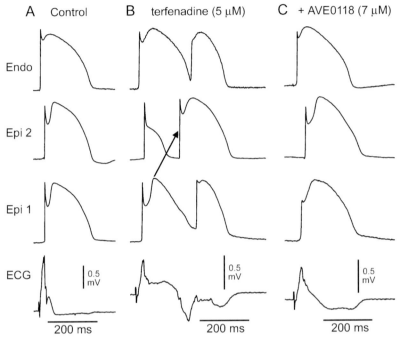

Fig. 9a–c Effects of I_{to} blockade with AVE0118 to suppress phase 2 reentry induced by terfenadine in an arterially perfused canine RV wedge preparation. **a** Control, BCL 800 ms. **b** Terfenadine (5 μM) induces ST segment elevation as a result of heterogeneous loss of the epicardial action potential dome, leading to phase 2 reentry, which triggers a closely coupled extrasystole (BCL = 800 ms). **c** Addition of AVE0118 (7 μM) prevents loss of the epicardial action potential dome and phase 2 reentry-induced arrhythmias (BCL = 800 ms)

approach will oppose I_{to} and should prevent the development of both the substrate (transmural dispersion of repolarization) and trigger (phase 2 reentry) for the Brugada syndrome.

Acknowledgements Supported by grants HL47678 NHLBI (CA) and grants from the American Heart Association (JF and CA) and NYS and Florida Grand Lodges Free and Accepted Masons.

References

Alings M, Wilde A (1999) "Brugada" syndrome: clinical data and suggested pathophysiological mechanism. Circulation 99:666–673

Alings M, Dekker L, Sadee A, Wilde A (2001) Quinidine induced electrocardiographic normalization in two patients with Brugada syndrome. Pacing Clin Electrophysiol 24:1420–1422

Antzelevitch C (2001a) The Brugada syndrome: ionic basis and arrhythmia mechanisms. J Cardiovasc Electrophysiol 12:268–272

Antzelevitch C (2001b) The Brugada syndrome. Diagnostic criteria and cellular mechanisms. Eur Heart J 22:356–363

Antzelevitch C, Brugada R (2002) Fever and the Brugada syndrome. Pacing Clin Electrophysiol 25:1537–1539

Antzelevitch C, Shimizu W (2002) Cellular mechanisms underlying the long QT syndrome. Curr Opin Cardiol 17:43–51

Antzelevitch C, Sicouri S, Litovsky SH, Lukas A, Krishnan SC, Di Diego JM, Gintant GA, Liu DW (1991) Heterogeneity within the ventricular wall: electrophysiology and pharmacology of epicardial, endocardial and M cells. Circ Res 69:1427–1449

Antzelevitch C, Brugada P, Brugada J, Brugada R, Nademanee K, Towbin JA (1999a) The Brugada syndrome. In: Camm AJ (ed) Clinical approaches to tachyarrhythmias. Futura Publishing Company, Armonk, pp 1–99

Antzelevitch C, Shimizu W, Yan GX, Sicouri S, Weissenburger J, Nesterenko VV, Burashnikov A, Di Diego JM, Saffitz JE, Thomas GP (1999b) The M cell: Its contribution to the ECG and to normal and abnormal electrical function of the heart. J Cardiovasc Electrophysiol 10:1124–1152

Antzelevitch C, Brugada P, Brugada J, Brugada R, Shimizu W, Gussak I, Perez Riera AR (2002) Brugada syndrome: a decade of progress. Circ Res 91:1114–1118

Antzelevitch C, Brugada P, Borggrefe M, Brugada J, Brugada R, Corrado D, Gussak I, LeMarec H, Nademanee K, Riera ARP, Tan H, Shimizu W, Schulze-Bahr E, Wilde A (2005) Brugada syndrome. Report of the Second Consensus Conference. Circulation 111:659–670

Araki T, Konno T, Itoh H, Ino H, Shimizu M (2003) Brugada syndrome with ventricular tachycardia and fibrillation related to hypokalemia. Circ J 67:93–95

Ayerza MR, de Zutter M, Goethals M, Wellens F, Geelen P, Brugada P (2002) Heart transplantation as last resort against Brugada syndrome. J Cardiovasc Electrophysiol 13:943–944

Babaliaros VC, Hurst JW (2002) Tricyclic antidepressants and the Brugada syndrome: an example of Brugada waves appearing after the administration of desipramine. Clin Cardiol 25:395–398

Balser JR (2001) The cardiac sodium channel: gating function and molecular pharmacology. J Mol Cell Cardiol 33:599–613

Belardinelli L, Antzelevitch C, Vos MA (2003) Assessing predictors of drug-induced torsade de pointes. Trends Pharmacol Sci 24:619–625

Belhassen B, Viskin S (2004) Pharmacologic approach to therapy of Brugada syndrome: quinidine as an alternative to ICD therapy? In: Antzelevitch C, Brugada P, Brugada J, Brugada R (eds) The Brugada syndrome: from bench to bedside. Blackwell Futura, Oxford, pp 202–211

Belhassen B, Viskin S, Fish R, Glick A, Setbon I, Eldar M (1999) Effects of electrophysiologic-guided therapy with Class IA antiarrhythmic drugs on the long-term outcome of patients with idiopathic ventricular fibrillation with or without the Brugada syndrome [see comments]. J Cardiovasc Electrophysiol 10:1301–1312

Belhassen B, Viskin S, Antzelevitch C (2002) The Brugada syndrome: is ICD the only therapeutic option? Pacing Clin Electrophysiol 25:1634–1640

Bezzina C, Veldkamp MW, van Den Berg MP, Postma AV, Rook MB, Viersma JW, Van Langen IM, Tan-Sindhunata G, Bink-Boelkens MT, Der Hout AH, Mannens MM, Wilde AA (1999) A single Na(+) channel mutation causing both long-QT and Brugada syndromes. Circ Res 85:1206–1213

Bolognesi R, Tsialtas D, Vasini P, Conti M, Manca C (1997) Abnormal ventricular repolarization mimicking myocardial infarction after heterocyclic antidepressant overdose. Am J Cardiol 79:242–245

Bordachar P, Reuter S, Garrigue S, Cai X, Hocini M, Jais P, Haissaguerre M, Clementy J (2004) Incidence, clinical implications and prognosis of atrial arrhythmias in Brugada syndrome. Eur Heart J 25:879–884

Brugada J, Brugada R, Brugada P (1998) Right bundle-branch block and ST-segment elevation in leads V1 through V3. A marker for sudden death in patients without demonstrable structural heart disease. Circulation 97:457–460

Brugada J, Brugada R, Brugada P (2000a) Pharmacological and device approach to therapy of inherited cardiac diseases associated with cardiac arrhythmias and sudden death. J Electrocardiol 33 Suppl:41–47

Brugada P, Brugada R, Brugada J, Geelen P (1999) Use of the prophylactic implantable cardioverter defibrillator for patients with normal hearts. Am J Cardiol 83:98D–100D

Brugada P, Brugada J, Brugada R (2000b) Arrhythmia induction by antiarrhythmic drugs. Pacing Clin Electrophysiol 23:291–292

Brugada P, Brugada R, Antzelevitch C, Nademanee K, Towbin J, Brugada J (2003) The Brugada syndrome. In: Gussak I, Antzelevitch C (eds) Cardiac repolarization. bridging basic and clinical sciences. Humana Press, Totowa, pp 427–446

Brugada P, Bartholomay E, Mont L, Brugada R, Brugada J (2004) Treatment of Brugada syndrome with an implantable cardioverter defibrillator. In: Antzelevitch C, Brugada P, Brugada J, Brugada R (eds) The Brugada syndrome: from bench to bedside. Blackwell Futura, Oxford, pp 194–201

Brugada R, Brugada J, Antzelevitch C, Kirsch GE, Potenza D, Towbin JA, Brugada P (2000c) Sodium channel blockers identify risk for sudden death in patients with ST-segment elevation and right bundle branch block but structurally normal hearts. Circulation 101:510–515

Chen Q, Kirsch GE, Zhang D, Brugada R, Brugada J, Brugada P, Potenza D, Moya A, Borggrefe M, Breithardt G, Ortiz-Lopez R, Wang Z, Antzelevitch C, O'Brien RE, Schultze-Bahr E, Keating MT, Towbin JA, Wang Q (1998) Genetic basis and molecular mechanisms for idiopathic ventricular fibrillation. Nature 392:293–296

Chinushi M, Aizawa Y, Ogawa Y, Shiba M, Takahashi K (1997) Discrepant drug action of disopyramide on ECG abnormalities and induction of ventricular arrhythmias in a patient with Brugada syndrome. J Electrocardiol 30:133–136

Di Diego JM, Cordeiro JM, Goodrow RJ, Fish JM, Zygmunt AC, Perez GJ, Scornik FS, Antzelevitch C (2002) Ionic and cellular basis for the predominance of the Brugada syndrome phenotype in males. Circulation 106:2004–2011

Dumaine R, Towbin JA, Brugada P, Vatta M, Nesterenko VV, Nesterenko DV, Brugada J, Brugada R, Antzelevitch C (1999) Ionic mechanisms responsible for the electrocardiographic phenotype of the Brugada syndrome are temperature dependent. Circ Res 85:803–809

Dzielinska Z, Bilinska ZT, Szumowski L, Grzybowski J, Michalak E, Przybylski A, Lubiszewska B, Walczak F, Ruzyllo W (2004) [Recurrent ventricular fibrillation during a febrile illness as the first manifestation of Brugada syndrome—a case report]. Kardiol Pol 61:269–273

Eckardt L, Kirchhof P, Johna R, Haverkamp W, Breithardt G, Borggrefe M (2001) Wolff-Parkinson-White syndrome associated with Brugada syndrome. Pacing Clin Electrophysiol 24:1423–1424

Fish JM, Antzelevitch C (2004) Role of sodium and calcium channel block in unmasking the Brugada syndrome. Heart Rhythm 1:210–217

Fish JM, Extramiana F, Antzelevitch C (2004a) AVE0118, an Ito and IKur blocker, suppresses VT/VF in an experimental model of the Brugada syndrome. Circulation 110:III-193 (abstr)

Fish JM, Extramiana F, Antzelevitch C (2004b) Tedisamil abolishes the arrhythmogenic substrate responsible for VT/VF in an experimental model of the Brugada syndrome. Heart Rhythm 1:S158 (abstr)

Fujiki A, Usui M, Nagasawa H, Mizumaki K, Hayashi H, Inoue H (1999) ST segment elevation in the right precordial leads induced with class IC antiarrhythmic drugs: insight into the mechanism of Brugada syndrome [see comments]. J Cardiovasc Electrophysiol 10:214–218

Gasparini M, Priori SG, Mantica M, Napolitano C, Galimberti P, Ceriotti C, Simonini S (2003) Flecainide test in Brugada syndrome: a reproducible but risky tool. Pacing Clin Electrophysiol 26:338–341

Goldgran-Toledano D, Sideris G, Kevorkian JP (2002) Overdose of cyclic antidepressants and the Brugada syndrome. N Engl J Med 346:1591–1592

Gonzalez Rebollo G, Madrid H, Carcia A, Garcia de Casto A, Moro AM (2000) Recurrent ventricular fibrillation during a febrile illness in a patient with the Brugada syndrome. Rev Esp Cardiol 53:755–757

Gussak I, Antzelevitch C, Bjerregaard P, Towbin JA, Chaitman BR (1999) The Brugada syndrome: clinical, electrophysiological and genetic aspects. J Am Coll Cardiol 33:5–15

Haissaguerre M, Extramiana F, Hocini M, Cauchemez B, Jais P, Cabrera JA, Farre G, Leenhardt A, Sanders P, Scavee C, Hsu LF, Weerasooriya R, Shah DC, Frank R, Maury P, Delay M, Garrigue S, Clementy J (2003) Mapping and ablation of ventricular fibrillation associated with long-QT and Brugada syndromes. Circulation 108:925–928

Hermida JS, Denjoy I, Clerc J, Extramiana F, Jarry G, Milliez P, Guicheney P, Di Fusco S, Rey JL, Cauchemez B, Leenhardt A (2004) Hydroquinidine therapy in Brugada syndrome. J Am Coll Cardiol 43:1853–1860

Hong K, Berruezo-Sanchez A, Poungvarin N, Oliva A, Vatta M, Brugada J, Brugada P, Towbin JA, Dumaine R, Pinero-Galvez C, Antzelevitch C, Brugada R (2004) Phenotypic characterization of a large European family with Brugada syndrome displaying a sudden unexpected death syndrome mutation in SCN5A. J Cardiovasc Electrophysiol 15:64–69

Horigome H, Shigeta O, Kuga K, Isobe T, Sakakibara Y, Yamaguchi I, Matsui A (2003) Ventricular fibrillation during anesthesia in association with J waves in the left precordial leads in a child with coarctation of the aorta. J Electrocardiol 36:339–343

Kalla H, Yan GX, Marinchak R (2000) Ventricular fibrillation in a patient with prominent J (Osborn) waves and ST segment elevation in the inferior electrocardiographic leads: a Brugada syndrome variant? J Cardiovasc Electrophysiol 11:95–98

Kasanuki H, Ohnishi S, Ohtuka M, Matsuda N, Nirei T, Isogai R, Shoda M, Toyoshima Y, Hosoda S (1997) Idiopathic ventricular fibrillation induced with vagal activity in patients without obvious heart disease. Circulation 95:2277–2285

Kies P, Wichter T, Schafers M, Paul M, Schafers KP, Eckardt L, Stegger L, Schulze-Bahr E, Rimoldi O, Breithardt G, Schober O, Camici PG (2004) Abnormal myocardial presynaptic norepinephrine recycling in patients with Brugada syndrome1. Circulation 110:3017–3022

Krishnan SC, Antzelevitch C (1991) Sodium channel blockade produces opposite electrophysiologic effects in canine ventricular epicardium and endocardium. Circ Res 69:277–291

Krishnan SC, Antzelevitch C (1993) Flecainide-induced arrhythmia in canine ventricular epicardium: phase 2 reentry? Circulation 87:562–572

Krishnan SC, Josephson ME (1998) ST segment elevation induced by class IC antiarrhythmic agents: underlying electrophysiologic mechanisms and insights into drug-induced proarrhythmia. J Cardiovasc Electrophysiol 9:1167–1172

Kum L, Fung JWH, Chan WWL, Chan GK, Chan YS, Sanderson JE (2002) Brugada syndrome unmasked by febrile illness. Pacing Clin Electrophysiol 25:1660–1661

Kurita T, Shimizu W, Inagaki M, Suyama K, Taguchi A, Satomi K, Aihara N, Kamakura S, Kobayashi J, Kosakai Y (2002) The electrophysiologic mechanism of ST-segment elevation in Brugada syndrome. J Am Coll Cardiol 40:330–334

Litovsky SH, Antzelevitch C (1990) Differences in the electrophysiological response of canine ventricular subendocardium and subepicardium to acetylcholine and isoproterenol. A direct effect of acetylcholine in ventricular myocardium. Circ Res 67:615–627

Littmann L, Monroe MH, Svenson RH (2000) Brugada-type electrocardiographic pattern induced by cocaine. Mayo Clin Proc 75:845–849

Lukas A, Antzelevitch C (1996) Phase 2 reentry as a mechanism of initiation of circus movement reentry in canine epicardium exposed to simulated ischemia. The antiarrhythmic effects of 4-aminopyridine. Cardiovasc Res 32:593–603

Madle A, Kratochvil Z, Polivkova A (2002) [The Brugada syndrome]. Vnitr Lek 48:255–258

Matana A, Goldner V, Stanic K, Mavric Z, Zaputovic L, Matana Z (2000) Unmasking effect of propafenone on the concealed form of the Brugada phenomenon. Pacing Clin Electrophysiol 23:416–418

Matsuo K, Shimizu W, Kurita T, Inagaki M, Aihara N, Kamakura S (1998) Dynamic changes of 12-lead electrocardiograms in a patient with Brugada syndrome. J Cardiovasc Electrophysiol 9:508–512

Miyasaka Y, Tsuji H, Yamada K, Tokunaga S, Saito D, Imuro Y, Matsumoto N, Iwasaka T (2001) Prevalence and mortality of the Brugada-type electrocardiogram in one city in Japan. J Am Coll Cardiol 38:771–774

Miyazaki T, Mitamura H, Miyoshi S, Soejima K, Aizawa Y, Ogawa S (1996) Autonomic and antiarrhythmic drug modulation of ST segment elevation in patients with Brugada syndrome. J Am Coll Cardiol 27:1061–1070

Mizumaki K, Fujiki A, Tsuneda T, Sakabe M, Nishida K, Sugao M, Inoue H (2004) Vagal activity modulates spontaneous augmentation of ST elevation in daily life of patients with Brugada syndrome. J Cardiovasc Electrophysiol 15:667–673

Mok NS, Chan NY, Chi-Suen CA (2004) Successful use of quinidine in treatment of electrical storm in Brugada syndrome. Pacing Clin Electrophysiol 27:821–823

Morita H, Kusano-Fukushima K, Nagase S, Fujimoto Y, Hisamatsu K, Fujio H, Haraoka K, Kobayashi M, Morita ST, Nakamura K, Emori T, Matsubara H, Hina K, Kita T, Fukatani M, Ohe T (2002) Atrial fibrillation and atrial vulnerability in patients with Brugada syndrome. J Am Coll Cardiol 40:1437

Morita H, Fukushima-Kusano K, Nagase S, Miyaji K, Hiramatsu S, Banba K, Nishii N, Watanabe A, Kakishita M, Takenaka-Morita S, Nakamura K, Saito H, Emori T, Ohe T (2004) Sinus node function in patients with Brugada-type ECG. Circ J 68:473–476

Nademanee K, Veerakul G, Nimmannit S, Chaowakul V, Bhuripanyo K, Likittanasombat K, Tunsanga K, Kuasirikul S, Malasit P, Tansupasawadikul S, Tatsanavivat P (1997) Arrhythmogenic marker for the sudden unexplained death syndrome in Thai men. Circulation 96:2595–2600

Nimmannit S, Malasit P, Chaovakul V, Susaengrat W, Vasuvattakul S, Nilwarangkur S (1991) Pathogenesis of sudden unexplained nocturnal death (lai tai) and endemic distal renal tubular acidosis. Lancet 338:930–932

Noda T, Shimizu W, Taguchi A, Satomi K, Suyama K, Kurita T, Aihara N, Kamakura S (2002) ST-segment elevation and ventricular fibrillation without coronary spasm by intracoronary injection of acetylcholine and/or ergonovine maleate in patients with Brugada syndrome. J Am Coll Cardiol 40:1841–1847

Nogami A, Nakao M, Kubota S, Sugiyasu A, Doi H, Yokoyama K, Yumoto K, Tamaki T, Kato K, Hosokawa N, Sagai H, Nakamura H, Nitta J, Yamauchi Y, Aonuma K (2003) Enhancement of J-ST-segment elevation by the glucose and insulin test in Brugada syndrome. Pacing Clin Electrophysiol 26:332–337

Ortega-Carnicer J, Bertos-Polo J, Gutierrez-Tirado C (2001) Aborted sudden death, transient Brugada pattern, and wide QRS dysrhythmias after massive cocaine ingestion. J Electrocardiol 34:345–349

Ortega-Carnicer J, Benezet J, Ceres F (2003) Fever-induced ST-segment elevation and T-wave alternans in a patient with Brugada syndrome. Resuscitation 57:315–317

Pastor A, Nunez A, Cantale C, Cosio FG (2001) Asymptomatic Brugada syndrome case unmasked during dimenhydrinate infusion. J Cardiovasc Electrophysiol 12:1192–1194

Pitzalis MV, Anaclerio M, Iacoviello M, Forleo C, Guida P, Troccoli R, Massari F, Mastropasqua F, Sorrentino S, Manghisi A, Rizzon P (2003) QT-interval prolongation in right precordial leads: an additional electrocardiographic hallmark of Brugada syndrome. J Am Coll Cardiol 42:1632–1637

Porres JM, Brugada J, Urbistondo V, Garcia F, Reviejo K, Marco P (2002) Fever unmasking the Brugada syndrome. Pacing Clin Electrophysiol 25:1646–1648

Potet F, Mabo P, Le Coq G, Probst V, Schott JJ, Airaud F, Guihard G, Daubert JC, Escande D, Le Marec H (2003) Novel Brugada SCN5A mutation leading to ST segment elevation in the inferior or the right precordial leads. J Cardiovasc Electrophysiol 14:200–203

Priori SG, Napolitano C, Gasparini M, Pappone C, Della BP, Brignole M, Giordano U, Giovannini T, Menozzi C, Bloise R, Crotti L, Terreni L, Schwartz PJ (2000) Clinical and genetic heterogeneity of right bundle branch block and ST-segment elevation syndrome: a prospective evaluation of 52 families. Circulation 102:2509–2515

Priori SG, Napolitano C, Gasparini M, Pappone C, Della BP, Giordano U, Bloise R, Giustetto C, De Nardis R, Grillo M, Ronchetti E, Faggiano G, Nastoli J (2002) Natural history of Brugada syndrome: insights for risk stratification and management. Circulation 105:1342–1347

Proclemer A, Facchin D, Feruglio GA, Nucifora R (1993) Recurrent ventricular fibrillation, right bundle-branch block and persistent ST segment elevation in V1–V3: a new arrhythmia syndrome? A clinical case report [see comments]. G Ital Cardiol 23:1211–1218

Rolf S, Bruns HJ, Wichter T, Kirchhof P, Ribbing M, Wasmer K, Paul M, Breithardt G, Haverkamp W, Eckardt L (2003) The ajmaline challenge in Brugada syndrome: diagnostic impact, safety, and recommended protocol. Eur Heart J 24:1104–1112

Rouleau F, Asfar P, Boulet S, Dube L, Dupuis JM, Alquier P, Victor J (2001) Transient ST segment elevation in right precordial leads induced by psychotropic drugs: relationship to the Brugada syndrome. J Cardiovasc Electrophysiol 12:61–65

Sangwatanaroj S, Prechawat S, Sunsaneewitayakul B, Sitthisook S, Tosukhowong P, Tungsanga K (2001) New electrocardiographic leads and the procainamide test for the detection of the Brugada sign in sudden unexplained death syndrome survivors and their relatives. Eur Heart J 22:2290–2296

Saura D, Garcia-Alberola A, Carrillo P, Pascual D, Martinez-Sanchez J, Valdes M (2002) Brugada-like electrocardiographic pattern induced by fever. Pacing Clin Electrophysiol 25:856–859

Shimizu W (2004) Acquired forms of Brugada syndrome. In: Antzelevitch C, Brugada P, Brugada J, Brugada R (eds) The Brugada syndrome: from bench to bedside. Blackwell Futura, Oxford, pp 166–177

Shimizu W, Antzelevitch C, Suyama K, Kurita T, Taguchi A, Aihara N, Takaki H, Sunagawa K, Kamakura S (2000a) Effect of sodium channel blockers on ST segment, QRS duration, and corrected QT interval in patients with Brugada syndrome. J Cardiovasc Electrophysiol 11:1320–1329

Shimizu W, Matsuo K, Takagi M, Tanabe Y, Aiba T, Taguchi A, Suyama K, Kurita T, Aihara N, Kamakura S (2000b) Body surface distribution and response to drugs of ST segment elevation in Brugada syndrome: clinical implication of eighty-seven-lead body surface potential mapping and its application to twelve-lead electrocardiograms. J Cardiovasc Electrophysiol 11:396–404

Shimizu W, Aiba T, Kurita T, Kamakura S (2001) Paradoxic abbreviation of repolarization in epicardium of the right ventricular outflow tract during augmentation of Brugada-type ST segment elevation. J Cardiovasc Electrophysiol 12:1418–1421

Smits JP, Eckardt L, Probst V, Bezzina CR, Schott JJ, Remme CA, Haverkamp W, Breithardt G, Escande D, Schulze-Bahr E, LeMarec H, Wilde AA (2002) Genotype-phenotype relation-ship in Brugada syndrome: electrocardiographic features differentiate SCN5A-related patients from non-SCN5A-related patients. J Am Coll Cardiol 40:350–356

Suzuki H, Torigoe K, Numata O, Yazaki S (2000) Infant case with a malignant form of Brugada syndrome. J Cardiovasc Electrophysiol 11:1277–1280

Tada H, Nogami A, Shimizu W, Naito S, Nakatsugawa M, Oshima S, Taniguchi K (2000) ST segment and T wave alternans in a patient with Brugada syndrome. Pacing Clin Electrophysiol 23:413–415

Tada H, Sticherling C, Oral H, Morady F (2001) Brugada syndrome mimicked by tricyclic antidepressant overdose. J Cardiovasc Electrophysiol 12:275

Takehara N, Makita N, Kawabe J, Sato N, Kawamura Y, Kitabatake A, Kikuchi K (2004) A cardiac sodium channel mutation identified in Brugada syndrome associated with atrial standstill. J Intern Med 255:137–142

Takenaka S, Emori T, Koyama S, Morita H, Fukushima K, Ohe T (1999) Asymptomatic form of Brugada syndrome. Pacing Clin Electrophysiol 22:1261–1263

Tan HL, Bezzina CR, Smits JP, Verkerk AO, Wilde AA (2003) Genetic control of sodium channel function. Cardiovasc Res 57:961–973

Tanaka H, Kinoshita O, Uchikawa S, Kasai H, Nakamura M, Izawa A, Yokoseki O, Kitabayashi H, Takahashi W, Yazaki Y, Watanabe N, Imamura H, Kubo K (2001) Suc-cessful prevention of recurrent ventricular fibrillation by intravenous isoproterenol in a patient with Brugada syndrome. Pacing Clin Electrophysiol 24:1293–1294

Tsuchiya T, Ashikaga K, Honda T, Arita M (2002) Prevention of ventricular fibrillation by cilostazol, an oral phosphodiesterase inhibitor, in a patient with Brugada syndrome. J Cardiovasc Electrophysiol 13:698–701

van Den Berg MP, Wilde AA, Viersma TJW, Brouwer J, Haaksma J, van der Hout AH, Stolte-Dijkstra I, Bezzina TCR, Van Langen IM, Beaufort-Krol GC, Cornel JH, Crijns HJ (2001) Possible bradycardic mode of death and successful pacemaker treatment in a large family with features of long QT syndrome type 3 and Brugada syndrome. J Cardiovasc Electrophysiol 12:630–636

Vatta M, Dumaine R, Varghese G, Richard TA, Shimizu W, Aihara N, Nademanee K, Bru-gada R, Brugada J, Veerakul G, Li H, Bowles NE, Brugada P, Antzelevitch C, Towbin JA (2002) Genetic and biophysical basis of sudden unexplained nocturnal death syndrome (SUNDS), a disease allelic to Brugada syndrome. Hum Mol Genet 11:337–345

Weiss R, Barmada MM, Nguyen T, Seibel JS, Cavlovich D, Kornblit CA, Angelilli A, Villanueva F, McNamara DM, London B (2002) Clinical and molecular heterogeneity in the Brugada syndrome. A novel gene locus on chromosome 3. Circulation 105:707–713

Wichter T, Matheja P, Eckardt L, Kies P, Schafers K, Schulze-Bahr E, Haverkamp W, Borggrefe M, Schober O, Breithardt G, Schafers M (2002) Cardiac autonomic dysfunction in Brugada syndrome. Circulation 105:702–706

Wilde AA, Antzelevitch C, Borggrefe M, Brugada J, Brugada R, Brugada P, Corrado D, Hauer RN, Kass RS, Nademanee K, Priori SG, Towbin JA (2002a) Proposed diagnostic criteria for the Brugada syndrome: consensus report. Circulation 106:2514–2519

Wilde AA, Antzelevitch C, Borggrefe M, Brugada J, Brugada R, Brugada P, Corrado D, Hauer RN, Kass RS, Nademanee K, Priori SG, Towbin JA (2002b) Proposed diagnostic criteria for the Brugada syndrome: consensus report. Eur Heart J 23:1648–1654

Yan GX, Antzelevitch C (1996) Cellular basis for the electrocardiographic J wave. Circulation 93:372–379

Yan GX, Antzelevitch C (1999) Cellular basis for the Brugada Syndrome and other mechanisms of arrhythmogenesis associated with ST segment elevation. Circulation 100:1660–1666

HEP (2006) 171:331–347
© Springer-Verlag Berlin Heidelberg 2006

Molecular Basis of Isolated Cardiac Conduction Disease

P. C. Viswanathan · J. R. Balser (✉)

Vanderbilt University Medical Center, 560 Preston Research Building, 2220 Pierce Avenue,
Nashville TN, 37232-6602, USA
jeff.balser@vanderbilt.edu

Abstract Cardiac conduction disorders are among the most common rhythm disturbances causing disability in millions of people worldwide and necessitating pacemaker implantation. Isolated cardiac conduction disease (ICCD) can affect various regions within the heart, and therefore the clinical features also vary from case to case. Typically, it is characterized by progressive alteration of cardiac conduction through the atrioventricular node, His–Purkinje system, with right or left bundle branch block and QRS widening. In some instances, the disorder may progress to complete atrioventricular block, with syncope and even death. While the role of genetic factors in conduction disease has been suggested as early as the 1970s, it was only recently that specific genetic loci have been reported. Multiple mutations in the gene encoding for the cardiac voltage-gated sodium channel (*SCN5A*), which plays a fundamental role in the initiation, propagation, and maintenance of normal cardiac rhythm, have been linked to conduction disease, allowing for genotype–phenotype correlation. The electrophysiological characterization of heterologously expressed mutant Na$^+$ channels has revealed gating defects that consistently lead to a loss of channel function. However, studies have also revealed significant overlap between aberrant rhythm phenotypes, and single mutations have been identified that evoke multiple distinct rhythm disorders with common gating lesions. These new insights highlight the complexities involved in linking single mutations, ion-channel behavior, and cardiac rhythm but suggest that interplay between multiple factors could underlie the manifestation of the disease phenotype.

Keywords Na$^+$ channel · Mutation · Channelopathies · Polymorphism ·
Structural determinants · Antiarrhythmic · Proarrhythmic · Na$_V$1.5 · SCN5A · Activation ·
Inactivation · Recovery from inactivation · Long QT syndrome · Brugada syndrome ·
Conduction disorders · Arrhythmia · conduction system

1
Introduction

The challenge in describing the role of molecular mechanisms in cardiac conduction derives partly from the many tissues and structures that assemble to form the conduction pathway. The cardiac impulse originates in the sinoatrial (SA) node and spreads rapidly over the atria, converging on the atrioventricular (AV) node (Fig. 1). From the AV node the impulse travels through the specific ventricular conduction system that includes the His bundle, the right and left bundle branch and Purkinje fibers, and finally both ventricles to produce a synchronized excitation of the myocardium. While depolarization (or activation) of the atria forms the P wave in the electrocardiogram, depolarization of the ventricles forms the QRS complex in the electrocardiogram (Fig. 1). Repolarization of the atria coincides with ventricular depolarization and is therefore masked by the QRS complex. However, ventricular repolarization produces the T wave on the electrocardiogram. The time involved for the impulse to travel from the SA node to the AV node is referred to as the AV conduction time (PR interval on the electrocardiogram).

The basic biophysical properties underlying impulse propagation are common to nerve tissue, skeletal muscle, and cardiac muscle, and derive largely from the ionic channels that underlie cellular excitation and govern conduction throughout the heart. Cardiac conduction defects are among the most com-

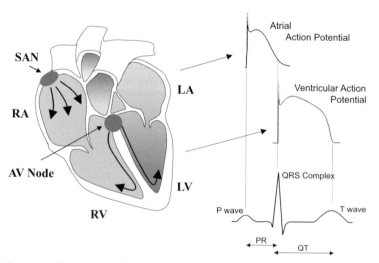

Fig. 1 Schematic of the electrical activity in the normal myocardium. Representative action potentials are shown from the atrium and ventricles with the corresponding body surface electrocardiogram. The P wave corresponds to atrial depolarization, the QRS complex corresponds to ventricular depolarization, and the T wave represents ventricular repolarization. (Reprinted with permission from Nattel 2002)

mon cardiac rhythm disturbances and are often characterized by progressive alteration of cardiac conduction through the His–Purkinje system with right or left bundle branch block and widening of the QRS complex. The disorder may progress to complete AV block, with syncope and in some cases sudden death. Figure 2 shows representative electrocardiograms of isolated cardiac conduction disease. Note the marked QRS widening and P-Q interval prolongation in panel A, while panel B illustrates a typical second-degree conduction block, but with normal QT and QRS duration. Changes in ion channel properties that govern excitability with or between cells are often invoked to explain slow or abnormal conduction of the cardiac impulse in discrete areas of the heart, as is the case during ischemia and hypoxia. As such, ischemia and hypoxia have been shown to change not only cellular excitability (Shaw and Rudy 1997) but have also been associated with changes in cell-to-cell coupling (Kleber et al. 1987).

With the advancement of molecular biology techniques, the identification and subsequent cloning of genes that encode various proteins, including pore-forming subunits of key ion channels that play a role in cardiac excitation, has progressed by leaps and bounds. Conduction disease was first genetically mapped to a group of four linked loci on chromosome 19q13.2–13.3 (Brink et al. 1995; de Meeus et al. 1995). While no gene has yet been identified in this region, this locus seems to be particularly rich in genes with known cardiac functions. For example, the proximity of this locus to one encoding myotonin

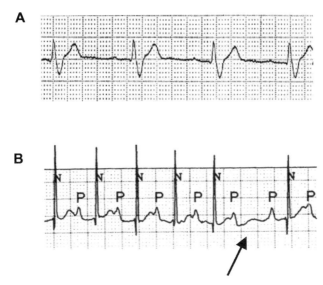

Fig. 2a,b Representative ECG traces from two patients with isolated conduction disease. Note the marked QRS widening and PQ interval prolongation in **a**, and second-degree conduction block (as indicated by the *arrow*) but normal QT and QRS durations in **b**

protein kinase (Gharehbaghi-Schnell et al. 1998), implicated in myotonic dystrophy (Phillips and Harper 1997), a disease with cardiac complications that include bundle branch blocks and intraventricular conduction disturbances, suggests a causal relationship. Subsequently, numerous studies have identified mutations in the gene encoding for cardiac voltage-gated sodium channel, SCN5A, on chromosome 3p21 (hNa$_V$1.5) (Schott et al. 1999).

In atrial and ventricular myocardium, and in the specific ventricular conduction system, the main current responsible for the initial phase of the action potential (AP) is carried by Na$^+$ ions through voltage-gated sodium channels. Therefore, Na$^+$ channels are molecular determinants of cardiac excitability and impulse propagation. Exceptions include the sinoatrial and antrioventricular nodal cells, where depolarization is a consequence of slow inward calcium currents. The cardiac sodium channel is a transmembrane protein composed of the main pore forming α-subunit (hNa$_V$1.5), and one or more subsidiary β-subunits (Catterall 2000; Balser 2001). The human β$_1$-subunit encoded by the SCN1B gene located on chromosome 19q13.1 is highly expressed in the heart, skeletal muscle, and brain. Coexpression of the α-subunit with the β$_1$-subunit recapitulates the characteristics of channels observed in vivo by modulating their gating and increasing the efficiency of their expression. Considering that Na$^+$ channels play a fundamental role in the initiation and maintenance of normal cardiac rhythm, association of inherited mutations in the Na$^+$ channel to isolated conduction diseases is not surprising. However, mutations in the SCN5A gene have also been associated with multiple life-threatening cardiac diseases ranging from tachyarrhythmias to bradyarrhythmias (Moric et al. 2003; Tan et al. 2003). The diseases include the congenital long QT syndrome (LQT3) (Wang et al. 1995), Brugada syndrome (BS) (Brugada and Brugada 1992; Alings and Wilde 1999), isolated cardiac conduction disease (ICCD) (Schott et al. 1999), sudden unexpected nocturnal death syndrome (SUNDS) (Vatta et al. 2002), and sudden infant death syndrome (SIDS) (Ackerman et al. 2001; Wedekind et al. 2001), constituting a spectrum of disease entities termed "sodium channelopathies." Although patients with SCN5A mutations linked to LQT3, BS, SUNDS, and SIDS may experience sudden, life-threatening arrhythmias, patients with isolated conduction disease exhibit heart rate slowing (bradycardia) that manifests clinically as syncope, or perhaps only as lightheadedness (Tan et al. 2001).

Electrophysiologic characterization of heterologously expressed mutant Na$^+$ channels have revealed functional defects that, in many cases, can explain the distinct phenotype associated with the rhythm disorders. However, recent studies have revealed significant overlap between aberrant rhythm phenotypes, and single mutations have been identified that evoke multiple rhythm disorders with a single lesion. These new insights enhance understanding of the structure–function relationships of Na$^+$ channels, and also highlight the complexities involved in linking single mutations, ion-channel behavior, and cardiac rhythm.

2
Sodium Channel Gating States: Linking Structure to Function

The α-subunit of the tetrodotoxin "insensitive" human cardiac Na^+ channel (hH1) is composed of four homologous domains, attached to one another by cytoplasmic linker sequences (Fig. 3a). Each domain consists of six transmembrane spanning segments connected by intracellular or extracellular sequences. Na^+ channels are dynamic molecules that undergo rapid structural rearrangements in response to changes in the transmembrane electric field, a process termed "gating." Over the last decade, site-directed mutagenesis, as well as spontaneous disease-causing mutations, have been used in

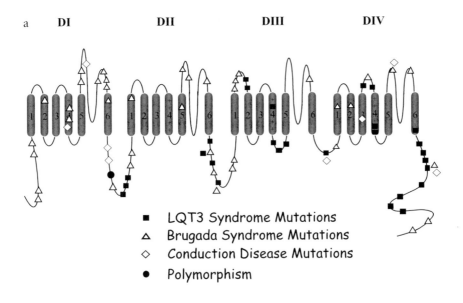

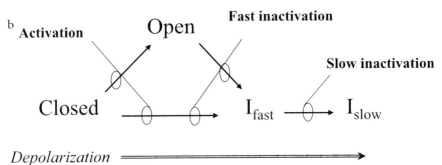

Fig. 3 a Predicted transmembrane topology of the α-subunit of the human cardiac Na^+ channel displaying locations of known disease-causing mutations. **b** Simplified scheme of the Na^+ channel conformational changes in response to membrane depolarization

concert with patch-clamp electrophysiologic measurements to define specific amino acids and structural elements involved in voltage-dependent gating function.

Upon membrane depolarization, Na^+ channels rapidly undergo conformational changes that lead to opening of the channel pore to allow Na^+ influx, a process termed "activation" (Fig. 3b). Simultaneously, depolarization triggers initiation of "fast inactivation" (I_{fast}) that terminates Na^+ influx. Inactivation differs qualitatively from channel closure in that inactivated channels do not normally open unless the membrane potential is hyperpolarized, often for a sustained period. In addition, Na^+ channels may inactivate without ever opening (so-called "closed-state" inactivation; Horn et al. 1981). With prolonged depolarizations, Na^+ channels progressively enter "slow inactivated" states (I_{slow}) with diverse lifetimes ranging from hundreds of milliseconds to many seconds (Cannon 1996; Balser 2001). Slow inactivation reduces cellular excitability, particularly in pathophysiologic conditions associated with prolonged membrane depolarization, such as epilepsy, neuromuscular diseases, or cardiac arrhythmias. It is clear that a great many single amino acid substitutions within the SCN5A coding region can evoke a broad spectrum of cardiac rhythm behavior by modulating these gating processes (Fig. 3a). At the same time, common sequence variants ("polymorphisms") in the Na^+ channel gene have also been implicated as risk factors in cardiac diseases (Viswanathan et al. 2003), as well as determinants of drug sensitivity (Splawski et al. 2002). Recent studies have shown that polymorphisms in the Na^+ channel gene can confer enhanced drug sensitivity promoting arrhythmias (Splawski et al. 2002), or even modulate the biophysical effects of disease-causing mutations (Viswanathan et al. 2003; Ye et al. 2003). Functional studies of mutations associated with cardiac diseases have provided us with a wealth of information that highlights the exquisite sensitivity of cardiac rhythm to Na^+ channel function.

Na^+ channel activation involves the concerted outward movement of all four charged S4 segments that leads to opening of the channel pore (Catterall 1988). Fast inactivation, like activation, is tied to the outward movement of the S4 sensors, but primarily those in domains III and IV. Consistent with this dual role for the S4-voltage sensor is the observation that activation and fast inactivation gating are tightly coupled and proceed almost simultaneously. Fast inactivation also critically involves the domain III–IV cytoplasmic linker, which may function as a lid that occludes the pore by binding to sites situated on or near the inner vestibule. Slow inactivation involves structural elements near the pore, particularly the P segments, external linker sequences between S5 and S6 segments in each domain that bend back into the membrane and line the outer pore. Hence, the mechanisms underlying slow inactivation in Na^+ channels might resemble slow, C-type inactivation in potassium channels. However, identification of mutations in other regions of the α-subunit of the channel, as well as site-directed mutations of externally directed residues that influence both fast and slow inactivation, suggests that both gating processes cannot be mapped to a single piece of the channel structure, but instead depend upon coupled interactions among several domains.

3
Electrophysiological Effects of Na$^+$ Channel Mutations

Progressive cardiac conduction defect (PCCD), also called Lenègre or Lev's disease, is one of the most common cardiac conduction diseases. PCCD is characterized by progressive alteration of conduction through the His–Purkinje system with right or left bundle branch block (LBBB or RBBB) and widening of the QRS complex, leading to complete AV block. Schott et al. (1999) first associated isolated conduction disease due to PCCD with mutations in *SCN5A*. Based on the sequence analysis, it was predicted that the resulting "sodium channel" would have large deleted segments, and would be entirely non-functional. Since the first report, multiple mutations have been identified in the cardiac sodium channel gene and linked to isolated conduction disease (Fig. 3a).

Electrophysiological characterization of *SCN5A* mutations in heterologous expression systems has provided us with a wealth of information that has allowed us to relate altered channel function to the observed phenotype. Functional studies of mutations associated with conduction disease have consistently revealed a loss of Na$^+$ channel function. Mutations in the gene have been found to affect key gating components that determine Na$^+$ current, such as activation, fast inactivation, closed-state inactivation, and even slow inactivation. While it would be reasonable to expect that any gating defect that causes reduced channel availability would evoke a conduction disturbance, in practice mutations often evoke multiple gating effects that, when considered together, may not reduce Na$^+$ current. For example, a positive shift of the voltage dependence of activation would increase the voltage difference between the resting membrane potential and the activation threshold, leading to reduced Na$^+$ current. At the same time, a positive shift in the voltage dependence of inactivation would tend to increase Na$^+$ channel availability, and thus increase Na$^+$ current. Therefore, a parallel shift in activation and inactivation (either positive or negative) may evoke little overall change in Na$^+$ channel activity. This was demonstrated recently when five members of a Dutch family carrying a mutation in the Na$^+$ channel I–II linker (G514C) exhibited isolated cardiac conduction disease requiring pacemaker therapy (Tan et al. 2001). Biophysical characterization of the G514C mutation revealed balanced changes in Na$^+$ channel gating function, with both activation and inactivation gating requiring stronger depolarizing membrane potentials (Fig. 4a, b). A parallel depolarizing shift in activation (loss of function) and inactivation (gain of function) might evoke little overall net change in Na$^+$ channel activity. However, in this case the activation shift predominated (by ∼3 mV), yielding a slight net decrease

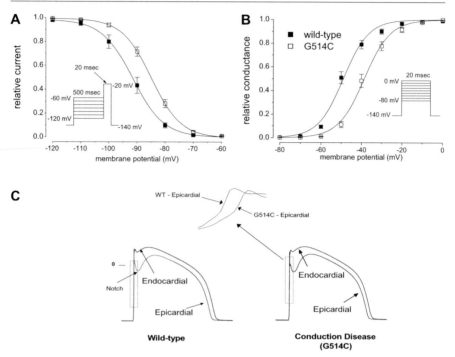

Fig. 4a–c Balanced gating effects resulting from G514C mutation. **a** The voltage dependence of channel inactivation assessed using the protocol in the *inset*. **b** The voltage-dependence of channel activation as evaluated using the protocol in the *inset*. A positive shift in inactivation would suggest an increase in the number of channels available to open at any given potential. In contrast, a positive shift in activation would suggest that channels are less likely to open at any given potential. **c** Simulated endocardial and epicardial action potentials using the Luo-Rudy cable model. Incorporation of gating defects associated with G514C into the model show a reduction in the upstroke velocity of the action potential (*inset* in panel *C*) without any change in action potential morphology. (Reprinted with permission from Tan et al. 2001)

in Na$^+$ channel function. In a computational model of cardiac conduction (Luo and Rudy 1994; Viswanathan and Rudy 1999), this loss of function was not sufficient to induce premature epicardial AP repolarization and Brugada syndrome (Fig. 4c), but did reduce AP upstroke velocity by 20%, an effect that slowed conduction and explained the observed phenotype (Table 1).

Although excessive slow inactivation leading to reduced Na$^+$ channel availability was previously associated with tachyarrhythmias and Brugada syndrome, recent studies have linked excessive slow inactivation to isolated conduction disease as well. A 2-year-old, with second-degree AV block carries a mutation, T512I, in the DI–DII cytoplasmic linker, and is also homozygous for a common polymorphism (H558R) present in the Na$^+$ channel DI–DII linker with a frequency of 20% (Yang et al. 2002). Studies showed that the

Table 1 Transverse conduction velocity (cm/s)

	Wild-type	G514C
Endocardial	22.8	20.0 (12%)
Epicardial	20.4	17.3 (15%)

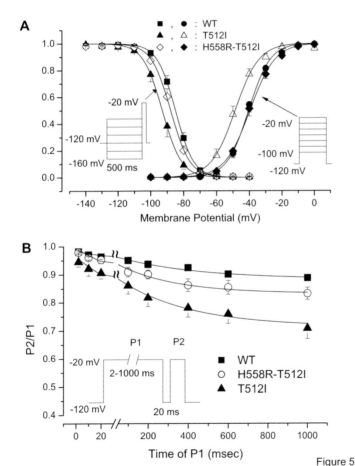

Figure 5

Fig. 5a,b Attenuation of the gating defects of a mutation (T512I) by a common polymorphism (H558R). **A** Voltage-dependence of activation and inactivation of wild-type, *T512I*, and *H558R-T512I* evaluated using the protocol shown in the *inset*. The polymorphism restores the hyperpolarizing shifts caused by the mutation. **b** Slow inactivation as evaluated using the protocol shown in the *inset*. Once again H558R attenuates the extent of slow inactivation caused by the mutation, *T512I* polymorphism alone had no effect on the channel, but had a modulatory effect on the gating lesions caused by the T512I mutation. Functional studies of the T512I mutation alone revealed shifts in the voltage-dependence of activation and inactivation and, more importantly, a significant enhancement of slow inactivation. However, in channels that carry both H558R and T512I, these defects are mitigated, although still maintain reduced function compared to normal channels (Fig. 5). In this case, it would seem that even a slight increase in slow inactivation may produce a cumulative loss of function, leading to conduction slowing.

4
Reduction in Na⁺ Current: A Common Mechanism Underlying Brugada Syndrome and Conduction Disease

Electrophysiological analysis of mutations linked to Brugada syndrome or isolated conduction disease has revealed defects that consistently lead to a reduction in Na^+ current (Veldkamp et al. 2000; Wang et al. 2000; Herfst et al. 2003; Probst et al. 2003). Whereas even a single change in gating function, considered alone, could drastically increase or decrease the Na^+ current, computational models of cardiac excitability equipped to consider the ensemble of these gating effects may predict only a mild "net" increase or decrease in Na^+ current (i.e., G514C, Fig. 4) (Tan et al. 2001). Furthermore, with the identification of several Na^+ channel mutations that alone can evoke multiple rhythm disturbances, such as 1795insD (Brugada and LQT3; Bezzina et al. 1999; Veldkamp et al. 2000), ΔK1500 (Brugada, LQT3, and conduction disease; Grant et al. 2002), it is becoming clear that the manifestation of a particular phenotype is the result of the complex interplay between gating defects, as well as other "unseen" regulatory factors.

Single nucleotide polymorphisms (SNPs), DNA sequence variations that are common in the population, have been implicated in phenotypic variability in physiology, pharmacology, and pathophysiology by altering gene function and susceptibility to disease. Studies have linked gene polymorphisms to elevated risk for cystic fibrosis (Hull and Thomson 1998), Alzheimer's disease (Roses 1998), certain forms of cancer (El-Omar et al. 2000) or even heart disease (Roses 2000). In addition to their role in disease, polymorphisms are also thought to confer sensitivity or resistance to drug therapy, as well as proarrhythmic risk from drug therapy (Splawski et al. 2002). Recently, a polymorphism in SCN5A (S1102Y) was identified in individuals with African descent and implicated in an elevated risk for proarrhythmia with drug therapy (Splawski et al. 2002). Electrophysiological and computational analyses predict negligible effects on AP properties as a result of the polymorphism. But surprisingly, the polymorphism increased AP duration and the susceptibility to the development of arrhythmogenic early afterdepolarizations in the background of reduced outward potassium current as a result of drug block. Other studies have identified a more common polymorphism, H558R, in the I–II linker of the Na$^+$ channel (20% allelic frequency; Yang et al. 2002). As discussed above, the H558R polymorphism had no effect of wild-type Na$^+$ channel current, but attenuated the gating defects caused by an intragenic mutation identified in a patient with isolated conduction defect. Another study also reported a modulatory role of this polymorphism when present in tandem with a mutation linked to the long QT syndrome (Ye et al. 2003).

Functional studies of disease-causing mutations have identified gating themes common to Brugada syndrome and isolated conduction disease. Reduction in Na$^+$ current, irrespective of the underlying mechanism, evokes BS, conduction disease, or both. Widening of the QRS and an increased P-R interval, indicative of conduction slowing, has indeed been observed in the original report describing BS (Brugada and Brugada 1992), and in other Brugada kindreds (Kyndt et al. 2001; Potet et al. 2003). However, predictable effects related to reduced Na$^+$ channel function may be more the exception than the rule. For example, a 50% reduction in Na$^+$ current in an *SCN5A+/−* mouse leads to only minor conduction abnormalities (Papadatos et al. 2002), while certain mutations evoke complete loss of Na$^+$ channel function, yet the phenotype is only mild (Smits et al. 2002; Herfst et al. 2003; Probst et al. 2003). Hence, the phenotypic predominance of BS or conduction disease cannot be predicted by the degree of reduction in Na$^+$ channel current alone. It is important to note that the electrophysiological characterization of *SCN5A* mutations associated with BS or conduction disease have been carried out in heterologous expression systems that do not necessarily recapitulate in vivo conditions. As such, while the severity of the Na$^+$ channel functional defect may parallel the ECG phenotype in some cases, other unrecognized factors are certain to play a role in vivo. Such factors may include humoral regulation, auxiliary subunits, chaperone proteins, anchoring proteins, and transcriptional regulation. A recent study identified a new BS locus, distinct from *SCN5A,* and associated with progressive conduction disease (Weiss et al. 2002). While this locus has not yet been associated with any ion channel or protein, the finding supports the notion that non-*SCN5A* gene products are likely to play a role in the manifestation of the conduction phenotype.

5
Loss of Na$^+$ Channel Function: Phenotypic Variability in Conduction Disease?

Cardiac conduction defects are characterized by progressive alteration of cardiac conduction through the His–Purkinje system with right or left bundle branch block and widening of the QRS complex. The disorder may progress to complete AV block, with syncope and in some cases sudden death. Inherited mutations in the Na$^+$ channel have been associated with isolated AV conduction slowing in some cases, and with "pan-conduction" slowing throughout the atria and ventricles, including right or left bundle branch block, in other cases. Although functional studies have consistently shown that these mutations reduce Na^+ current, in most cases it is still unclear as to how this links to distinct phenotypes. Additional studies will be required to clarify the causality between conduction block phenotypes and channel gating lesions. Heterogeneous AP properties in the myocardium could be a major factor in linking the biophysical observations and the distinct conduction phenotypes.

Nonetheless, some general patterns have emerged. Mutations identified in patients with conduction defects have revealed three distinct functional lesions: shifts in voltage-dependence of activation and inactivation, enhanced slow inactivation, and entirely non-functional channels. Mutations causing enhanced slow inactivation have generally been associated with AV conduction slowing (prolonged PR intervals), but no intraventricular or intra-atrial conduction defect (normal P wave and QRS duration). In contrast, mutations leading to shifts in voltage dependence of activation and inactivation have often been associated with slow AV conduction, as well as delayed conduction throughout the atria and ventricles—including broad P waves, P-R interval prolongation, and widening of the QRS complex. Mutations that preferentially delay recovery from inactivation (via enhancing slow inactivation) could disproportionately affect cells with longer inherent AP duration (Purkinje cells).

Hence, it is possible that excess slow inactivation, which would cause mutant Na^+ channels to recover from inactivation more slowly than normal during diastole, could slow AV conduction without altering atrial or ventricular conduction. In contrast, as in the case of G514C, mutations targeting the channel activation process would have similar effects regardless of AP duration, and may thus affect the myocardium more uniformly, as was observed. Consistent with this idea, it is noteworthy that H558R entirely eliminated the T512I effect on activation gating, but only partly corrected the slow inactivation defect. As such, greater accumulation of Na^+ channel slow inactivation upon successive stimuli in Purkinje cells, with their longer AP duration and smaller consequent diastolic interval, could lead to proportionally greater loss of Na^+ channel function in these cells and thereby produce the isolated AV conduction delay observed. Moreover, a premature stimulus could also further compromise the Purkinje diastolic interval and lead to a dramatic loss of Na^+ current, and even result in early repolarization and conduction block.

This is illustrated in Fig. 6, which shows superimposed APs computed using the Luo-Rudy mathematical AP model. Panel a shows APs computed from a ventricular cell (in this case an epicardial cell) during three conditions: control (thin line), enhanced slow inactivation (thick line), and a positive shift in the voltage-dependence of activation (dotted line). Panel b shows APs computed from a Purkinje cell during the same three conditions. To simulate the longer AP characteristic of the Purkinje cell, the expression of the outward delayed rectifier potassium currents and the inward calcium currents was reduced.

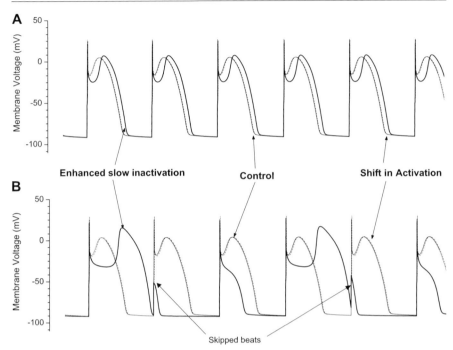

Fig. 6a,b Simulated action potentials obtained using the Luo-Rudy model of action potential. **a** Train of epicardial cell action potentials computed during three conditions (control, positive shift in activation, and enhanced slow inactivation). All three conditions had minimal effect on the shape or duration of the action potential. **b** Train of Purkinje cell action potentials during the three conditions. While activation shift had minimal effect (APs overlap with control), enhanced slow inactivation resulted in abnormal AP prolongation and skipped beats

While enhanced slow inactivation prolonged the action potential duration (APD) modestly, the other two interventions did not have any effect on the ventricular cell AP (thin line and dotted line are almost indistinguishable in panel a). In contrast, Purkinje cell APs became irregular during conditions of enhanced slow inactivation (thick line in panel b), occasionally exhibiting complete loss of the AP dome (skipped beats as indicated by the arrow), but not during a positive shift in activation (dotted trace is again indistinguishable from control in panel b).

6
Therapeutic Intervention: Pharmacologic Versus Implantable Devices

Although major advances have been made in the prevention, diagnosis, and treatment of cardiovascular diseases, 62 million people in the United States currently have these diseases. Furthermore, cardiac conduction disorders cause disability in millions of people worldwide. However, major advances have been made in device-based therapy of cardiac rhythm and conduction (Schron and Domanski 2003). In fact, progressive cardiac conduction disease represents the major worldwide cause for pacemaker implantation. Pacemakers and implantable cardioverter–defibrillators continue to be the mainstay therapy for not only a variety of primary conduction disorders, but also to decrease mortality and morbidity related to secondary conduction disturbances and arrhythmias in patients with coronary heart disease.

Since a recurrent theme in sodium channel-linked ICCD is a loss of channel function, it maybe logical to expect that sodium channel "openers" might have a therapeutic benefit by accelerating conduction throughout the myocardium. Based on the examples presented, it would also be logical to expect drugs that target voltage dependence of activation and/or inactivation might have a beneficial effect on sodium channel availability. In this regard, in vitro studies have shown that glucocorticoid steroids can reverse some of the gating defects caused by G514C in vitro, while analysis of ECGs of G514C carriers indicate that steroid therapy using prednisolone improves not only atrial, but also ventricular conduction (Tan et al. 2001). Whether steroids target either directly or indirectly, particular features of channel function require further study.

Conduction of impulses in the myocardium is also dependent on cable properties, that is, the degree of "connectivity" between cardiac myocytes and other conduction elements, such as the Purkinje fibers. Electrical cell-to-cell coupling through gap junctions is an important determinant of AP propagation, and alterations in gap-junction properties can significantly modulate conduction velocity in the myocardium (Saffitz et al. 1995). Recent studies have shown significant heterogeneities in the distribution of connexin 43, the principal ventricular gap-junction protein, leading to reduced conduction velocities and altered excitability due to changes in the upstroke velocity of the AP (Poelzing et al. 2004). Studies have also identified SCN5A mutations co-segregating with a rare connexin 40 genotype in familial atrial standstill, thereby providing yet another locus for conduction disease mutations (Groenewegen et al. 2003). While these studies clarify the importance of screening gap-junction proteins for variants in ICCD, they also suggest that interventions that might enhance coupling between cells could potentially mitigate conduction slowing in these cases. In this regard, studies have shown that pharmacological modulation of gap junctions can enhance cardiac conduction and diminish heterogeneous AP repolarization in experimental models of slow conduction (Eloff et al. 2003). These studies highlight the possibility of providing a genotype-specific rationale for particular therapies in patients with altered conduction.

References

Ackerman MJ, Siu BL, Sturner WQ, Tester DJ, Valdivia CR, Makielski JC, Towbin JA (2001) Postmortem molecular analysis of SCN5A defects in sudden infant death syndrome. JAMA 286:2264–2269

Alings M, Wilde A (1999) "Brugada" syndrome: clinical data and suggested pathophysiological mechanism. Circulation 99:666–673

Balser JR (2001) The cardiac sodium channel: gating function and molecular pharmacology. J Mol Cell Cardiol 33:599–613

Bezzina CR, Veldkamp MW, van der Berg MP, Postma AV, Rook MB, Viersma JW, van Langen IM, Tan-Sindhunata G, Bink-Boelkens MTE, van der Hout AH, Mannens MM, Wilde AA (1999) A single sodium channel mutation causing both long QT and Brugada syndrome. Circulation 100:I494

Brink PA, Ferreira A, Moolman JC, Weymar HW, van der Merwe PL, Corfield VA (1995) Gene for progressive familial heart block type I maps to chromosome 19q13. Circulation 91:1633–1640

Brugada P, Brugada J (1992) Right bundle branch block, persistent ST segment elevation and sudden cardiac death: a distinct clinical and electrocardiographic syndrome. A multicenter report. J Am Coll Cardiol 20:1391–1396

Cannon SC (1996) Slow inactivation of sodium channels: more than just a laboratory curiosity. Biophys J 71:5–7

Catterall WA (1988) Structure and function of voltage-sensitive ion channels. Science 242:50–61

Catterall WA (2000) From ionic currents to molecular mechanisms: the structure and function of voltage-gated sodium channels. Neuron 26:13–25

de Meeus A, Stephan E, Debrus S, Jean MK, Loiselet J, Weissenbach J, Demaille J, Bouvagnet P (1995) An isolated cardiac conduction disease maps to chromosome 19q. Circ Res 77:735–740

El-Omar EM, Carrington M, Chow WH, McColl KE, Bream JH, Young HA, Herrera J, Lissowska J, Yuan CC, Rothman N, Lanyon G, Martin M, Fraumeni JF Jr, Rabkin CS (2000) Interleukin-1 polymorphisms associated with increased risk of gastric cancer. Nature 404:398–402

Eloff BC, Gilat E, Wan X, Rosenbaum DS (2003) Pharmacological modulation of cardiac gap junctions to enhance cardiac conduction: evidence supporting a novel target for antiarrhythmic therapy. Circulation 108:3157–3163

Gharehbaghi-Schnell EB, Finsterer J, Korschineck I, Mamoli B, Binder BR (1998) Genotype-phenotype correlation in myotonic dystrophy. Clin Genet 53:20–26

Grant AO, Carboni MP, Neplioueva V, Starmer CF, Memmi M, Napolitano C, Priori S (2002) Long QT syndrome, Brugada syndrome, and conduction disease are linked to a single sodium channel mutation. J Clin Invest 110:1201–1209

Groenewegen WA, Firouzi M, Bezzina CR, Vliex S, van Langen IM, Sandkuijl L, Smits JP, Hulsbeek M, Rook MB, Jongsma HJ, Wilde AA (2003) A cardiac sodium channel mutation cosegregates with a rare connexin40 genotype in familial atrial standstill. Circ Res 92:14–22

Herfst LJ, Potet F, Bezzina CR, Groenewegen WA, Le Marec H, Hoorntje TM, Demolombe S, Baro I, Escande D, Jongsma HJ, Wilde AA, Rook MB (2003) Na+ channel mutation leading to loss of function and non-progressive cardiac conduction defects. J Mol Cell Cardiol 35:549–557

Horn R, Patlak JB, Stevens CF (1981) Sodium channels need not open before they inactivate. Nature 291:426–427

Hull J, Thomson AH (1998) Contribution of genetic factors other than CFTR to disease severity in cystic fibrosis. Thorax 53:1018–1021

Kleber AG, Riegger CB, Janse MJ (1987) Electrical uncoupling and increase of extracellular resistance after induction of ischemia in isolated, arterially perfused rabbit papillary muscle. Circ Res 61:271–279

Kyndt F, Probst V, Potet F, Demolombe S, Chevallier JC, Baro I, Moisan JP, Boisseau P, Schott JJ, Escande D, Le Marec H (2001) Novel SCN5A mutation leading either to isolated cardiac conduction defect or Brugada syndrome in a large French family. Circulation 104:3081–3086

Luo CH, Rudy Y (1994) A dynamic model of the cardiac ventricular action potential. I Simulations of ionic currents and concentration changes. Circ Res 74:1071–1096

Moric E, Herbert E, Trusz-Gluza M, Filipecki A, Mazurek U, Wilczok T (2003) The implications of genetic mutations in the sodium channel gene (SCN5A). Europace 5:325–334

Nattel S (2002) New ideas about atrial fibrillation 50 years on. Nature 415:219–220

Papadatos GA, Wallerstein PM, Head CE, Ratcliff R, Brady PA, Benndorf K, Saumarez RC, Trezise AE, Huang CL, Vandenberg JI, Colledge WH, Grace AA (2002) Slowed conduction and ventricular tachycardia after targeted disruption of the cardiac sodium channel gene SCN5A. Proc Natl Acad Sci U S A 99:6210–6215

Phillips MF, Harper PS (1997) Cardiac disease in myotonic dystrophy. Cardiovasc Res 33:13–22

Poelzing S, Akar FG, Baron E, Rosenbaum DS (2004) Heterogeneous connexin43 expression produces electrophysiological heterogeneities across ventricular wall. Am J Physiol Heart Circ Physiol 286:H2001–2009

Potet F, Mabo P, Le Coq G, Probst V, Schott JJ, Aireau F, Guihard G, Daubert JC, Escande D, Le Marec H (2003) Novel Brugada SCN5A mutation leading to ST segment elevation in the inferior or the right precordial leads. J Cardiovasc Electrophysiol 14:1–4

Probst V, Kyndt F, Potet F, Trochu JN, Mialet G, Demolombe S, Schott JJ, Baro I, Escande D, Le Marec H (2003) Haploinsufficiency in combination with aging causes SCN5A-linked hereditary Lenegre disease. J Am Coll Cardiol 41:643–652

Roses AD (1998) Apolipoprotein E and Alzheimer's disease. The tip of the susceptibility iceberg. Ann N Y Acad Sci 855:738–743

Roses AD (2000) Genetic susceptibility to cardiovascular diseases. Am Heart J 140:S45–47

Saffitz JE, Davis LM, Darrow BJ, Kanter HL, Laing JG, Beyer EC (1995) The molecular basis of anisotropy: role of gap junctions. J Cardiovasc Electrophysiol 6:498–510

Schott JJ, Alshinawi C, Kyndt F, Probst V, Hoorntje TM, Hulsbeek M, Wilde AA, Escande D, Mannens MM, Le Marec H (1999) Cardiac conduction defects associate with mutations in SCN5A. Nat Genet 23:20–21

Schron EB, Domanski MJ (2003) Implantable devices benefit patients with cardiovascular diseases. J Cardiovasc Nurs 18:337–342

Shaw RM, Rudy Y (1997) Electrophysiologic effects of acute myocardial ischemia: a theoretical study of altered cell excitability and action potential duration [see comment]. Cardiovasc Res 35:256–272

Smits JP, Eckardt L, Probst V, Bezzina CR, Schott JJ, Remme CA, Haverkamp W, Breithardt G, Escande D, Schulze-Bahr E, LeMarec H, Wilde AA (2002) Genotype-phenotype relationship in Brugada syndrome: electrocardiographic features differentiate SCN5A-related patients from non-SCN5A-related patients. J Am Coll Cardiol 40:350–356

Splawski I, Timothy KW, Tateyama M, Clancy CE, Malhotra A, Beggs AH, Cappuccio FP, Sagnella GA, Kass RS, Keating MT (2002) Variant of SCN5A sodium channel implicated in risk of cardiac arrhythmia. Science 297:1333–1336

Tan HL, Bink-Boelkens MT, Bezzina CR, Viswanathan PC, Beaufort-Krol GC, van Tintelen PJ, van den Berg MP, Wilde AA, Balser JR (2001) A sodium-channel mutation causes isolated cardiac conduction disease. Nature 409:1043–1047

Tan HL, Bezzina CR, Smits JP, Verkerk AO, Wilde AA (2003) Genetic control of sodium channel function. Cardiovasc Res 57:961–973

Vatta M, Dumaine R, Varghese G, Richard TA, Shimizu W, Aihara N, Nademanee K, Brugada R, Brugada J, Veerakul G, Li H, Bowles NE, Brugada P, Antzelevitch C, Towbin JA (2002) Genetic and biophysical basis of sudden unexplained nocturnal death syndrome (SUNDS), a disease allelic to Brugada syndrome. Hum Mol Genet 11:337–345

Veldkamp MW, Viswanathan PC, Bezzina C, Baartscheer A, Wilde AA, Balser JR (2000) Two distinct congenital arrhythmias evoked by a multidysfunctional Na(+) channel. Circ Res 86:E91–E97

Viswanathan PC, Rudy Y (1999) Cellular arrhythmogenic effects of the long QT syndrome in the heterogeneous myocardium. Circulation 101:1192–1198

Viswanathan PC, Benson DW, Balser JR (2003) A common SCN5A polymorphism modulates the biophysical effects of an SCN5A mutation. J Clin Invest 111:341–346

Wang DW, Makita N, Kitabatake A, Balser JR, George AL Jr (2000) Enhanced Na(+) channel intermediate inactivation in Brugada syndrome. Circ Res 87:E37–E43

Wang Q, Shen J, Li Z, Timothy K, Vincent GM, Priori SG, Schwartz PJ, Keating MT (1995) Cardiac sodium channel mutations in patients with long QT syndrome, an inherited cardiac arrhythmia. Hum Mol Genet 4:1603–1607

Wedekind H, Smits JP, Schulze-Bahr E, Arnold R, Veldkamp MW, Bajanowski T, Borggrefe M, Brinkmann B, Warnecke I, Funke H, Bhuiyan ZA, Wilde AA, Breithardt G, Haverkamp W (2001) De novo mutation in the SCN5A gene associated with early onset of sudden infant death [see comment]. Circulation 104:1158–1164

Weiss R, Barmada MM, Nguyen T, Seibel JS, Cavlovich D, Kornblit CA, Angelilli A, Villanueva F, McNamara DM, London B (2002) Clinical and molecular heterogeneity in the Brugada syndrome: a novel gene locus on chromosome 3. Circulation 105:707–713

Yang P, Kanki H, Drolet B, Yang T, Wei J, Viswanathan PC, Hohnloser S, Shimizu W, Schwartz PJ, Stanton M, Murray KT, Norris K, George AL Jr, Roden DM (2002) Allelic variants in long QT disease genes in patients with drug-associated torsade de pointes. Circulation 105:1943–1948

Ye B, Valdivia CR, Ackerman MJ, Makielski JC (2003) A common human SCN5A polymorphism modifies expression of an arrhythmia causing mutation. Physiol Genomics 12:187–193

HEP (2006) 171:349–355
© Springer-Verlag Berlin Heidelberg 2006

hERG Trafficking and Pharmacological Rescue of LQTS-2 Mutant Channels

G. A. Robertson[1] (✉) · C. T. January[2]

[1]Dept. of Physiology, University of Wisconsin-Madison, 601 Science Drive,
Madison WI, 53711, USA
robertson@physiology.wisc.edu

[2]Medicine (Cardiovascular), H6/354 CSC, University of Wisconsin Medical School,
600 Highland Avenue, 53792-1618 WI, Madison, USA

Abstract The *human ether-a-go-go-related gene* (*hERG*) encodes an ion channel subunit underlying I_{Kr}, a potassium current required for the normal repolarization of ventricular cells in the human heart. Mutations in *hERG* cause long QT syndrome (LQTS) by disrupting I_{Kr}, increasing cardiac excitability and, in some cases, triggering catastrophic torsades de pointes arrhythmias and sudden death. More than 200 putative disease-causing mutations in *hERG* have been identified in affected families to date, but the mechanisms by which these mutations cause disease are not well understood. Of the mutations studied, most disrupt protein maturation and reduce the numbers of hERG channels at the membrane. Some trafficking-defective mutants can be rescued by pharmacological agents or temperature. Here we review evidence for rescue of mutant hERG subunits expressed in heterologous systems and discuss the potential for therapeutic approaches to correcting I_{Kr} defects associated with LQTS.

Keywords K^+ channel · hERG · LQTS · LQT-2 · Mutation · Channelopathies ·
Antiarrhythmic · Proarrhythmic · I_{kr} · Trafficking defects · Glycosylation ·
Torsades de pointes · RXR · Golgi · Golgi-resident protein GM130 · G601S · N470D ·
R752W · F805C · Fexofenadine · hERG1b · hERG1a · Rescue · Heteromultimer

1
Introduction

The *human ether-a-go-go related gene* (*hERG*) was first identified in the human
hippocampus based on its similarity to *Drosophila ether-a-go-go* (Warmke and
Ganetzky 1994), a potassium channel gene regulating membrane excitability
at the neuromuscular junction (Ganetzky and Wu 1983). A candidate-gene
approach led to the identification of mutations in *hERG* in families with type 2
inherited long QT syndrome (LQTS-2) (Curran et al. 1995), an autosomal-
dominant disease associated with ventricular arrhythmias and sudden death
(Roden 1993). Shortly thereafter, hERG subunits were shown to be primary
constituents of cardiac I_{Kr} channels, thus explaining the underlying cause of
disease as a disruption of this repolarizing current (Sanguinetti et al. 1995;
Trudeau et al. 1995). Recent evidence indicates that I_{Kr} channels are hetero-
multimers (Jones et al. 2004), comprising the original subunit, now termed
hERG1a, and hERG1b, a subunit encoded by an alternate transcript of the
hERG gene (Lees-Miller et al. 1997; London et al. 1997). The subunits are
identical except for the N-terminal region, which in hERG1b is much shorter
and contains a unique stretch of 36 amino acids. To date, no hERG1b-specific
mutations have been associated with LQTS-2.

2
hERG Trafficking

Mutations in *hERG* are thought to cause disease by altering I_{Kr} functional properties (Keating and Sanguinetti 1996) and by reducing channel number at the
surface via "trafficking defects" (Delisle et al. 2004). Although only a fraction
of the more than 200 potential disease-causing mutations in *hERG* have been
analyzed, most of those studied in heterologous expression systems lead to
reduced surface membrane expression of channels, lower current magnitudes,
and failure of mutant subunits to exit the endoplasmic reticulum (ER) and
become maturely glycosylated (Zhou et al. 1998a, 1999; Furutani et al. 1999;
Ficker et al. 2000a,b; January et al. 2000).

   The normal maturation process can be monitored by the appearance of two
glycoforms reflecting progressive glycosylation in HEK-293 cells (Zhou et al.
1998b; Gong et al. 2002). hERG channels are initially core-glycosylated in the
ER, producing a 135-kDa band on Western blots that is reduced in size by
endoglycosidase (Endo) H. Additional glycosylation takes place in the Golgi,
rendering the species that appear as the mature, Endo H resistant 155-kDa band
on Western blots. The time course of maturation can be measured by pulse-
chase metabolic labeling using ^{35}S and observing the time course of appearance
of the 155-kDa band captured on a phosphoimager (Gong et al. 2002). At
37°C, channels reach maturity in about 24 h. As virtually all the mature band

visible on a Western blot is sensitive to degradation by extracellularly applied proteases, such as proteinase K (Zhou et al. 1998a), transport from the Golgi to the plasma membrane must be very rapid. Many LQTS-2 mutants expressed heterologously are characterized by an abundance of the lower, immature band with little or no protein maturation. In contrast to wildtype channels, which exhibit prominent immunostaining at the membrane, the mutant channels accumulate in the ER (Zhou et al. 1998a, 1999; Ficker et al. 2000c).

Although we can measure the maturation that reports the arrival of hERG subunits to the Golgi apparatus, we know little about the interactions that characterize their travels along the way. The hERG carboxy terminus carries an arginine-rich signal (RXR) that causes the subunits to be retained in the ER, but so far this is known to operate only when downstream sequences are truncated, thus presumably exposing the RXR to the ER retrieval machinery (Kupershmidt et al. 1998). How or whether the RXR sequence functions in normal hERG trafficking is unknown, but it is reasonable to hypothesize that there is an interaction with the coat protein I (COPI) machinery responsible for retrieval of misfolded or non-oligomerized subunits escaping from the ER. In ATP-gated potassium (KATP) channels, the RXR motif together with a neighboring phosphorylation site serves as a binding site for COPI, and also for 14-3-3 γ, ζ, and ε isoforms expressed in the heart. The 14-3-3 proteins compete with COPI proteins for the RXR binding site, but only when the subunits are phosphorylated and oligomerized (Yuan et al. 2003). By detecting the multimeric state of the KATP subunits, 14-3-3 thus competes with COPI for the complex and promotes its exit from the ER. hERG is known to interact with 14-3-3, though studies to date have focused on interactions mediating functional effects at the plasma membrane (Kagan et al. 2002).

Upon entry to the Golgi, hERG interacts with the Golgi-resident protein GM130 (Roti Roti et al. 2002). Anchored to the Golgi membrane by an interaction with GRASP-65, GM130 tethers COPII vesicles arriving from the ER-Golgi intermediate compartment (ERGIC) via an interaction with p115 (Nakamura et al. 1997; Marra et al. 2001; Moyer et al. 2001). GM130 co-immunoprecipitates with both immature and mature hERG, suggesting it may accompany hERG from the *cis* to the medial Golgi, where the final glycosylation marking maturation occurs. Overexpression of GM130 in *Xenopus* oocytes reduces hERG current amplitude, consistent with a role as a trafficking checkpoint (Roti Roti et al. 2002). Further characterization of GM130's role in hERG trafficking is currently under way.

3
Trafficking Defects and Rescue of Mutant Phenotypes

The defects underlying the failure of LQTS-2 mutants to mature are poorly understood. LQTS-2 mutations are found throughout the hERG protein, including the cytosolic amino terminus, the transmembrane domains, and throughout the long, cytosolic carboxy terminus (Delisle et al. 2004). Possible mechanisms preventing maturation include folding or assembly defects, failure to be appropriately processed in the Golgi, loss of checkpoint protein interactions, or mistargeting to degradative pathways rather than to the plasma membrane (Ellgaard and Helenius 2003). Mutations with a dominant-negative phenotype may cause protein misfolding but do not disrupt oligomerization with wildtype subunits, which are rendered dysfunctional by association with the mutants. In contrast, loss-of-function mutations, may signal defects in oligomerization, as wildtype subunits form functional channels unhindered by coexpressed mutant subunits. Both classes of mutant proteins are unlikely to proceed beyond the ER, following instead an expedited path to degradation.

Perhaps surprisingly, our understanding of these underlying defects may be illuminated by the even more mysterious phenomenon of rescue. The plasma membrane expression of some LQTS-2 mutants in heterologous systems can be restored by reducing temperature or applying hERG channel blockers (Zhou et al. 1999). Other compounds, such as fexofenadine (Rajamani et al. 2002), a derivative of the hERG blocker terfenadine (Suessbrich et al. 1996), and thapsigargin (Delisle et al. 2003), a calcium pump inhibitor that diminishes calcium-dependent chaperone protein activity, have also been shown to rescue LQTS-2 mutations. Each of these interventions likely mediates rescue by a different mechanism. Reduced temperature is thought to stabilize folding intermediates, whereas channel blockers, which bind to the internal pore vestibule where the four subunits interact, may stabilize oligomeric integrity. Thapsigargin inhibits the sarcoplasmic/ER Ca^{++}-ATPase, resulting in a reduced lumenal Ca^{++} concentration in the ER (Inesi and Sagara 1992). For mutant cystic fibrosis transmembrane regulator (CFTR) channels, it has been proposed that Ca^{++}-dependent chaperones, which handcuff improperly folded proteins while they await degradation, lose their grip as Ca^{++} levels drop and allow the errant channels to escape to the plasma membrane (Egan et al. 2002; Delisle et al. 2003).

At least four hERG mutations, G601S, N470D, R752W and F805C can be rescued by reduced temperature, consistent with folding defects (Zhou et al. 1999; Ficker et al. 2000c; Delisle et al. 2003). G601S and R752W subunits exhibit enhanced binding to the chaperone proteins Hc70 and Hsp90, accompanied by an increase in degradation, suggesting the mutants cannot be coaxed by the normal, physiological mechanisms into the correct conformation for export (Ficker et al. 2003). G601S and F805C can be rescued by thapsigargin but not other inhibitors of the sarcoplasmic/ER Ca^{++}-ATPase, suggesting a mechanism distinct from that for CFTR mutant rescue (Delisle et al. 2003).

Interestingly, of these three mutants, G601S is perhaps the most compliant of all, as it is rescued by all approaches utilized so far. In contrast, F805C is rescued only by temperature and thapsigargin (Delisle et al. 2003), and N470D by temperature and pore blockers (Zhou et al. 1999; Rajamani et al. 2002).

Thus, even among mutants characterized as folding-defective, the molecular mechanisms of disease must be quite diverse. R752W is unlikely even to form oligomers, as it exhibits a loss-of-function rather than a dominant-negative phenotype (Ficker et al. 2003), whereas G601S and N470D oligomerize effectively and respond to the stabilizing effects of reduced temperature on folding or the binding of drugs to the pore (Zhou et al. 1999; Rajamani et al. 2002), which may reinforce the oligomeric structure required for ER export.

4
Therapeutic Potential for Rescue

Pharmacological or chemical rescue strategies have a therapeutic potential only if the rescued I_{Kr} channels are sufficiently functional to support normal cardiac repolarization. Most examples of rescue have occurred with hERG channel blockers, which carry the risk for acquired LQTS. This is not so for fexofenadine, a derivative of the hERG blocker terfenadine. Fexofenadine mediates rescue of G601S and N470D at an IC_{50} 300-fold lower than that for drug block, indicating for the first time that rescue and restoration of normal I_{Kr} function can potentially be decoupled from the risk for LQTS and torsades de pointes (Rajamani et al. 2002). At the surface, rescued G601S and N470D channels exhibit normal gating and permeation (Furutani et al. 1999; Zhou et al. 1999).

5
Conclusions

Recent advances indicate that hERG channels with LQTS mutations may be rescued pharmacologically, opening the door for therapeutic intervention in the disease process. One compound, fexofenadine, can rescue certain mutants without the deleterious effects of channel block and associated risk of acquired LQTS. Mutants with relatively mild folding defects are likely the best candidates for rescue, as they seem to function normally upon reaching the plasma membrane. These studies underscore the importance of determining the specific mutation carried by a patient and evaluating the corresponding mutant phenotype and its receptiveness to rescue in heterologous systems.

There is also a need to understand in greater detail the mechanisms of hERG subunit folding and assembly, as well as the protein–protein interactions in the trafficking pathway. Disruption of any of these events may lead to disease, and all represent potential targets for therapeutic rescue. Heterologous expression systems used to evaluare LQTS mutants should incorporate wildtype subunits as well as hERG1b subunits to better mimic native I_{Kr} channels. Mutations introduced into heteromeric hERG1a/1b channels may confer different mutant phenotypes and responses to rescue agents compared with hERG1a homomeric mutant channels. Ultimately, this information will contribute to a rational and personalized approach to therapeutic treatment of patients with long QT syndrome.

References

Curran ME, Splawski I, Timothy KW, Vincent GM, Green ED, Keating MT (1995) A molecular basis for cardiac arrhythmia: HERG mutations cause long QT syndrome. Cell 80:795–803

Delisle BP, Anderson CL, Balijepalli RC, Anson BD, Kamp TJ, January CT (2003) Thapsigargin selectively rescues the trafficking defective LQT2 channels G601S and F805C. J Biol Chem 278:35749–35754

Delisle BP, Anson BD, Rajamani S, January CT (2004) Biology of cardiac arrhythmias: ion channel protein trafficking. Circ Res 94:1418–1428

Egan ME, Glockner-Pagel J, Ambrose C, Cahill PA, Pappoe L, Balamuth N, Cho E, Canny S, Wagner CA, Geibel J, Caplan MJ (2002) Calcium-pump inhibitors induce functional surface expression of Delta F508-CFTR protein in cystic fibrosis epithelial cells. Nat Med 8:485–492

Ellgaard L, Helenius A (2003) Quality control in the endoplasmic reticulum. Nat Rev Mol Cell Biol 4:181–191

Ficker E, Dennis AT, Obejero-Paz CA, Castaldo P, Taglialatela M, Brown AM (2000a) Retention in the endoplasmic reticulum as a mechanism of dominant-negative current suppression in human long QT syndrome. J Mol Cell Cardiol 32:2327–2337

Ficker E, Thomas D, Viswanathan PC, Dennis AT, Priori SG, Napolitano C, Memmi M, Wible BA, Kaufman ES, Iyengar S, Schwartz PJ, Rudy Y, Brown AM (2000b) Novel characteristics of a misprocessed mutant HERG channel linked to hereditary long QT syndrome. Am J Physiol Heart Circ Physiol 279:H1748–1756

Ficker E, Dennis AT, Wang L, Brown AM (2003) Role of the cytosolic chaperones Hsp70 and Hsp90 in maturation of the cardiac potassium channel HERG. Circ Res 92:e87–100

Ficker EK, Thomas D, Viswanathan P, Rudy Y, Brown AM (2000c) Rescue of a misprocessed mutant HERG channel linked to hereditary long QT syndrome. Biophys J 78:342A

Furutani M, Trudeau MC, Hagiwara N, Seki A, Gong Q, Zhou Z, Imamura S, Nagashima H, Kasanuki H, Takao A, Momma K, January CT, Robertson GA, Matsuoka R (1999) Novel mechanism associated with an inherited cardiac arrhythmia: defective protein trafficking by the mutant HERG (G601S) potassium channel. Circulation 99:2290–2294

Gong Q, Anderson CL, January CT, Zhou Z (2002) Role of glycosylation in cell surface expression and stability of HERG potassium channels. Am J Physiol Heart Circ Physiol 283:H77–84

Inesi G, Sagara Y (1992) Thapsigargin, a high affinity and global inhibitor of intracellular Ca^{2+} transport ATPases. Arch Biochem Biophys 298:313–317

January CT, Gong Q, Zhou Z (2000) Long QT syndrome: cellular basis and arrhythmia mechanism in LQT2. J Cardiovasc Electrophysiol 11:1413–1418

Jones EM, Roti Roti EC, Wang J, Delfosse SA, Robertson GA (2004) Cardiac IKr channels minimally comprise hERG 1a and 1b subunits. J Biol Chem 279:44690–44694

Kagan A, Melman YF, Krumerman A, McDonald TV (2002) 14-3-3 amplifies and prolongs adrenergic stimulation of HERG K$^+$ channel activity. EMBO J 21:1889–1898

Keating MT, Sanguinetti MC (1996) Molecular genetic insights into cardiovascular disease. Science 272:681–685

Kupershmidt S, Snyders DJ, Raes A, Roden DM (1998) A K^+ channel splice variant common in human heart lacks a C-terminal domain required for expression of rapidly activating delayed rectifier current. J Biol Chem 273:27231–27235

Lees-Miller JP, Kondo C, Wang L, Duff HJ (1997) Electrophysiological characterization of an alternatively processed ERG K^+ channel in mouse and human hearts. Circ Res 81:719–726

London B, Trudeau MC, Newton KP, Beyer AK, Copeland NG, Gilbert DJ, Jenkins NA, Satler CA, Robertson GA (1997) Two isoforms of the mouse ether-a-go-go-related gene coassemble to form channels with properties similar to the rapidly activating component of the cardiac delayed rectifier K^+ current. Circ Res 81:870–878

Marra P, Maffucci T, Daniele T, Tullio GD, Ikehara Y, Chan EKL, Luini A, Beznoussenko G, Mironov A, DeMatteis MA (2001) The GM130 and GRASP65 golgi proteins cycle through and define a subdomain of the intermediate compartment. Nat Cell Biol 3:1101–1114

Moyer BD, Allan BB, Balch WE (2001) Rab1 interaction with a GM130 effector complex regulates COPII vesicle cis-Golgi tethering. Traffic 2:268–276

Nakamura N, Lowe M, Levine TP, Rabouille C, Warren G (1997) The vesicle docking protein p115 binds GM130, a cis-Golgi matrix protein, in a mitotically regulated manner. Cell 89:445–455

Rajamani S, Anderson CL, Anson BD, January CT (2002) Pharmacological rescue of human K(+) channel long-QT2 mutations: human ether-a-go-go-related gene rescue without block. Circulation 105:2830–2835

Roden DM (1993) Torsade de pointes. Clin Cardiol 16:683–686

Roti Roti EC, Myers CD, Ayers RA, Boatman DE, Delfosse SA, Chan EK, Ackerman MJ, January CT, Robertson GA (2002) Interaction with GM130 during HERG ion channel trafficking. Disruption by type 2 congenital long QT syndrome mutations. J Biol Chem 277:47779–47785

Sanguinetti MC, Jiang C, Curran ME, Keating MT (1995) A mechanistic link between an inherited and an acquired cardiac arrhythmia: HERG encodes the IKr potassium channel. Cell 81:299–307

Suessbrich H, Waldegger S, Lang F, Busch AE (1996) Blockade of HERG channels expressed in Xenopus oocytes by the histamine receptor antagonists terfenadine and astemizole. FEBS Lett 385:77–80

Trudeau MC, Warmke JW, Ganetzky B, Robertson GA (1995) HERG, a human inward rectifier in the voltage-gated potassium channel family. Science 269:92–95

Warmke JW, Ganetzky B (1994) A family of potassium channel genes related to eag in Drosophila and mammals. Proc Natl Acad Sci U S A 91:3438–3442

Yuan H, Michelsen K, Schwappach B (2003) 14-3-3 dimers probe the assembly status of multimeric membrane proteins. Curr Biol 13:638–646

Zhou Z, Gong Q, Epstein ML, January CT (1998a) HERG channel dysfunction in human long QT syndrome. Intracellular transport and functional defects. J Biol Chem 273:21061–21066

Zhou Z, Gong Q, January CT (1999) Correction of defective protein trafficking of a mutant HERG potassium channel in human long QT syndrome. Pharmacological and temperature effects. J Biol Chem 274:31123–31126

Subject Index

Printing: Krips bv, Meppel
Binding: Stürtz, Würzburg